Lecture Notes in Artificial Intelligence   1095

Subseries of Lecture Notes in Computer Science
Edited by J. G. Carbonell and J. Siekmann

Lecture Notes in Computer Science

Edited by G. Goos, J. Hartmanis and J. van Leeuwen

Springer

Berlin
Heidelberg
New York
Barcelona
Budapest
Hong Kong
London
Milan
Paris
Santa Clara
Singapore
Tokyo

W. McCune   R. Padmanabhan

Automated Deduction in Equational Logic and Cubic Curves

 Springer

Series Editors
Jaime G. Carbonell, Carnegie Mellon University, Pittsburgh, PA, USA
Jörg Siekmann, University of Saarland, Saarbrücken, Germany

Authors

W. McCune
Mathematics and Computer Science Division,Argonne National Laboratory
Argonne, Illinois 60439, USA

R. Padmanabhan
Department of Mathematics, University of Manitoba
Winnipeg, Manitoba R3T 2N2, Canada

Cataloging-in-Publication Data applied for

Die Deutsche Bibliothek - CIP-Einheitsaufnahme

MacCune, William:
Automated deduction in equational logic and cubic curves / W.
Mc Cune ; R. Padmanabhan. - Berlin ; Heidelberg ; New York
; Barcelona ; Budapest ; Hong Kong ; London ; Milan ; Paris ;
Santa Clara ; Singapore ; Tokyo : Springer, 1996
  (Lecture notes in computer science ; Vol. 1095 : Lecture notes in
  artificial intelligence)
  ISBN 3-540-61398-6
NE: Padmanabhan, R.:; GT

CR Subject Classification (1991): I.2.3, F.4.1, I.3.5
1991 Mathematics Subject Classification: 03B35, 03C05, 14Q05

ISBN 3-540-61398-6 Springer-Verlag Berlin Heidelberg New York

© Springer-Verlag Berlin Heidelberg 1996
Printed in Germany

Typesetting: Camera ready by author
SPIN 10513209   06/3142 – 5 4 3 2 1 0   Printed on acid-free paper

Preface

The aim of this monograph is to demonstrate that automated deduction is starting to become a practical tool for working mathematicians. It contains a set of problems and theorems that arose in correspondence between the authors starting early in 1993. One of us (Padmanabhan), a mathematician working in universal algebra and geometry, contacted the other (McCune), a computer scientist working in automated deduction, after reading a survey article by Larry Wos on applications of automated reasoning [80]. Padmanabhan sent McCune a few first-order theorems and asked whether Argonne's theorem prover Otter could prove them. Otter succeeded and, in a few cases, found slightly better results. The collaboration quickly took off.

We worked mostly in four areas: (1) equational proofs of theorems that had been proved previously with higher-order arguments, (2) a new equational inference rule for problems about cubic curves in algebraic geometry, (3) a conjecture about cancellative semigroups, proving many theorems that support the conjecture, and (4) equational bases with particular properties such as single axioms and independent self-dual bases. Some of the results presented here have also appeared (or will also appear) elsewhere in more detail.

The intended audience of this monograph is both mathematicians and computer scientists. We include many new results, and we hope that mathematicians working in equational logic, universal algebra, or algebraic geometry will gain some understanding of the current capabilities of automated deduction (and, of course, we hope that readers will find new practical uses for automated deduction). Computer scientists working in automated reasoning will find a large and varied source of theorems and problems that will be useful in designing and evaluating automated theorem-proving programs and strategies.

Otter (version 3.0.4) and MACE (version 1.2.0) are the two computer programs that made this work possible. Both programs are in the public domain and are available by anonymous FTP and through the World Wide Web (WWW). See either of

```
ftp://info.mcs.anl.gov/pub/Otter/README
http://www.mcs.anl.gov/home/mccune/ar/
```

for information on obtaining the programs. The primary documentation for the programs is [39] and [37, 38]; these are included with the programs when obtained from the above network locations. We also have the WWW document

`http://www.mcs.anl.gov/home/mccune/ar/monograph/`

associated with this work; it points to all of the Otter and MACE input files and proofs to which we refer, and it is particularly useful if the reader wishes to see precisely the search strategy we used or to experiment with related theorems.

Special thanks go to Larry Wos and to Ross Overbeek. Larry introduced the notions of strategy and simplification to automated deduction and invented the inference rule paramodulation for equality; all three of these concepts are at the center of this work. Also, Larry simplified several of our Otter proofs (in one case from 816 steps to 99!) so that they could be included in these pages. Ross is due a lot of the credit for Otter's basic design and high performance, because in building Otter, McCune borrowed so heavily from Ross's earlier theorem provers and ideas. Special thanks also go to Dr. David Kelly (of the University of Manitoba) and Dr. Stanley Burris (of the University of Waterloo) for encouraging Padmanabhan to get in touch with the Argonne group. Padmanabhan thanks Dr. Lynn Margarett Batten and Dr. Peter McClure, successive heads of the Department of Mathematics at the University of Manitoba, for creating a pleasant atmosphere conducive to creative research, without which this project could not have been completed so smoothly. We also thank Dr. Harry Lakser for compiling a Macintosh version of Otter, which enabled Padmanabhan to experiment with some of his conjectures on a Mac. And we are deeply indebted to Gail Pieper, who read several versions of the manuscript and made many improvements in the presentation.

The cubic curves in most of the figures were drawn with data generated by the Pisces software [77] developed at The Geometry Center of the University of Minnesota.

McCune was supported by the Mathematical, Information, and Computational Sciences Division subprogram of the Office of Computational and Technology Research, U.S. Department of Energy, under Contract W-31-109-Eng-38. Padmanabhan was supported by the University of Manitoba and by operating grant #A8215 from NSERC of Canada.

Table of Contents

List of Figures

List of Tables

1. Introduction

This monograph contains a collection of theorems and problems in equational logic and in logics close to equational logic. Some are easy, some are difficult, and some are still open. We have attacked nearly all of the problems with the automated deduction system Otter, and many of the resulting proofs are presented in the following chapters. Several previously open problems were solved by Otter, and many more new problems arose and were solved by Otter during the course of this work.

About two-thirds of the theorems and problems are in several areas of universal algebra, including lattice theory, Boolean algebra, groups, quasigroups, loops, and semigroups. These are about axiomatizations of algebras, relationships between various algebras, investigations of new algebras, finding proofs simpler than those previously known, finding new equational descriptions satisfying certain preassigned syntactic properties, and finding first-order proofs where previous proofs have been model theoretic or higher order. The remaining one-third of the problems are related to algebraic geometry. These use an equational inference rule, implemented for the first time in Otter, that captures a generalization principle of cubic curves in the complex projective plane. The rule can be used to construct short proofs of some interesting and useful theorems that ordinarily require higher-order notions from algebraic geometry.

In this work, the program Otter was used to *find* proofs rather than to check proofs supplied by the user or to complete proofs outlined by the user. Otter is programmable in the sense that the user can specify inference rules and supply parameters to restrict and direct the search for proofs, and we freely used these features. But we did not use any knowledge about proofs or about the algebras under study when giving problems to Otter; in most cases we did not know a proof of the theorem in question. We also made some use of a complementary program called MACE, which searches for small finite models and counterexamples. Although not particularly powerful, MACE made several important contributions to this work.[1]

---

[1] We view Otter and MACE as assistants, and we anthropomorphize freely about them. In fact, we considered making Otter a coauthor.

1.1 Algebras and Equational Logic

The material in this section follows Tarski's 1968 paper on equational logic
[72]. An *algebra* is a system $\langle A; O_1, \ldots, O_n \rangle$ consisting of a nonempty set A
and a finite sequence of finitary operations from A to A. The sequence of
arities of $O_1, \ldots, O_n$ is the *type* of the algebra. For example, a group might
be $\langle G; \cdot,', e \rangle$ with type $\langle 2, 1, 0 \rangle$. If the type of an algebra or class of algebras
is clear from the context, for example, if it is specified by a set of equations,
we may omit mention of the type. To specify classes of algebras and to prove
theorems about them, we can use equational logic.

Equational logic.[2] A logic typically is specified by a language of formulas
and rules of inference over formulas. The language of an equational logic
consists of a set of function symbols of fixed arity (including constants) and
an unlimited supply of variables. (Otter distinguishes variables from constants
with the following rule: the symbol is a variable if and only if it starts with a
member of $\{u, v, w, x, y, z\}$; we use the same convention here.) Formulas are
simply equations in which all variables are interpreted as being universally
quantified. From another view, equational logic is that part of first-order
predicate logic with equality in which universally quantified equations are the
only formulas. In the terminology of automated theorem proving (Sec. 2.1),
an equation is a unit equality clause.

The *equational theory* of a class of algebras is the set of equations that
are satisfied in all algebras of the class. A *variety* is a class of algebras that
consists of all models of some set of equations. A *basis* for a theory is a set of
equations from which all equations of the theory can be derived by the rules
of inference. A basis is *independent* if none of its members can be derived
from the remaining equations. An n-basis is a basis with n elements.

There has been great interest in simple bases and bases with particular
properties (see Tarski's paper [72] for an excellent survey), and many of our
theorems are in those areas. For example, we study (1) one-based theories,
including the existence of simple single axioms, (2) self-dual independent
bases, with several different notions of the dual of an equation, (3) cardinalities of independent bases, including self-dual bases, and (4) bases in terms
of nonstandard operations.

A standard set of inference rules for equational logic is Birkhoff's, listed
in Table 1.1. Birkhoff's rules are sound and deduction complete [4]; that is,

---

[2] We exclude from consideration first-order systems, such as the finite first-order
axiomatizations of von Neumann-Gödel-Bernays set theory [13] or the Tarski-
Givant equational formulation of set theory [74], in which all of classical mathematics can be cast. In such systems, the terms of the logic can refer to all
objects of classical mathematics, in particular, to "higher-order" objects such
as relations and functions. Throughout the present work, we use first-order or
equational logic to express and reason about algebras, and terms of the logic
refer to nothing more than elements of algebras.

an equation E is a model-theoretic consequence of a set Σ of equations if and only if E can be derived from Σ by Birkhoff's rules.

Table 1.1. Birkhoff's Inference Rules for Equational Logic

$\dfrac{}{t = t}$	Reflexivity
$\dfrac{t_1 = t_2}{t_2 = t_1}$	Symmetry
$\dfrac{t_1 = t_2,\; t_2 = t_3}{t_1 = t_3}$	Transitivity
$\dfrac{f(x) = g(x)}{f(t) = g(t)}$	Substitution
$\dfrac{t_1 = t_2}{f[t_1] = f[t_2]}$	Replacement

Although of theoretical interest, Birkhoff's rules have not been useful in searching automatically for proofs of theorems in equational logic because no one has found an effective way to control the use of the substitution and replacement rules. Instead, we use L. Wos's inference rule *paramodulation* [67], which combines substitution with a different kind of replacement, in particular, replacing a term with an equal term. See Table 1.2, where σ is a most general substitution of terms for variables such that $t_1\sigma$ is identical to $s\sigma$, and $P[s]$ is an equation or negated equation containing a term s. (Paramodulation is ordinarily defined for first-order predicate logic with equality; see Sec. 2.1.)

Table 1.2. Wos's Paramodulation for Equational Logic

$\dfrac{t_1 = t_2,\; P[s] \;\text{(with } t_1\sigma \equiv s\sigma)}{(P[t_2])\sigma}$	Paramodulation

Paramodulation is not deduction complete, because there is no substitution rule; for example, $f(t, t) = t$ cannot be derived from $f(x, x) = x$. It is, however, *refutation* complete; that is, if $\Sigma \models \{t_1 = t_2\}$, then $\Sigma \cup \{t_1' \neq t_2'\} \vdash \{t \neq t\}$ for some term t, where $t_1' \neq t_2'$ is the *denial* of $t_1 = t_2$, obtained by replacing all variables with new (Skolem) constants.

1.2 Outside of Equational Logic

We deviate from equational logic in several ways; all but the last one listed here are still within first-order predicate logic with equality. First, we occasionally include deduction rules, for example, left cancellation, $x \cdot y = x \cdot z \rightarrow y = z$, as axioms. In some cases, for example the theory of cancellative semigroups, the rules are necessary, because there is no equational axiomatization. In other cases it leads to more effective searches for proofs; for example, quasigroups can be defined equationally, but because left and right cancellation hold for quasigroups, we may include them as deduction rules. Second, we prove some theorems involving existential quantification. For example, to prove that an operation is a quasigroup, the straightforward way is to show left and right cancellation and the existence of left and right solutions (see Thms. MFL-4, MFL-5, and MFL-6, starting on p. 187). Third, we occasionally make simple transformations to a representation in predicate logic. For example, for Thm. CS-3 (p. 97), to prove $t_1 = t_2$, we assert $P(t_1')$, deny $P(t_2')$, then derive a contradiction. The purpose of the transformation in this case is to achieve a particular highly constrained bidirectional search strategy.

Finally, many of the theorems in this work involve the new inference rule (gL), which is based on a property of cubic curves in the complex projective plane. We refer to (gL) as a first-order equational rule because it derives equations from equations; however, it is outside of ordinary equational logic because it cannot be axiomatized with equations. See Sec. 2.2.4 for details of Otter's implementation of (gL), and see Sec. 3.1 for the mathematical presentation of the rule.

1.3 First-Order and Higher-Order Proofs

Birkhoff's subdirect decomposition theorem, a basic structure theorem in universal algebra, says that every algebra in an equational class K is a subdirect product of subdirectly irreducible members of K (proved in [4]; see [5, p. 193] for applications to universal algebras). In order to prove that a set Σ of equations of a given type implies another equation of the same type, it is enough if we verify the validity of the implication in subdirectly irreducible algebras satisfying Σ.

Birkhoff's theorem plays a central role in equational logic. However, in using it and other theorems of the same kind, one calls on second-order tools such as Zorn's lemma and transfinite induction. Thus, one area of research in equational logic is to take results that were proved with the aid of higher-order theorems and try to obtain proofs that lie strictly within the realm of first-order logic with equality. (By the completeness of equational logic, if an equational statement is a theorem, there must exist an equational proof of it.)

Why invest in re-proving such theorems? First, we simply like to find proofs without using unnecessary machinery. Second, we are better at automating searches in lower logics. When faced with an open equational problem about which we have little insight, our only practical choice is an automated first-order search. Experience with known theorems gives us insight into open problems. Third, it is currently more difficult to rigorously formalize higher-order notions; as formal verification becomes more important throughout mathematics, science, and engineering, lower-level proofs may be regarded as more trustworthy because the inference rules are simpler. See Sec. 6.2 for a set of examples related to the M_5-N_5 lattice theory argument, Frink's theorem in Sec. 6.5.1, and Prob. HBCK-1 in Appendix B.

MED-1 is an example of a theorem whose previously known proofs are higher order. First, we list a proof found by Otter after several failed attempts with various search strategies (see p. 24 for proof notation); then, we sketch a higher-order proof.

Theorem MED-1. Cancellative median algebras.

The type is $\langle 2, 0 \rangle$ with constant e.

$$\left\{ \begin{array}{l} \cdot \text{ is cancellative} \\ (xy)(zu) = (xz)(yu) \\ ee = e \end{array} \right\} \Rightarrow \{(x(yz))((uv)w) = (x(uz))((yv)w)\}.$$

Proof (found by Otter 3.0.4 on gyro[3] at 142.15 seconds).

1	$x = x$	
2	$x \cdot y = z,\ x \cdot u = z\ \rightarrow\ y = u$	
3	$x \cdot y = z,\ u \cdot y = z\ \rightarrow\ x = u$	
4	$(x \cdot y) \cdot (z \cdot u) = (x \cdot z) \cdot (y \cdot u)$	
5	$e \cdot e = e$	
7	$(A \cdot (D \cdot C)) \cdot ((B \cdot E) \cdot F) = (A \cdot (B \cdot C)) \cdot ((D \cdot E) \cdot F)\ \rightarrow\ \square$	
8	$((x \cdot y) \cdot (z \cdot u)) \cdot (v \cdot w) = ((x \cdot z) \cdot v) \cdot ((y \cdot u) \cdot w)$	$[4 \rightarrow 4]$
9	$(x \cdot y) \cdot ((z \cdot u) \cdot (v \cdot w)) = (x \cdot (z \cdot v)) \cdot (y \cdot (u \cdot w))$	$[4 \rightarrow 4]$
10	$((x \cdot y) \cdot z) \cdot ((u \cdot v) \cdot w) = ((x \cdot u) \cdot (y \cdot v)) \cdot (z \cdot w)$	$[\text{flip } 8]$
13,12	$(x \cdot e) \cdot (y \cdot e) = (x \cdot y) \cdot e$	$[5 \rightarrow 4, \text{flip}]$
15,14	$(e \cdot x) \cdot (e \cdot y) = e \cdot (x \cdot y)$	$[5 \rightarrow 4, \text{flip}]$
16	$(e \cdot x) \cdot e = e \cdot (x \cdot e)$	$[5 \rightarrow 12, \text{flip}]$
27	$((e \cdot x) \cdot y) \cdot (e \cdot z) = (e \cdot (x \cdot e)) \cdot (y \cdot z)$	$[16 \rightarrow 4, \text{flip}]$
30	$(x \cdot (e \cdot y)) \cdot (z \cdot (e \cdot u)) = (x \cdot z) \cdot (e \cdot (y \cdot u))$	$[14 \rightarrow 4, \text{flip}]$
33,32	$((e \cdot x) \cdot y) \cdot ((e \cdot z) \cdot u) = (e \cdot (x \cdot z)) \cdot (y \cdot u)$	$[14 \rightarrow 4, \text{flip}]$
74	$((x \cdot y) \cdot (z \cdot u)) \cdot e = ((x \cdot z) \cdot (y \cdot u)) \cdot e$	$[5 \rightarrow 8 :13]$
81	$(e \cdot (x \cdot e)) \cdot ((e \cdot y) \cdot z) = e \cdot ((x \cdot y) \cdot z)$	$[14 \rightarrow 27 :15, \text{flip}]$

---

[3] gyro is a 486 DX2/66 computer with 40 megabytes of RAM running the Linux (a UNIX clone) operating system. It runs Otter and MACE about 1.4 times faster than a Sun SPARCstation 2.

147	$(x \cdot ((y \cdot z) \cdot (u \cdot v))) \cdot e = (x \cdot ((y \cdot u) \cdot (z \cdot v))) \cdot e$	$[74 \to 12 : 13]$
193	$(e \cdot (x \cdot y)) \cdot ((z \cdot u) \cdot v) = (e \cdot (z \cdot y)) \cdot ((x \cdot u) \cdot v)$	$[14 \to 10 : 33]$
238,237	$(x \cdot (y \cdot e)) \cdot ((e \cdot z) \cdot (u \cdot v)) = (x \cdot (y \cdot u)) \cdot (e \cdot (z \cdot v))$	$[9 \to 30]$
306	$x \cdot ((y \cdot z) \cdot (u \cdot v)) = x \cdot ((y \cdot u) \cdot (z \cdot v))$	$[3,1,147]$
369	$e \cdot ((x \cdot (y \cdot z)) \cdot ((u \cdot v) \cdot w)) = e \cdot ((x \cdot (y \cdot v)) \cdot ((u \cdot z) \cdot w))$	
		$[193 \to 81 : 238,15]$
485	$(x \cdot (y \cdot z)) \cdot ((u \cdot v) \cdot w) = (x \cdot (u \cdot z)) \cdot ((y \cdot v) \cdot w)$	$[2,306,369]$
486	□	$[485,7]$

Higher-order proof. Every cancellative median algebra A satisfies Ore's quotient condition (Thm. MED-7, p. 106). If A has an idempotent element, it can be embedded in a median quasigroup Q [69]. By a well-known representation theorem of A. A. Albert, D. C. Murdoch, S. K. Stein, T. Evans, and others [1], the quasigroup multiplication $x \cdot y$ in Q is given by $x \cdot y = \alpha x + \beta y$ for some Abelian group structure $+$ in Q with α and β being commuting automorphisms of the group. In such an algebra, the identity on the right-hand side is obviously valid.

1.4 Previous Applications of Automated Deduction

We list here some highlights of previous work related to the automated discovery of new results in mathematics and logic.

- *SAM's Lemma* [16]. In 1966, a theorem about modular lattices was proved with the interactive theorem prover Semi-Automated Mathematics (SAM V). While trying to guide SAM V to the proof of a known theorem, the user noticed that an equation had been derived that led directly to the answer of a related open question. This is widely regarded as the first case of a new result in mathematics being found with help from an automated theorem prover. (See Thm. LT-2, p. 109.)
- *Ternary Boolean Algebra* [78]. Several dependencies were found by S. Winker in the original axiomatization of ternary Boolean algebra. Although the results are based on small finite models, a theorem prover with equational capabilities (NIUTP [33]) was used to analyze various conjectures. The work [78] predates the existence of fast model searching programs such as MACE, which would have been helpful.
- *Equivalential Calculus.* Several single axioms of minimum length were previously known. Many of the other theorems of the same length were shown by J. Peterson [62] to be too weak to be single axioms, and one was found by J. A. Kalman to be a single axiom [22]. Then, the seven remaining minimal candidates were classified by members of the Argonne group [85]: two were found to be single axioms, and five were shown (by hand) to be too weak by using insight from failed theorem prover searches.

- *Robbins Algebra* [79]. The famous Robbins problem, whether an algebra with a commutative and associative operation satisfying the Robbins axiom is necessarily a Boolean algebra, is still open. S. Winker and L. Wos attacked the problem by finding very weak equations that force such an algebra to be Boolean (see Sec. 6.5.2).
- *Fixed Point Combinators* [41, 81]. A set of special-purpose equational strategies was developed to answer questions on the existence of fixed points and fixed point combinators in fragments of combinatory logic; many new results were obtained through their use.
- *Nonassociative Rings* [76]. A special-purpose theorem prover for the Z-module reasoning method was used by T.-C. Wang and R. Stevens to answer several open questions in nonassociative rings.
- *Group Calculi* [36]. This work is on nonequational axiomatizations of group theory in terms of division, with inference rules substitution and detachment (modus ponens). The first known single axioms were found for two variants of the theory. This was the first use of a theorem-proving program to generate large sets of candidate axioms, then run separate searches with each candidate. Similar methods were used in other successful studies, including the present work and that summarized in the next point.
- *Single Axioms for Groups.* There has been a lot of recent work in this area, almost all of it relying heavily on equational provers. Short single axioms for groups and for Abelian groups, in terms of various operations, were presented in [35]. In [28], K. Kunen presented new single axioms for groups in terms of product and inverse, and proved that these axioms have the fewest possible number of variables and that the ones in [35] are the shortest possible. In [42], schemas for single axioms for odd exponent groups, in terms of {product} and {product,identity}, were presented, and in [18], J. Hart and Kunen formalized and generalized those results.
- *Quasigroup Existence Problems.* Many quasigroup existence problems posed by F. Bennett and others have been answered by J. Slaney, M. Stickel, M. Fujita, H. Zhang, and McCune with model-searching programs similar to MACE [70, 37, 3]. Also, some of the quasigroup problems have been simplified with an equational theorem prover [71].

1.5 Organization

Chapter 2 contains introductions to Otter and MACE. The programs apply to full first-order predicate logic with equality, but we limit the discussion to the parts close to equational logic. Automated deduction is reviewed, Otter's algorithms are sketched, the implementation of the new inference rule (gL) for algebraic geometry is discussed, and some relevant Otter strategies are presented. Example input files, Otter proofs, and MACE models are listed, some advice on using the programs is given, and soundness of the programs is discussed.

Chapters 3 and 4, of central importance in this work, are on problems related to algebraic geometry. The equational inference rule (gL), defined by Padmanabhan and installed in Otter for this work, is presented from a mathematical perspective. Several important results in classical algebraic geometry are proved by Otter, quite elegantly, with (gL), and several improvements and new results are presented. Chapter 3 focuses on uniqueness theorems, and Chapter 4 on compatibility of various algebras with (gL), for example, on algebras that cannot be defined on a cubic curve.

Chapter 5 is about a conjecture on the relationship between groups and cancellative semigroups. Specifically, if a particular kind of statement can be proved for groups, then it holds also for cancellative semigroups. Few think the conjecture plausible at first, but no counterexample has been found. We present many examples of theorems, proved first by Otter, that support the conjecture.

Chapter 6, on problems in lattices and related algebras, is a return to pure equational logic. We first give a few Otter proofs of classical results about modularity and distributivity; then we present new single axioms, found using some new reduction methods, for lattices and for weakly associative lattices. Next, we generalize to quasilattices several previously known theorems about lattices. Finally, we look at Boolean algebra, giving Otter proofs of some previously known theorems, including some about the famous Robbins algebra problem, and presenting a single axiom for ternary Boolean algebra.

Chapter 7 is about self-dual axiom systems for groups and for lattices. This work is motivated by Tarski's interpolation theorem on spectra of independent bases. We apply the additional syntactic constraint of self-duality and use Otter and MACE to test specific conjectures in support of the main results, which will be reported in detail elsewhere.

Finally, in Chapter 8, we present Otter proofs of some classical theorems about Moufang loops, some new and simple bases for Moufang loops, a single axiom for inverse loops in terms of product and inverse, and a schema for single axioms for subvarieties of inverse loops, including Moufang loops.

Choice of Material. Our two objectives in writing this monograph have been to demonstrate that programs such as Otter and MACE can be valuable assistants to mathematicians, and to present new results in mathematics that have been obtained by them or with their help. As a consequence, it contains a mixed bag of theorems, counterexamples, conjectures, and problems. We have included new results, both significant and supportive, along with some well-known theorems for perspective and for comparison with the new results. Although much of this work is about group theory, lattices, and Boolean algebra, we have not included the basic theorems in those areas cited so often in the automated deduction literature and used so often as test problems for theorem provers. Some of our results are presented elsewhere [59, 57, 40, 58, 60]. We have included them here as well to indicate how the programs were used to obtain them and to give a unified and complete picture of our collaboration on this project.

2. Otter and MACE

It is like having a window on the thought processes of a powerful but very different kind of mathematician.

J. R. Guard et al. [16]

Otter [39] is a program that searches for proofs, and MACE [37, 38] is a program that searches for small finite models or counterexamples. Both apply to statements in first-order logic with equality. Otter is more powerful and flexible at its task than MACE is at its task, and Otter is more difficult to use. When searching for a proof with Otter, the user typically formulates a search strategy and makes several attempts, modifying the strategy along the way. When searching for a counterexample with MACE, the user simply supplies a statement of the conjecture; if the conjecture is not too complex, and if small finite models exist, MACE will find them. For many of our conjectures (quasigroup problems in particular), we know that counterexamples, if they exist, must be infinite; MACE is useless in such cases. But the two programs nicely complement one another in many other cases.

The following descriptions of the two programs are informal. See the manuals and [83] for more formal and detailed presentations. First, we define some terms of automated theorem proving.

2.1 Definitions

Our definitions lean toward Otter and equational theorem proving.

- A *term* is a variable, a constant, or the application of an n-ary function symbol to n terms.
- An *atom* is the application of an n-ary predicate symbol to n terms. Nearly all atoms in this work are equalities.
- A *literal* is either an atom or the negation of an atom.
- A *clause* is a disjunction of literals. The variables in a clause are implicitly universally quantified.
- The *length* of a term, literal, or clause is the symbol count, excluding punctuation. In particular, variables and constants have length 1, the length of a term or atom is the sum of the lengths of the arguments plus 1, and the

length of a clause is the sum of the lengths of its atoms (negation symbols
are not counted).

- A *unit clause* contains exactly one literal. Most of the clauses in this work
  are positive unit equality clauses, that is, equations.
- The *empty clause* contains no literals and represents a contradiction. It is
  written as □ or \$F.
- A *Horn clause* has at most one positive literal. All of the clauses associated
  with this work are Horn clauses.
- Two terms or atoms *unify* if there is a substitution of terms for variables
  that makes the terms or atoms identical. When we speak of a *unifying
  substitution*, we refer to the most general unifying substitution.
- An *equational deduction rule* is a nonunit Horn clause containing at least
  one negative equality (the antecedents) and one positive equality (the consequent). It is frequently written as an implication,

$$s_1 = t_1, \cdots, s_n = t_n \rightarrow s = t,$$

  and is typically used by Otter with the inference rule hyperresolution.
- *Hyperresolution* (for equational deduction rules) takes an equational deduction rule with n antecedents and n equality unit clauses and derives, if
  possible, the appropriate instance of the consequent.
- *Paramodulation* (for unit equality clauses) is an inference rule that combines variable instantiation (by unification) and equality substitution into
  one step. Consider $t = s$, and let $P[t']$ be a positive or negative equality
  containing a term t'. If t and t' unify, with unifying substitution σ, then
  paramodulation *from* $t = s$ *into* $P[t']$ yields the paramodulant $(P[s])\sigma$. For
  example, from $uu' = e$ into $x(x'y) = y$, we can derive $xe = x''$ by unifying
  uu' and $x'y$.
- *Unit conflict* is a trivial inference rule that derives the empty clause (a
  contradiction) from two unifying unit clauses of opposite sign. All Otter
  proofs are by contradiction.
- A *demodulator* is a positive equality unit, $t_1 = t_2$, that is used as a rewrite
  rule to rewrite instances of t_1 to the corresponding instances of t_2.
- *Demodulation* is the process of repeatedly rewriting a clause with a set
  of demodulators until no more rewriting steps can be applied. A desir-
  able property of a set of demodulators is that demodulation with the set
  terminates on all terms to which it is applied.
- *Forward demodulation* is demodulation applied to newly inferred clauses.
  Otter always performs forward demodulation when demodulators are
  present.
- *Back demodulation* occurs when a newly inferred equality is added to the
  set of demodulators. All clauses that can be rewritten with the new demodulator are demodulated with the (extended) set of demodulators. Back
  demodulation is optional in Otter.

- Clause C *subsumes* clause D if there exists a substitution σ such that $(C)\sigma \subseteq D$. In such a situation, $C \Rightarrow D$, and D can usually be discarded without affecting completeness of the proof procedure. For this work, Otter always applied *forward subsumption*, discarding newly inferred clauses that are subsumed by clauses already in memory, and *back subsumption*, discarding all other clauses in memory subsumed by a newly retained clause.

- *Skolemization* is a process of eliminating existentially quantified variables from a formula by replacing them with *Skolem constants* and *Skolem functions*. The resulting formula is equiconsistent with the original formula. In this work, Skolemization is used mainly to assert the denials of conclusions. For example, when proving commutativity, the conclusion is $\forall x \forall y (xy = yx)$, its negation is $\exists x \exists y (xy \neq yx)$, and the Skolemized negation is $AB \neq BA$, where A and B are new symbols, that is, Skolem constants. In other words, to deny the conclusion, we simply assert that there are two elements, A and B, that do not commute.

- *Knuth-Bendix completion* is a method for attempting to transform a set of equalities into a *complete set of reductions*, that is, a set of rewrite rules that produces canonical forms. Although we are not concerned in this work with complete sets of reductions, the mechanics of some of our search strategies are similar to Knuth-Bendix completion. See Sec. 2.2.3.

2.2 Otter

Otter is a resolution/paramodulation theorem-proving program that applies to statements in first-order logic with equality. Otter operates on clauses; if the statements supplied by the user are not clauses, Otter immediately translates them with Skolemization (to eliminate existential quantifiers) and conjunctive normal form (CNF) conversion. This work focuses mainly on equational logic; since equations are clauses already, CNF conversion is not necessary. We search for proofs by contradiction, and Skolemization is used here mainly to deny the conclusion of the conjecture.

Otter is not interactive; the user prepares an input file that contains a denial of the conjecture and a specification of a search strategy, then runs Otter and receives an output file. (Otter has a primitive interactive mode, but it was not used in this work.) If the output file does not contain a proof, the user examines the output, modifies the input, and tries again for a proof. Most trivial theorems are proved within the first few attempts, and most difficult theorems require some iteration, resulting in the development of a specialized search strategy. For this work, we did not use Otter as a proof checker; that is, the strategies were not designed to lead Otter to particular proofs. Rather, the strategies were designed to guide the program toward a particular *type* of proof or to avoid search paths that appear to be fruitless or redundant.

Because of the iterative way in which Otter typically is used, we do not think of it as an automatic theorem prover. However, Otter does have a mode, the *autonomous mode*, which allows the user to supply simply a denial of the conjecture; the program does a simple syntactic analysis of the conjecture, decides on a simple strategy, and searches for a proof. Although not used for the main body of this work, the autonomous mode can be used to prove many of the theorems we present.[1]

The real, practical strength of Otter is its ability to quickly explore large search spaces. Strategies are used mainly to restrict the search rather than to guide it, resulting in a search less focused than with some of the other automated theorem provers currently in use, which use heuristics to carefully plan and control the search for a proof.

The theorems we present fall mostly into two classes: ordinary equational theorems, and equational theorems with the rule (gL). In some cases we add deduction rules also, such as cancellation or various closure conditions (Sec. 4.4). The strategies we use for ordinary equational theorems (both with and without deduction rules) are simpler, more uniform, and better understood than our strategies for (gL). Before going into detail about particular strategies in Secs. 2.2.3 and 2.2.4, we present some general features of Otter.

2.2.1 The Main Loop

Otter maintains three primary lists of clauses.

usable. These clauses are able to participate in the search through application of inference rules.

sos (set of support). These clauses are waiting to participate in the search through application of inference rules. A subset (possibly empty) of sos occurs also in the list demodulators; members of that subset can participate in the search through rewriting (demodulation).

demodulators. All of these clauses are unit equalities. They are applied as rewrite rules to all inferred clauses.

In the input file, the user typically partitions the denial of the conjecture into usable and sos; if any equalities are present, they may be inserted into demodulators also. Otter's highest-level operation is a simple loop that drives the search:

Repeat
1. Select a clause, the *given clause*, from sos.
2. Move the given clause from sos to usable.
3. Make inferences from the given clause by using other clauses in usable. With each inferred clause, rewrite it with demodulators and decide whether it should be retained; if so, append it to sos,

---

[1] See Appendix C.

and if it should be a rewrite rule, append it to `demodulators`.

Until a proof is found or a resource is exhausted.

The five main determinants of the search are (1) the initial partition into `usable` and `sos`, (2) the method for selecting the given clause, (3) the inference rules that are applied, (4) the rules for retention of inferred clauses, and (5) the rules for deciding whether inferred equalities are to be rewrite rules. These are considered to be part of the search strategy and are covered in some detail in the following subsection.

2.2.2 General Strategies

As mentioned above, use of Otter frequently involves iteration through a sequence of searches, and part of the iteration is toward achieving a *well-behaved search*. The notion of a well-behaved search is not precise, and recognizing one is largely a matter of experience in finding proofs with Otter. However, we can list here some ways in which searches can be ill behaved. One determines the behavior of a search by examining the output file.

- The `sos` list grows very rapidly, which wastes time and memory, because most `sos` clauses will never participate in the search.
- The program focuses on many trivial variants of clauses, because they are smaller than other, more useful clauses.
- The program is not using, or not using to the desired extent, some of the inference rules that are enabled.
- A bidirectional search is desired, but the given clauses that are selected force the search to be mostly in one direction.
- The program wastes time making checks or performing other operations that have little or no effect on the search.

To adjust the behavior of a search, one changes the specification of one or more strategies. The rest of this subsection summarizes Otter's most important general strategies.

Set of Support

One of the earliest strategies for automated theorem proving was the set of support strategy [84], which is designed to prevent lines of reasoning. To use the set of support strategy, the user partitions the input clauses into two sets: S (clauses with support) and T (those without support). The strategy requires that all lines of reasoning start with a member of S. That is, no inference is drawn entirely from clauses in T.

Otter's main loop can be viewed as a simple implementation of the set of support strategy. The initial `sos` list is S, and the initial `usable` list is T. Because the clauses in the initial `usable` list never occur in the `sos` list, and because each inference is started with the given clause, which is selected from the sos list, no inference is made in which all of the participants are in the initial usable list. To do without the set of support strategy, the user simply places all input clauses in the sos list.

The primary motivation for the set of support strategy is to focus the reasoning on the problem at hand instead of on the general theory. Many theorems formulate naturally as axioms of a theory, special hypotheses, and a conclusion. The standard recommendation for partitioning the input into T and S is to place the axioms of the theory into T and clauses corresponding to the special hypotheses and denial of the conclusion into S. The result is that all lines of reasoning start with the hypotheses and the conclusion.

However, our experience has shown that the set of support strategy is not particularly effective for two classes of theory: (1) equational and nearly equational theories, and (2) theories with few axioms and few concepts. Since the theories on which we focus in this work fall into both classes, for the most part we do without the set of support strategy.

Selecting the Given Clause

Selection of the given clause has a great effect on the character and outcome of the search; it is the next path to explore. Since new clauses are appended to the sos list, always selecting the first sos clause leads to breadth-first search (occasionally useful), and always taking the last leads to depth-first search (rarely useful). Otter's default method, however, is to select the best clause in the sos list as the given clause. For example, if a simple equation is derived, it is usually best to use it right away. The default measure of "best", and the one used for this work, is simply the length of the clause. Alternatively, Otter's weighting mechanism can be used to specify rules for assigning weights to clauses.

For most of the theorems and conjectures in this work, we use a selection method that is a combination of best-first and breadth-first. One of Otter's parameters, pick_given_ratio, can be used to specify the ratio. A value of n means n parts best-first to 1 part breadth-first. That is, through n iterations of the main loop, the best clause is taken; then, in the next iteration, the first is used, and so on. With this method, Otter focuses mainly on the small clauses, while occasionally allowing a larger, richer clause to participate in the search to mix things up.

Demodulation (Rewriting)

The list demodulators contains a set of equations that are used as rewrite rules. When a new clause is inferred, one of the first processes applied to it is forward demodulation, in which the clause is rewritten, as much as possible, with the members of demodulators. The primary motivation for demodulation is to reduce redundancy by canonicalizing clauses. For example, if the theory has a left identity, $ex = x$, it is usually wasteful to retain both $(xy)z = x(yz)$ and $(xy)(ez) = x(yz)$; with $ex = x$ as a demodulator, the second will be demodulated to an equality identical to the first and will be deleted.

Equations can become demodulators in three ways: (1) demodulators can be specified in the input file, (2) input equations in sos or usable can be copied into demodulators, and (3) derived equalities can be copied into demodulators. Choices (2) and (3), which apply only if certain flags are set, can also cause the new demodulator to rewrite previously retained clauses, through back demodulation.

Retaining Inferred Clauses

After a newly inferred clause is demodulated and possibly rewritten in other ways, Otter must decide whether it is to be retained. Forward subsumption is nearly always applied; that is, if the new clause is subsumed by any of Otter's existing clauses, it is is discarded. Also, the user can assign limits on several measures, including number of variables, number of literals, and the weight of the clause. A limit on the weight of retained clauses is nearly always used for difficult theorems.

The default (which we use unless stated otherwise) weight of a new clause is its length. If there is no limit on the weight of retained clauses, or if the limit is too high, the sos list will grow very rapidly; most of the sos clauses will simply be wasting space, with little chance of participating in the search. Also, if back demodulation is enabled, when newly derived equalities are adjoined as demodulators, a lot of time will be wasted attempting to rewrite all of the sos clauses. If the limit is too low, the sos list will be exhausted, and the search will terminate. The typical strategy, and the one used in this work, is to start with a low limit and make several test runs, adjusting the limit until a well-behaved search is achieved. (This iterative process explains the apparently arbitrary weight limits specified in many of the input files associated with this work.)

Forward, Backward, and Bidirectional Searches

Most of the theorems and conjectures in this work fit the pattern

$$E \cup D \Rightarrow C,$$

where E and C are sets of equalities, D is a set of deduction rules, and $\Rightarrow$ may include an extension to the theory.

In a forward search, one reasons from $E \cup D$ and derives additional equalities, succeeding if all members of C are derived. An easy (but possibly incomplete) way to have Otter conduct a forward search is to place the denials of the conclusions C in an auxiliary list called passive. Clauses in passive do not participate actively in the search, but if a derived equality conflicts with a member of `passive`, a proof will be reported and printed. This method is incomplete, however, if the members of C are complex enough to be rewritten by a demodulator. For example, suppose e is a constant, the denial of $(ex)x = yy$ is in `passive`, and the equality $ex = x$ is derived and adjoined as a demodulator. If $(ex)x = yy$ is derived later, it will be demodulated to $xx = yy$ before the unit conflict check with `passive`, and the proof will be missed. Therefore, if the user suspects that the denial might be rewritten by a derived equation, the denial should not be placed in `passive`.

In a backward search, one reasons from the denial of C (assume for simplicity that C is a singleton), using members of E and D to derive negated equalities, until a contradiction with a member of E is derived. To have Otter conduct a backward search, one can simply start the search with $E \cup D$ in list `usable` and the denial of C in `sos`. All derived clauses will be negative. (If D is nonempty, its members will have to be used to derive negative clauses with the inference rule UR-resolution [83].) However, *purely backward searches are ineffective for nearly all theories in which equality plays a dominant role*; deriving and using equalities are simply too important.

In a bidirectional search, one reasons both forward and backward. To achieve a bidirectional search with Otter, one can place a subset of E in the list `sos`; both positive and negative equalities can then be derived. However, achieving the desired balance between forward and backward search can be difficult, because Otter uses a single method for selecting the given clause, and both positive and negative clauses must be selected.

Most of the searches and most of the proofs in this work are either forward or mostly-forward bidirectional. If C was a simple singleton equality, its denial was usually placed in the list `passive`, resulting in a forward search (proof); otherwise, its denial was placed in list `usable`, resulting in a forward or bidirectional search (proof).

Problem Formulation

Although problem formulation is not really a search strategy, from the practical view, it falls in with the strategies presented in this section.

One usually has choices in the statement of the theorem or conjecture, and Otter can be sensitive to these choices. We list four examples. First, given an equational theory E, which basis for E should be used? We usually find it best to use a basis that includes simple equations.

Second, should dependent equations be included? Consider group theory in terms of product, inverse, and identity e. Three logically equivalent alternatives are the following.

$$\left\{ \begin{array}{l} ex = x \\ x'x = e \\ (xy)z = x(yz) \end{array} \right\} \left\{ \begin{array}{l} ex = x \\ xe = x \\ x'x = e \\ xx' = e \\ (xy)z = x(yz) \end{array} \right\} \left\{ \begin{array}{ll} ex = x & e' = e \\ xe = x & x'' = x \\ x'x = e & (xy)' = y'x' \\ xx' = e & x(x'y) = y \\ (xy)z = x(yz) & x'(xy) = y \end{array} \right\}$$

In addition, should cancellation be included as a deduction rule? We don't have simple answers; a good choice usually depends on the situation and requires some experimentation to achieve a well-behaved search.

Third, consider the following two forms of the deduction rule for left cancellation:

$$xy = xz \;\rightarrow\; y = z$$
$$xy = u, \; xz = u \;\rightarrow\; y = z$$

The first form, although more natural, is usually less effective than the second. Both rules apply to the equation $ab = ac$ (assuming the presence of $x = x$ for use with the second rule), but only the second rule applies to the pair $\{ab = d, ac = d\}$. However, the second rule causes much more redundancy in the sequence of derived equations.

Finally, if the denial of the conclusion is complex, it is sometimes advantageous to introduce new terms (usually constants) to, in effect, abbreviate complex terms. We call this the *naming strategy*, and it has been long advocated by R. Overbeek in the context of hyperresolution [33, 32]. For example, we might reformulate a complex ground denial $\alpha \neq \beta$ as $\alpha = C \;\&\; \beta = D \;\&\; C \neq D$, where C and D are new constants. This reformulation tends to prevent some inferences within the named terms, because the subterms are hidden, and to favor some inferences containing named terms, because those clauses are simpler.

2.2.3 Equational Problems

Many of our proof searches in equational theories use strategies that are similar to the Knuth-Bendix completion method [27]. Because the method is well known, we review it only briefly here and state how it differs from our Otter strategies.

Knuth-Bendix Completion

Knuth-Bendix completion is a method for attempting to transform a set of equations E into an equivalent set of rewrite rules R such that rewriting with R terminates on all terms and produces a canonical form. Rewriting with R is a decision procedure for equality of terms in the theory of E. (Thus, the existence of R is undecidable.) The set R is called a *complete set of reductions*. The method uses an ordering, say $\succ$, on terms to orient equations into rewrite rules and to guarantee termination of rewriting; each member of R satisfies *left-side $\succ$ right-side*. The following statements summarize the method.

1. All equations, whether members of E or derived, must be orientable into rules as *left-side $\succ$ right-side*; otherwise the method fails.
2. New equations are inferred by paramodulation from left sides of rules into left sides of rules, then simplified with the current set of rules.
3. When a new (simplified) equation is derived and oriented into a rule, it is used to rewrite all known rules.
4. The method succeeds if no new paramodulants (left into left) can be derived.

If the method fails, a different term ordering may lead to success.

The Otter strategy we typically use for equational searches is quite similar to the Knuth-Bendix method; the main differences are (1) our goal is to prove theorems, (2) instead of failing when a nonorientable equation is derived, we continue, using it for paramodulation, but not for rewriting, (3) we typically impose a limit on the size of retained clauses, and (4) we can paramodulate into negative clauses as well as positive ones, resulting in a bidirectional search.

Historical remark: The roots of our current equational rules and strategies are in the methods used by Larry Wos and Ross Overbeek in the late 1960s and early 1970s. Those methods were developed independently from the Knuth-Bendix method (at about the same time) and were aimed at proving theorems rather than at finding a complete set of reductions. We tell Otter to apply the strategy by setting a flag called `knuth_bendix`; hence, this flag name is somewhat misleading.

2.2.4 Equational Problems with the Rule (gL)

This section (which can be skipped on first reading of this monograph) contains a description of Otter's implementation of the derivation rule (gL) and some practical information on its use. The rule (gL), which builds in a key property of algebras over projective curves, is presented mathematically in Sec. 3.1.

The rule (gL) generalizes equations: it replaces terms with variables. From the operational point of view of Otter, (gL) is implemented in two ways: as an inference rule and as a rewrite rule. Application of (gL) requires the equality to have identical terms in two positions. When (gL) is used as an inference rule, the identical terms occur implicitly through the use of unification (analogous to the inference rules resolution and paramodulation). Let $F[a_1, x]$ represent a term that contains a subterm a_1 at a particular position, with x representing everything else in the term. Suppose we have $F[a_1, x] = F[a_2, y]$, (i.e., a_1 and a_2 are in corresponding positions and occur in the same nest of symbols), with a_1 and a_2 unifiable. By (gL) we infer $F[z, x'] = F[z, y']$, where z is a new variable and x' and y' are the appropriate instances of x and y. For example, from

$$f(f(x,y), f(z, \underbrace{f(x,z)})) = f(u, f(y, \underbrace{u})),$$
$$ a_1)) = f(u, f(y,} a_2$$

we can derive

$$f(f(x,y), f(z,w)) = f(f(x,z), f(y,w))$$

with (gL) by unifying $f(x,z)$ and u, then introducing the new variable w.

When used as a rewrite rule, (gL) is applied to all pairs of identical terms in appropriate positions, in analogy to ordinary rewriting. The equality $F[a,x] = F[a,y]$ is rewritten by (gL) to $F[z,x] = F[z,y]$, where z is a new variable. For example, from

$$f(e,e,e,e) = f(e,e,x,f(e,e,x,e)),$$

we can derive

$$f(y,z,e,e) = f(y,z,x,f(e,e,x,e))$$

by two applications of (gL) as a rewrite rule: to the first occurrences of e on each side, then to the second occurrences of e.

The user has the options of applying (gL) as an inference rule, as a rewrite rule before ordinary rewriting, as a rewrite rule after ordinary rewriting, or any combination of these. We usually apply it as an inference rule and as a rewrite rule before ordinary rewriting.

Strategies for Use with (gL)

Otter's implementation of (gL) can be difficult to work with because it does not fit well with our ordinary strategies for equational deduction (activated with the Otter flag knuth_bendix). The main reason is that in many cases, the positions at which (gL) is effectively and usefully applied occur in *simplifiable* terms. With our ordinary strategies, the simplifiable terms are eliminated by demodulation before (gL) can be applied. Consider, for example, the theorem

$$\{ex = xe = x\} \;\;=\!(gL)\!\Rightarrow\;\; \{xy = yx\}$$

and the following proof (found by Otter 3.0.4 on gyro at 22.26 seconds).

2	$e \cdot x = x$	
3	$x \cdot e = x$	
4	$B \cdot A \neq A \cdot B$	
5	$(e \cdot e) \cdot x = x$	$[2 \to 2]$
7	$(x \cdot e) \cdot e = x$	$[3 \to 3]$
47	$(x \cdot e) \cdot e = (e \cdot e) \cdot x$	$[5 \to 7]$
812	$(x \cdot y) \cdot e = (e \cdot y) \cdot x$	$[(gL)\ 47]$
827	$(e \cdot x) \cdot y = y \cdot x$	$[3 \to 812,\ \text{flip}]$
899	$x \cdot y = y \cdot x$	$[2 \to 827]$
900	$\square$	$[899,4]$

Note the following points about the proof. (The notation for the justification of each step is explained on p. 24.)

- The positions in clause 47 at which (gL) applies (to derive clause 812) occur in terms that can be simplified by clauses 2 and 3; in fact, if clauses 2 and 3 become demodulators (as in our ordinary strategies), everything that Otter derives is demodulated to $x = x$, and the search terminates without finding a proof.
- The proof is quite simple, but Otter required 22 seconds to find it on a fast computer. (A proof this simple but without (gL) would be found immediately with our ordinary equational strategies.)
- The derivation of clauses 5, 7, and 47 involve paramodulation from a variable, a process that is never allowed with our ordinary equational strategies but is frequently required with (gL).

Our basic strategy for use with (gL) allows unrestricted paramodulation from both sides of equalities and from and into variables, and it does not use any demodulation. Such a strategy (which produced the preceding proof) usually results in a very redundant search, mostly because of the lack of demodulation, and it takes a long time to find proofs.

In many cases, however, we *can* use demodulation with our basic (gL) strategy. (Such cases usually have a richer set of axioms than the preceding example.) Hence, when faced with a new (gL) conjecture, we typically start with full demodulation as in our ordinary equational strategy, then iteratively restrict demodulation (in successive Otter runs) until a well-behaved search is achieved.

Our search strategies for (gL) are still quite primitive; there is a lot of room for research in this area.

2.2.5 Conjectures with Deduction Rules

Some of the theories we deal with are specified by axiom sets that contain equational deduction rules (implications) as well as equalities. An example is left cancellation (in its more useful form):

$$xy = u, \ xz = u \ \rightarrow \ y = z.$$

We use the following Otter flags whenever equational deduction rules are present.

```
set(hyper_res). set(unit_deletion).
set(para_from_units_only). set(para_into_units_only).
set(output_sequent).
```

The flag hyper_res turns on the inference rule hyperresolution, flag unit_deletion says to delete an antecedent when justified by a derived equation, the paramodulation flags disallow paramodulation from or into nonunit clauses (the deduction rules), and `output_sequent` causes deduction rules to be printed as implications instead of as disjunctions of literals.

If the equational theorem or conjecture we are trying to prove, say $E \Rightarrow C$, has *multiple goals* (i.e., if C is not a singleton), the denial of C is a nonunit negative clause. For example, if we are trying to prove that E is commutative and associative, the denial might be

$$AB \neq BA \mid (AB)C \neq A(BC),$$

where A, B, and C are Skolem constants. This can be viewed as the deduction rule

$$AB = BA, \ (AB)C = A(BC) \ \rightarrow \ \Box,$$

where $\Box$ represents falsehood. If commutativity and associativity are derived, then hyperresolution with that deduction rule gives a contradiction. Also, the goals can be rewritten with derived demodulators (and individual goals can be removed from the denial with the *unit deletion* process) before hyperresolution produces a contradiction.

2.2.6 Running Otter

For this work we used four standard sets of strategies, encoded as four partial input files, for our four basic types of problem—ordinary equational and (gL), both with and without deduction rules (see the examples in Sec. 2.2.7). These are good starting points for anyone wishing to apply Otter to these types of problem. We made major deviations from these strategies in very few cases. Most of our experimentation involved fiddling with various combinations of just a few parameters and strategies until a proof was found—and, if the proof was very long, until a shorter proof was found. In most cases, we adjusted the parameter `max_weight`, the limit on the length of retained equations, and in some cases we assigned higher priority to clauses containing Skolem constants from the denial; for the (gL) problems, we usually had to adjust the demodulation strategy. The easy theorems fell in one or two searches, and the difficult ones required ten, twenty, or more attempts, usually of one to five minutes; if a well-behaved search was achieved, we would let Otter run for a day or more.

Many of our problems required *finding* theorems, usually finding single axioms or other equational bases with particular properties. Since we don't know how to do this directly with Otter, which expects a specific first-order conjecture, we generate large sets candidate bases and give each to Otter for a separate search. (As far as we know, this technique was first used in [34].) Typical examples are (1) 10,000 single axiom candidates and (2) several hundred absorption equations, each to be tried with some set of absorption equations. The candidate sets are generated in several ways, including (1) enumerating clauses and selecting a subset, (2) using Otter to generate part of a theory and selecting a subset, and (3) transforming, possibly with Otter, an existing set of candidates. Such sets of Otter jobs are run automatically, with a trivial driver program, for a few seconds or a few minutes each, with a standard search strategy.

2.2.7 Example Input Files and Proofs

This section contains two examples of Otter input files and proofs, with some explanation of the input language and proof format.

Ordinary Equational Logic

Consider the basic theorem that groups satisfying $xx = e$ are commutative. The following annotated input file causes Otter to find a proof.

```
%%%%%%%%%%%%%%%%%%%%%  Basic options

op(400, xfx, [*,+,^,v,/,\,#]).  % Declare some infix operations.
op(300, yf, @).                 % Postfix operation.

clear(print_kept).       % Don't print retained clauses.
clear(print_new_demod).  % Don't print new demodulators.
clear(print_back_demod). % Don't print back demodulated clauses.

assign(pick_given_ratio, 4).  % 4 best-first : 1 breadth-first.
assign(max_mem, 20000).       % Use at most 20 megabytes of RAM.

%%%%%%%%%%%%%%%%%%%%%  Standard for equational problems

set(knuth_bendix).

%%%%%%%%%%%%%%%%%%%%%  Modifications to strategy

%%%%%%%%%%%%%%%%%%%%%  Clauses

list(usable).              % The usable list.
x = x.
end_of_list.

list(sos).                 % The sos list.
e * x = x.                 % Left identity.
x@ * x = e.                % Left inverse.
(x * y) * z = x * (y * z). % Associativity.
x * x = e.                 % The hypothesis.
end_of_list.

list(passive).             % The passive list.
A * B != B * A.            % Denial of commutativity.
end_of_list.
```

Note the following points about the input file.

- The first part of the input file, through "Modifications to strategy", is our standard header for ordinary equational problems. This theorem is easy enough that no modifications are required. For many theorems, the only addition to the strategy is the parameter max_weight (as explained in Sec. 2.2.2).
- Otter requires the clause x = x to be present when the inference rule paramodulation is used. (The operational reason is so that a contradiction can be found when $t \neq t$, for some t, is derived. The main reason is historical, due to the original definition of paramodulation.)
- Otter's rule for distinguishing variables from constants is that variables start with (lower case) u–z. Our convention for naming Skolem constants is to use A, B, C, ...
- For Otter, we use != instead of $\neq$.
- The denial of the conclusion is placed in the passive list, because we wish a forward proof; in this case it is safe in the passive list, because no derived clause can simplify it (Sec. 2.2.2).

If we run Otter with the preceding input file, it quickly finds a proof. We use a simple program to extract the proof from the output file and translate it into the following form.

Proof (found by Otter 3.0.4 on gyro at 0.14 seconds).

1	$A \cdot B \neq B \cdot A$	
4,3	$e \cdot x = x$	
7	$(x \cdot y) \cdot z = x \cdot (y \cdot z)$	
9	$x \cdot x = e$	
11	$x \cdot (x \cdot y) = y$	$[9 \rightarrow 7 : 4, \text{flip}]$
15	$x \cdot (y \cdot (x \cdot y)) = e$	$[9 \rightarrow 7, \text{flip}]$
18,17	$x \cdot e = x$	$[9 \rightarrow 11]$
25	$x \cdot (y \cdot x) = y$	$[15 \rightarrow 11 : 18, \text{flip}]$
29	$x \cdot y = y \cdot x$	$[25 \rightarrow 11]$
30	□	$[29,1]$

The clause numbers in the proof are not sequential, because they reflect the sequence of *retained* clauses. Some clauses are identified with two numbers; the first is cited when the clause is used as a demodulator, and the second when it is used with an inference rule (paramodulation or hyperresolution).

Each derived clause has a justification. The notation "$m \rightarrow n$" indicates paramodulation from m into n; "$: i, j, k, \ldots$" indicates rewriting with the demodulators $i, j, k, \ldots$; and "flip" indicates that equality was reversed (usually so that the complex side occurs on the left).

An Example with (gL)

Consider the theorem that a cancellative semigroup satisfying (gL) must be commutative. Otter finds a proof with the following input file.

```
%%%%%%%%%%%%%%%%%%%%%%  Basic options

op(400, xfx, [*,+,^,v,/,\,#]).  % Declare some infix operations.
op(300, yf, @).                 % Postfix operation.

clear(print_kept).        % Don't print retained clauses.
clear(print_new_demod).   % Don't print new demodulators.
clear(print_back_demod).  % Don't print back demodulated clauses.

assign(pick_given_ratio, 4).    % 4 best-first : 1 breadth-first.
assign(max_mem, 20000).         % Use at most 20 megabytes of RAM.

%%%%%%%%%%%%%%%%%%%%%%%  Standard for (gL) problems

set(geometric_rule).      % Apply (gL) as an inference rule.
set(geometric_rewrite_before).  % Apply (gL) as a rewrite rule
                                % before ordinary demodulation.
set(para_from).           % Paramodulate from the given clause.
set(para_into).           % Paramodulate into the given clause.
set(para_from_vars).      % Allow paramodulation into variables.
set(para_into_vars).      % Allow paramodulation from variables.
set(order_eq).            % Orient equalities.
set(back_demod).          % Apply back demodulation.
set(process_input).       % Process input clauses as if derived.
set(lrpo).                % Orient equalities with LRPO procedure.

%%%%%%%%%%%%%%%%%%%%%%%  Standard options for hyperresolution

set(output_sequent).         % Output nonunit cls. as implications.
set(hyper_res).              % Apply the inf. rule hyperresolution.
set(order_history).          % List given cl. first in just. list.
set(unit_deletion).          % Existing units simplify new nonunits.
set(para_from_units_only).   % Disable paramodulation from nonunits.
set(para_into_units_only).   % Disable paramodulation into nonunits.

%%%%%%%%%%%%%%%%%%%%%%%  Modifications to strategy

clear(dynamic_demod).        % Disable demodulation.

%%%%%%%%%%%%%%%%%%%%%%%  Clauses

list(usable).
x = x.
x * y != u | x * z != u | y = z.  % left cancellation
y * x != u | z * x != u | y = z.  % right cancellation
end_of_list.

list(sos).
```

```
(x * y) * z = x * (y * z).
end_of_list.

list(passive).
B * A != A * B.
end_of_list.
```

The first part of the input file, through "**Modifications to strategy**", is standard for (gL) searches with deduction rules. In this case, demodulation was found to constrain the search too much, so it was disabled. The preceding input file leads to the following proof.

Proof (found by Otter 3.0.4 on gyro at 26.99 seconds).

1	$B \cdot A = A \cdot B \rightarrow \square$	
3	$x \cdot y = z,\ x \cdot u = z \rightarrow y = u$	
4	$x \cdot y = z,\ u \cdot y = z \rightarrow x = u$	
5	$(x \cdot y) \cdot z = x \cdot (y \cdot z)$	
7	$((x \cdot (y \cdot z)) \cdot u) \cdot v = ((x \cdot y) \cdot z) \cdot (u \cdot v)$	$[5 \rightarrow 5]$
11	$(x \cdot (y \cdot z)) \cdot u = (x \cdot y) \cdot (z \cdot u)$	$[5 \rightarrow 5]$
1687	$((x \cdot (y \cdot z)) \cdot u) \cdot v = ((x \cdot y) \cdot u) \cdot (z \cdot v)$	$[(gL)\ 7]$
2076	$(x \cdot (y \cdot z)) \cdot u = (x \cdot z) \cdot (y \cdot u)$	$[11 \rightarrow 1687 : (gL)]$
2123	$(x \cdot y) \cdot (z \cdot u) = (x \cdot z) \cdot (y \cdot u)$	$[7 \rightarrow 1687 : (gL)]$
2887	$x \cdot (y \cdot z) = x \cdot (z \cdot y)$	$[4,11,2076]$
4307	$x \cdot y = y \cdot x$	$[3,2123,2887]$
4308	$\square$	$[4307,1]$

The justification "$[(gL)$" indicates the use of (gL) as an inference rule, and "$:(gL)$" indicates its use as a rewrite rule. Hyperresolution with a deduction rule is indicated by a sequence of clause identifiers at the beginning of the justification; for example, clauses 2887 and 4307 were derived by hyperresolution. Note also that the flag output_sequent causes the denial, clause 1, to be written as an implication. (The preceding theorem is Thm. CS-GL-1 in Sec. 4.1.4; it is a significant new result that demonstrates the machine-oriented nature of (gL)—the proof is short, but it is not intuitive and not one that would likely be found by a mathematician.)

2.2.8 Soundness of Otter

Otter is a large and complex program, with many optimizations and experimental features, coded in a low-level programming language (C). It has many hacks and kludges, and certainly some unknown bugs. Rather than to try to formally verify Otter, our approach to the problem of soundness is to have another program check Otter's proofs. This approach is gaining acceptance in the automated deduction community [2, 63], because verification of large low-level programs will not be practical within the next few years (decades?), but we are already starting to rely on automatically generated proofs.

If Otter's flag `build_proof_object` is set, and if it finds a proof, a very detailed *proof object* will be output along with the ordinary proof. The proof object is not meant for human consumption; its purpose is to be read and checked by an independent program, the *checker*. The proof object has enough detail and its steps are at such a low level that the checker can be a very simple program; for example, it doesn't have to handle term unification or matching, because variable substitutions are explicit and done as separate steps. We list here the proof object corresponding to the five-step proof above that $x^2 = e$ groups are commutative.

```
(
(1 (input) ((not (= (* (A) (B)) (* (B) (A)))))))
(2 (input) ((= (* (e) v0) v0)))
(3 (input) ((= (* (* v0 v1) v2) (* v0 (* v1 v2)))))
(4 (input) ((= (* v0 v0) (e))))
(5 (instantiate 4 ((v0 . v65))) ((= (* v65 v65) (e))))
(6 (instantiate 3 ((v0 . v65)(v1 . v65)(v2 . v66))) ((= (* (* v65 v65) v66) (* v65
    (* v65 v66)))))
(7 (paramod 5 (1 1) 6 (1 1 1)) ((= (* (e) v66) (* v65 (* v65 v66)))))
(8 (instantiate 2 ((v0 . v66))) ((= (* (e) v66) v66)))
(9 (paramod 8 (1 1) 7 (1 1)) ((= v66 (* v65 (* v65 v66)))))
(10 (flip 9 (1)) ((= (* v65 (* v65 v66)) v66)))
(11 (instantiate 10 ((v65 . v0)(v66 . v1))) ((= (* v0 (* v0 v1)) v1)))
(12 (instantiate 4 ((v0 . v64 v65))) ((= (* (* v64 v65) (* v64 v65)) (e))))
(13 (instantiate 3 ((v0 . v64)(v1 . v65)(v2 . (* v64 v65)))) ((= (* (* v64 v65)
    (* v64 v65)) (* v64 (* v65 (* v64 v65))))))
(14 (paramod 12 (1 1) 13 (1 1)) ((= (e) (* v64 (* v65 (* v64 v65))))))
(15 (flip 14 (1)) ((= (* v64 (* v65 (* v64 v65))) (e))))
(16 (instantiate 15 ((v64 . v0)(v65 . v1))) ((= (* v0 (* v1 (* v0 v1))) (e))))
(17 (instantiate 4 ((v0 . v65))) ((= (* v65 v65) (e))))
(18 (instantiate 11 ((v0 . v65)(v1 . v65))) ((= (* v65 (* v65 v65)) v65)))
(19 (paramod 17 (1 1) 18 (1 1 2)) ((= (* v65 (e)) v65)))
(20 (instantiate 19 ((v65 . v0))) ((= (* v0 (e)) v0)))
(21 (instantiate 16 ((v0 . v64))) ((= (* v64 (* v1 (* v64 v1))) (e))))
(22 (instantiate 11 ((v0 . v64)(v1 . (* v1 (* v64 v1))))) ((= (* v64 (* v64 (* v1
    (* v64 v1)))) (* v1 (* v64 v1)))))
(23 (paramod 21 (1 1) 22 (1 1 2)) ((= (* v64 (e)) (* v1 (* v64 v1)))))
(24 (instantiate 20 ((v0 . v64))) ((= (* v64 (e)) v64)))
(25 (paramod 24 (1 1) 23 (1 1)) ((= v64 (* v1 (* v64 v1)))))
(26 (flip 25 (1)) ((= (* v1 (* v64 v1)) v64)))
(27 (instantiate 26 ((v1 . v0)(v64 . v1))) ((= (* v0 (* v1 v0)) v1)))
(28 (instantiate 27 ((v0 . v64))) ((= (* v64 (* v1 v64)) v1)))
(29 (instantiate 11 ((v0 . v64)(v1 . (* v1 v64)))) ((= (* v64 (* v64 (* v1 v64)))
    (* v1 v64))))
(30 (paramod 28 (1 1) 29 (1 1 2)) ((= (* v64 v1) (* v1 v64))))
(31 (instantiate 30 ((v64 . v0))) ((= (* v0 v1) (* v1 v0))))
(32 (instantiate 31 ((v0 . (A))(v1 . (B)))) ((= (* (A) (B)) (* (B) (A)))))
(33 (resolve 1 (1) 32 (1)) ())
)
```

A checker for Otter proof objects has been written in the language of the Boyer-Moore logic [6, 7], and all of the non-(gL) Otter proofs in this work have been checked with it.[2] (The checker cannot yet handle (gL) inferences.)

---

[2] To be accepted by the proof checker, the proof objects must go through a round of translation, for example, to replace "*" and "=" with terms in the Boyer-Moore language.

Of course, we now have the question of the correctness of the checker. The goal of formally verifying the checker (say, with the Boyer-Moore prover) is certainly within reach, but that has not yet been attempted. However, it is just a few pages of high-level code, and it has been carefully read by an independent expert in the area.

2.3 MACE

The program MACE (Models And Counter-Examples) [37, 38] is used to search for finite models of formulas or sets of clauses. When the input is interpreted as the denial of a conjecture, any models found are counterexamples of the conjecture.

2.3.1 Use of MACE

MACE is easy to use. Its input language is the same as Otter's, and few decisions need to be made when using it; the search algorithm, unlike Otter's, is not adjustable by the user. To run MACE, the user takes the following steps.

1. Prepare a set of clauses. These are usually the same as the clauses prepared for Otter. If an equivalent set of simpler clauses is known, that should be used instead. (The key parameter is the length of the longest clause.) If very simple lemmas are known, they should be included as well.
2. Declare properties of operations. The two properties currently accepted by MACE are "quasigroup" for binary operations, and "bijection" for unary operations. These declarations speed the search; the clauses that specify the declared properties can be (and should be) omitted.
3. Assign constants. To reduce the number of isomorphic models in the search space, the user can assign constants in the clauses to elements of the domain. (If the domain has size n, its elements are always named $0, 1, \cdots, n - 1$.)
4. Specify a domain size. Each MACE search is for models of a fixed size. The user typically makes a sequence of searches, starting with a small size, then increasing it, until a model is found or until resources are exhausted. MACE is intended to be complete: if the search for a given domain size terminates without models, there should be none of that size.

MACE is quite limited, both in the size of clauses it can handle and in the size of domain it can search. The space it uses is proportional to n^{c+v}, where n is the domain size, c is the length of the longest clause, and v is the number of variables in the longest clause. If the clauses are very simple, MACE may be able to complete searches of sizes 10–12 in a reasonable time; examples of the longest clauses that MACE can handle are the following.

$$x \cdot y = u \cdot v, \; x \cdot w = z \cdot v \;\rightarrow\; u \cdot w = z \cdot y$$
$$x \cdot (((x \cdot y) \cdot z')' \cdot y)' = z$$

With such clauses, MACE can search for models of size at most 4 or 5.

An Example of a MACE Search

The following input file asks MACE to search for a noncommutative group.

```
op(400, xfx, [*,+,^,v,/,\,#]).  % Declare some infix operations.
op(300, yf, @).                 % Postfix operation.

list(usable).
0 * x = x.                      % left identity
x@ * x = 0.                     % left inverse
(x * y) * z = x * (y * z).      % associativity
1 * 2 != 2 * 1.                 % denial of commutativity
end_of_list.
```

By default, MACE recognizes integer constants as distinct elements of the domain. The group identity is assigned to element 0, and the noncommuting elements are 1 and 2. The user has reasoned a priori that the identity commutes with all other elements and that every element commutes with itself. Hence, no models are lost by using 1 and 2, which are distinct, to deny commutativity.

If we run a sequence of MACE searches specifying sizes $3, 4, \ldots$, the first model found has size 6, as expected, because the smallest noncommutative group is order 6. The following output is produced.

Model. The clauses have the following model (found by MACE 1.2.0 on gyro at 7.51 seconds).

```
*: | 0 1 2 3 4 5            @:  0 1 2 3 4 5
---+------------            ------------------
0 | 0 1 2 3 4 5                0 1 2 4 3 5
1 | 1 0 3 2 5 4
2 | 2 4 0 5 1 3
3 | 3 5 1 4 0 2
4 | 4 2 5 0 3 1
5 | 5 3 4 1 2 0
```

If we include the (dependent) clauses $\{x * 0 = x, x * g(x) = 0\}$, or if we declare * to be a quasigroup and @ to be a bijection, a model is found in slightly less time. If we use the ordinary constant e for the group identity and deny commutativity with $A * B \neq B * A$ instead of specifying the noncommuting elements with $1 * 2 \neq 2 * 1$, a model is found in about a minute, because MACE will try many isomorphic assignments for e, A, and B.

Although MACE is a primitive program, it has made several useful discoveries in this work. In some other cases where we cite its use, a mathematician familiar with the area would have found a model with little work. However, the program is useful to nonmathematicians, and model-searching programs are likely to become more powerful in the near future.

2.3.2 Soundness of MACE

MACE has not been formally verified in any way. Although models found by MACE can be checked without much difficulty (in fact, with a special Otter strategy), we have not done so in this work. A more difficult problem arises when the interest is in showing that there are no models of a given size [70]. In such cases the user has to trust MACE; but we have not faced that kind of problem in this work.

3. Algebras over Algebraic Curves

A beautiful result in classical algebraic geometry says that every compact complex manifold admits at most one algebraic structure and that every compact one-dimensional complex manifold has an algebraic structure and hence is unique. This is an application of the celebrated theorem of Chow, that the only complex analytic subsets of the complex projective space are algebraic varieties; see [50, Sec. 4B]. This theorem is an open-arm invitation by algebra to analysis, and vice versa, so that algebraic tools are automatically available to analytic subsets of projective space, and conversely, topological techniques are adaptable to geometrically defined algebras. In fact, the (gL)-implication considered in this chapter is one such important contribution to algebra by analysis and topology.

A good example of such an analytic subset of the complex projective plane is the so-called elliptic curves: these are compact complex Lie groups. The uniqueness of the group law over elliptic curves states that if $+$ and $\oplus$ are two group laws defined over an elliptic curve and if they have the same identity element, then $x + y = x \oplus y$. This is an easy consequence of the powerful rigidity lemma of projective curves (see [48, 49]).

3.1 What Is a Uniqueness Theorem?

Patterned after this uniqueness of group laws, we formulate the natural concept of a *uniqueness theorem*: Let $\mathcal{A}$ be a mathematical structure, F a class of functions defined on $\mathcal{A}$, and Σ a set of sentences in the first-order theory of a given type, say τ. A uniqueness theorem for the triple $\langle \mathcal{A}, F, \Sigma \rangle$ states that if f and g are two n-ary functions belonging to the class F and if the reducts $\langle \mathcal{A}; f \rangle$ and $\langle \mathcal{A}; g \rangle$ both satisfy the properties in Σ, then $f = g$. Examples are listed in Table 3.1.

3.1.1 The Rigidity Lemma

An elliptic curve, viewed as a plane algebraic curve in the complex projective plane, is given by a nonsingular cubic equation. In this chapter, we bring out the equational properties of the algebraic laws naturally definable on cubic

Table 3.1. Examples of Uniqueness of Algebraic Laws

Structure $\mathcal{A}$	Class of Functions F	Laws Σ	Status
Elliptic curves	morphisms	group laws	True
Elliptic curves	n-ary morphisms	n-ary Steiner laws	True [a]
Elliptic curves	n-ary morphisms	n-group laws	Open
Elliptic curves	ternary morphisms	Mal'cev laws	True
Groups	group words	group laws	False
Abelian groups	group words	group laws	True
R, real field	polynomials	group laws	True
R, real field	polynomials	Mal'cev + cancel.	True
R, real field	polynomials	Steiner laws	True
C, complex field	polynomials	Steiner laws	False
$R \times R$, the plane	ring polynomials	group laws	True
$R \times R$, the plane	ring polynomials	Mal'cev + cancel.	Open
$R \times R$, the plane	ring polynomials	Steiner laws	Open
$R \times R \times R$	ring polynomials	group laws	False [b]
Unit circle	continuous	group laws	True [c]
Unit circle	continuous	Mal'cev laws	Open
Join semilattice	binary functions	lattice meet	True
Join semilattice	binary functions	quasilattice meet	False [d]
$\{x+x=x, x+y=y+x\}$	binary functions	WA-lattice meet	True [e]
$\{x+x=x, x+y=y+x\}$	binary functions	TN-lattice meet	False [f]

[a] First proved by Otter for the cases $n = 5, 8$; see Sec. 3.5.

[b] For example, $(x_1, x_2, x_3) + (y_1, y_2, y_3) = (x_1 + y_1, x_2 + y_2 + x_1 + y_3, x_3 + y_3)$ is a noncommutative group law defined on $R \times R \times R$.

[c] A rule of inference analogous to (gL) is not known for continuous functions on circles.

[d] Example obtained by MACE; see Example QLT-7.

[e] First proof obtained by Otter; see Thm. WAL-2.

[f] Example obtained by MACE; see Example TNL-2.

curves (i.e., by means of synthetic constructions) from the point of view of equational logic. One may wonder what equational logic, a topic in universal algebra, has to do with cubic curves, a topic in classical algebraic geometry.

The subject of elliptic curves ignores interdisciplinary boundaries; it is a place where rationally defined quasigroup laws, certain combinatorial configurations, formal groups, number theory, incidence theorems, and coding theory all happily coexist. We say that an algebraic curve Γ admits an algebraic law (of arity n), if there exists an n-ary morphism $(x_1, x_2, \cdots, x_n)$ on the curve Γ; that is, $(x_1, x_2, \cdots, x_n)$ is the completion of a regular function on the product set $\Gamma_n = \Gamma \times \Gamma \times \cdots \times \Gamma$. The nonsingular cubic curves are rife with a number of such algebraic laws all of which are morphisms of the curve. For example, the famous Cayley-Bacharach theorem of classical algebraic geometry says that every algebraic curve induces a rational operation on cubic curves via a complete intersection cycle (e.g., see Fig. 3.7, p. 60, for the 5-ary conic process). Let us list a few equational properties enjoyed by such algebraic laws defined on nonsingular cubic curves.

1. Any two group laws sharing the same identity element are equal.
2. Any group law definable on an elliptic curve must be commutative.
3. Every quasigroup law definable on such a curve gives rise to a group law.
4. Every cancellative di-associative groupoid on an elliptic curve must be associative.
5. Every groupoid having a two-sided identity element is a group law.
6. Any two binary Steiner quasigroup laws having a common idempotent are equal; that is,

$$\left\{ \begin{array}{l} f(x, f(y, x)) = y, \ f(e, e) = e \\ g(x, g(y, x)) = y, \ g(e, e) = e \end{array} \right\} \ =\!(gL)\!\Rightarrow\ \{f(x, y) = g(x, y)\}.$$

7. Any two 5-ary Steiner quasigroup laws having a common idempotent are equal (see Lem. UAL-6, p. 57 below for details).

One of the purposes of this project is to develop a new geometric theorem-prover based on certain local-to-global principles of classical algebraic geometry so that we can formulate and prove properties like the above purely within the realm of first-order logic with equality. To this end we need additional context-sensitive rules of inference. One such rule is the famous rigidity lemma for the morphisms of projective curves, in particular for elliptic curves (see, e.g., [48, p. 104], [49, p. 43], or [68, p. 152]):

$$\exists y_0 z_0 \forall \mathbf{x}(f(\mathbf{x}, y_0) = z_0) \Rightarrow \forall \mathbf{x} z y(f(\mathbf{x}, y) = f(\mathbf{z}, y)). \qquad \text{(rigidity)}$$

That is, from local equality $f(\mathbf{x}, y_0) = z_0$ for some term f and some elements y_0, z_0, we can derive the global multivariable identity $f(\mathbf{x}, y) = f(\mathbf{z}, y)$, in which $\mathbf{x}$ and $\mathbf{z}$ can be vectors of variables.

A universal algebra $\langle A; F \rangle$ is called a (gL)-algebra if it satisfies the above rigidity implication for all operations f in its clone. Also, we write

$$\{\Sigma\} \ =\!(\text{rigidity})\!\Rightarrow\ \{\sigma\}$$

if σ is a formal consequence of $\{\Sigma, (\text{rigidity})\}$.

On an elementary level, the rigidity lemma is clearly valid for the so-called affine algebras: $p(x_1, x_2, \cdots, x_m) = n_i x_i + k$, where the n_i's are group homomorphisms of a group $\langle G; + \rangle$. The rigidity lemma says essentially that if an identity happens to be true in a dense open subset of a topological space, then it is globally valid.

In fact, this is the essence of almost all proofs of the associativity of the group law on cubics (see, e.g., [66, p. 36]). One first proves the associativity for distinct points (using, say the Bezout theorem), then notices that $+$ being a morphism is a continuous function, then appeals to continuity to extend the validity of the associativity to all points; that is, the two ternary continuous functions $(x + y) + z$ and $x + (y + z)$ agree on a dense open subset of the curve, and hence they agree everywhere. Such local-to-global principles have been well known in function theory and algebraic geometry for a long time.

3.1.2 Applications to Cubic Curves

Since nonsingular cubic curves (and hence, elliptic curves) over the complex projective plane (with n-ary morphisms as their algebraic operations) do satisfy this rigidity principle, any formal consequence valid under rigidity is true for these curves. As an example of the rigidity lemma, let us show a rather syntactic proof of the powerful four-variable median law $(xy)(zu) = (xz)(yu)$ for the classical binary morphism of chord-tangent construction (Fig. 3.2, p. 41) on nonsingular cubics from just the relatively weak two-variable Steiner quasigroup laws $\{x(yx) = y,\ (yz)z = y\}$ without any reference to the geometry or the topology of curves.

Theorem MED-2. Median law for Steiner quasigroups.

$$\left\{ \begin{array}{c} x(yx) = y \\ (yz)z = y \end{array} \right\} = (\text{rigidity}) \Rightarrow \{(xy)(zw) = (xz)(yw)\}.$$

Proof. Define the 5-ary composite operation f by

$$f(x, y, z, w, u) = ((xy)(zw))(u((xz)(yw))).$$

Now we have, by the law $x(yx) = y$, $f(x, y, y, w, u) = u$ for all x. Thus by the rigidity property, the 5-ary expression $f(x, y, z, w, u)$ does not depend upon x for all y, z, w, u. In particular, we have

$$
\begin{array}{lll}
f(x, y, z, w, u) & = f(x_1, y, z, w, u) & \forall x \forall x_1 \\
((xy)(zw))(u((xz)(yw))) & = ((x_1 y)(zw))(u((x_1 z)(yw))) & \forall x \forall x_1 \\
 & = (((yz)y)(zw))(u(((yz)z)(yw))) & \text{letting } x_1 = yz \\
 & = w(uw) & \text{by the Steiner laws} \\
 & = u & \\
 & = ((xz)(yw))(u((xz)(yw))), &
\end{array}
$$

and hence one right-cancellation of the common term $(u((xz)(yw)))$ immediately yields the desired median law $(xy)(zw) = (xz)(yw)$. Q.E.D.

Since a nonsingular cubic curve defined over an algebraically closed field is a projective curve and since the chord-tangent law of composition obviously satisfies the Steiner laws, we have the following.

Corollary. *The classical binary morphism of chord tangent construction defined on a nonsingular cubic curve is medial* (see Fig. 3.1).

Historical remark. This important identity for cubics usually is proved by using some sophisticated machinery such as the Bezout theorem, the Riemann-Roch theorem, or Weierstrass elliptic functions. It was first proved for plane cubic curves by I. M. S. Etherington using the Bezout theorem (see [11]). In [55], Padmanabhan gave a proof for elliptic curves over an arbitrary algebraically closed field. See also an elaborate proof of this identity by using the important concept of intersection multiplicities in [26, pp. 68–70]. The associativity of the classical group law is a simple consequence of this identity.

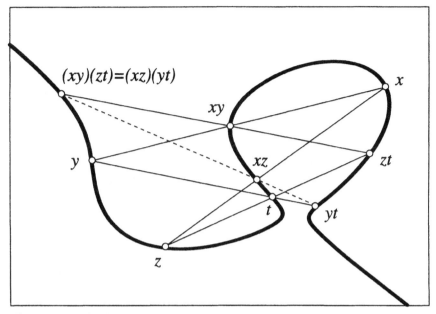

Fig. 3.1. The Median Law

Now let us connect this basic binary operation of chord-tangent construction with the classical group law on elliptic curves.

It turns out that the rigidity property is closely related to the so-called term condition, a universally quantified implication in universal algebra recently discovered while attempting to generalize the group theory concept of commutators to modular congruence varieties (for a history and comparison, see [65] and [44, Abelian algebras]). In fact, modulo nonsingular cubic curves over an algebraically closed field, these two properties turn out to be equivalent (see Thm. 3.3 below). Thus, hindsight suggests that the term condition, called (gL) here (for "geometric Logic") does not originate in universal algebra but comes from the classical algebraic geometry of complex projective varieties. After all, the uniqueness of group law for elliptic curves has been known for a long time.

The following three properties are of interest to us.

$$F(\mathbf{x}, b) = c \Rightarrow F(\mathbf{x}, z) = F(\mathbf{y}, z) \qquad \text{(rigidity)}$$
$$F(a, \mathbf{x}) = F(a, \mathbf{y}) \Rightarrow F(z, \mathbf{x}) = F(z, \mathbf{y}) \qquad \text{(weak-}gL)$$
$$F(a, b) = F(a, c) \Rightarrow F(z, b) = F(z, c) \qquad (gL)$$

Here, $\mathbf{x}$, and $\mathbf{y}$ are vectors of variables, and a, b, and c are terms not containing members of $\mathbf{x}$ or $\mathbf{y}$. It is clear that $(gL) \Rightarrow (\text{weak-}gL) \Rightarrow (\text{rigidity})$.

Theorem 3.1. *Let C be an elliptic curve in the complex projective plane. Then the morphisms of C satisfy the property* (weak-gL).

Proof. Without loss of generality we can assume that C is a nonsingular cubic curve, because it is well known that a biregular correspondence always exists between an elliptic curve and a smooth cubic in the complex projective plane (e.g., [50, Corollary 7.8]).

Let $F : C \times C \longrightarrow C$ be a binary morphism such that $F(a, x) = F(a, y)$ for some point a in C. Form the composite ternary morphism $f : C \times C \times C \longrightarrow C$ by the rule $f(x, y, z) = (F(z, x) \cdot e) \cdot F(z, y)$ where $\cdot$ is the synthetic binary morphism of chord-tangent construction on the cubic, and e is an inflection point which we choose to be the origin of the group law. Now,

$$\begin{aligned}
f(x, y, a) &= (F(a, x) \cdot e) \cdot F(a, y) \\
&= (F(a, x) \cdot e) \cdot F(a, x) \qquad \text{[since } F(a, x) = F(a, y)\text{]} \\
&= e \qquad\qquad\qquad\qquad\quad \text{[since } (u \cdot e) \cdot u = e\text{]}
\end{aligned}$$

Thus we have

$$f(x, y, a) = e \qquad\qquad\qquad [\forall x, y \in C]$$

and hence,

$$\begin{aligned}
f(x, y, z) &= f(u, v, z) \qquad\qquad\qquad \text{[by rigidity]} \\
&= f(e, e, z) \qquad\qquad\quad\; \text{[letting } v = u = e\text{]} \\
&= (F(z, e) \cdot e) \cdot F(z, e) \\
&= e.
\end{aligned}$$

In other words, $(F(z, x) \cdot e) \cdot F(z, x) = (F(z, x) \cdot e) \cdot F(z, y)$ and thus, after one left cancellation, we get the desired equality $F(z, x) = F(z, y)$ for all x, y, and z. Q.E.D.

Let us illustrate this deduction procedure with a typical example. Suppose we wish to find all binary morphisms $s(x, y)$ of an elliptic curve $\langle C; e \rangle$ satisfying the two equational identities $s(x, e) = x$ and $s(x, x) = e$. One obvious candidate for $s(x, y)$ is subtraction $x - y$, where the corresponding group addition has e as its identity. It turns out that this is the only possibility.

Theorem 3.2.

$$\left\{ \begin{array}{l} s(x, e) = x \\ s(x, x) = e \end{array} \right\} = \text{(rigidity)} \Rightarrow \{ s(x, y) = x - y \},$$

where $x - y$ is the usual subtraction on C corresponding to the group law $+$ having e as its identity element.

First Proof. Define $F(x, z) = s(x, s(x, z))$. We have $F(x, e) = s(x, s(x, e)) = s(x, x) = e$. Hence, by the rigidity property, we have $F(x, z) = F(y, z)$ for all x, y, and z. In particular, we have $F(x, z) = F(z, z) = s(z, (s(z, z)) = s(z, e) = z$. Thus we have $s(x, s(x, z)) = z$. Now define $m(x, y) = s(x, s(e, y))$. It is clear that $m(x, e) = x$ and $m(e, y) = s(e, s(e, y)) = y$. Appealing to [49, appendix to Sec. 4], we obtain the equality $m(x, y) = x + y$, the unique group law with e as its identity. Thus $s(x, s(e, y)) = x + y$. Now,

$$x + s(e, x) = s(x, s(e, s(e, x))) = s(x, x) = e,$$

and hence $s(e, x) = -x$. Finally, $x - y = x + (-y) = s(x, s(e, -y)) = s(x, s(e, s(e, y))) = s(x, y)$. Q.E.D.

Second Proof. See Thm. ABGT-5 (p. 64) where Otter shows that $\{s(x, e) = x, s(x, x) = e\}$ is a (gL)-basis for the equational theory of Abelian groups where $s(x, y) = x - y$.

Theorem 3.3. *Let C be a nonsingular cubic curve in the complex projective plane. Then the morphisms of C satisfy the rule (gL).*

Proof. Let F be an arbitrary binary morphism of a nonsingular cubic curve $\langle C; e \rangle$, and let $F(a, b) = F(a, c)$. Form the composite 4-ary morphism $h(x, y, z, u) = F(x, y) \cdot (u \cdot F(x, z))$. Now, $h(x, e, e, e) = F(x, e) \cdot (e \cdot F(x, e)) = e$; hence, by rigidity, $h(x, y, z, u) = h(w, y, z, u)$, and thus it does not depend upon x. Thus, in particular, we have the equality $F(x, y) \cdot (u \cdot F(x, z)) = F(a, y) \cdot (u \cdot F(a, z))$. So,

$$
\begin{aligned}
F(x, b) \cdot (u \cdot F(x, c)) &= F(a, b) \cdot (u \cdot F(a, c)) \\
&= F(a, c) \cdot (u \cdot F(a, c)) \\
&= u \\
&= F(x, c) \cdot (u \cdot F(x, c)).
\end{aligned}
$$

Since $\cdot$ is cancellative on both sides, we immediately get the desired equality $F(x, b) = F(x, c)$. Q.E.D.

The property (gL) is implemented as an inference rule in Otter (see Sec. 2.2.4 for details), and we write $=(gL)\Rightarrow$ for model-theoretic consequence under (gL). The rule derives an equation from an equation, so it is strictly within the language of equational logic, yet it captures the essence of the rigidity principle valid for complex projective curves. Using the rule, Otter can find formal equational proofs for the uniqueness of many geometrically constructed operations over nonsingular cubic curves and obtain elegant equational characterizations for these synthetic geometric constructions. It is this blend of universal algebra, algebraic geometry, and computer science on which we focus; it provides a rich source of examples for universal algebra and a new equational tool for algebraic geometry.

We now present in detail an example of a (gL)-derivation. One of the beauties of the group law on a nonsingular cubic curve C is that it is harmoniously blended with the geometry on the curve; in particular, $a + b + c = e$ if and only if there exists a line L such that $C \cap L = \{a, b, c\}$. Expressing this in our algebraic language, we have $a + b = e \cdot (a \cdot b)$. In [49, appendix to Sec. 4], D. Mumford and C. P. Ramanujam show that if $m(x, y)$ is a binary morphism of a complete variety and if m admits a two-sided identity, then it must be an Abelian group morphism. We capture a fragment of this deep theorem in the following implication.

$$
\left.
\begin{cases}
m(x, e) = x \\
m(e, x) = x \\
x \cdot (y \cdot x) = y \\
x \cdot y = y \cdot x
\end{cases}
\right\} =(gL)\Rightarrow \{m(x, y) = e \cdot (x \cdot y)\}.
$$

(Note that the hypotheses assert no connection at all between m and $\cdot$.)

Proof (found by Otter 3.0.4 on gyro at 8.07 seconds).

2	$m(x, e) = x$	
3	$m(e, x) = x$	
4	$x \cdot (y \cdot x) = y$	
5	$x \cdot y = y \cdot x$	
6	$m(A, B) \neq e \cdot (A \cdot B)$	
11	$m(e, x) \cdot (y \cdot x) = y$	[3 → 4]
17	$x \cdot m(e, y \cdot x) = y$	[3 → 4]
28,27	$m(e, x \cdot y) = y \cdot x$	[3 → 5]
29	$x \cdot (x \cdot y) = y$	[17 :28]
35	$m(x, y) \cdot (z \cdot y) = m(x, u) \cdot (z \cdot u)$	[11 → 11 :(gL)]
53	$x \cdot (y \cdot e) = m(x, z) \cdot (y \cdot z)$	[2 → 35]
106	$m(x, y) \cdot (m(x, z) \cdot y) = z$	[35 → 29]
180,179	$m(x, y) \cdot (z \cdot y) = x \cdot (e \cdot z)$	[5 → 53, flip]
194	$x \cdot (e \cdot m(x, y)) = y$	[106 :180]
225	$e \cdot m(x, y) = x \cdot y$	[194 → 29, flip]
231	$m(x, y) = e \cdot (x \cdot y)$	[225 → 29, flip]
233	$\square$	[231,6]

In Thm. ABGT-3 (p. 50), we show as well that each element has an inverse.

The logic of this project is now clear. First we transform the geometric concepts and statements associated with an elliptic curve into equivalent algebraic ones via a system of equational definitions. See Fig. 3.2 and Table 3.2.

Table 3.2. Equationally Definable Concepts in Cubic Curves

Geometry	Algebra
Points	elements
Chords	$\{(P, Q) \mid P, Q \text{ elements}\}$
Tangents	$\{(P, P) \mid P \text{ an element}\}$
P, Q, R collinear	$R = P \cdot Q$
P lies on the tangent at Q	$P = Q \cdot Q$
Inflection point	$\{P \mid P \cdot P = P\}$ (idempotent elements)
Singular point	$\{P \mid \forall Q, \ P \cdot Q = P\}$ (absorbing element)
Nonsingular curve	no absorbing elements
Conic	$\{(P, Q, R, S, T)\}$ (all 5-tuples)
Sextatic point P	$\{P \mid ((P \cdot P) \cdot (P \cdot P)) \cdot P = P\}$
Configuration theorem	equation or implication
Bitangents PQ	$2P + 2Q = 0$ under the induced group law (for elliptic quartics)

After transforming the geometric statements into the language of algebra, we formally derive the equational conditions that correspond to the required

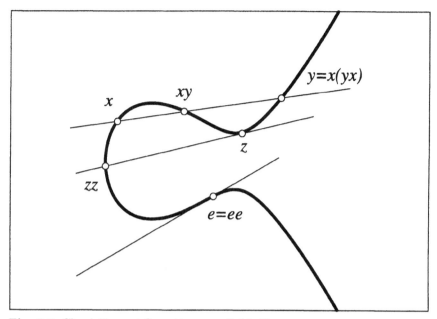

Fig. 3.2. Chord-Tangent Operation on a Cubic Curve

geometric conclusions. Let us start with a simple example of a well-known geometric property.

The following configuration theorem is a simple consequence of a Steiner law and (gL). The conclusion $S \cdot S = Q \cdot R$ means that the tangent at S and the chord QR meet the cubic at the same point. Thus, if QR meets the curve again at, say T, then the line ST is the tangent at S; see Fig. 3.3. This remains true even if the point T happens to be the point at infinity. In this case, the tangent at S is "parallel" to S for the denizens of the affine plane but we are still using only a ruler construction because in the projective plane the point at infinity is just like any other point in the plane; see Fig. 3.4.

Theorem GEO-1. Tangent construction.

$$\{x \cdot (y \cdot x) = y\} =\!(gL)\!\Rightarrow \left\{ \begin{bmatrix} A_1 \cdot A_2 = Q \\ B_1 \cdot B_2 = R \\ A_1 \cdot B_1 = S \\ A_2 \cdot B_2 = S \end{bmatrix} \rightarrow S \cdot S = Q \cdot R \right\}.$$

(In the statement of the theorem, $A_1, A_2, B_1, B_2, P, Q, R$ are variables; in the proof below, they are Skolem constants.)

Proof (found by Otter 3.0.4 on gyro at 27.17 seconds).

1	$S \cdot S \neq Q \cdot R$
3	$x \cdot (y \cdot x) = y$
4	$A_1 \cdot A_2 = Q$

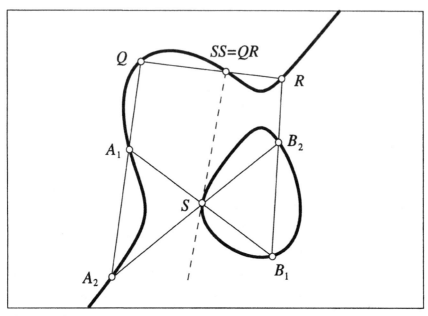

Fig. 3.3. Configuration I for Thm. GEO-1

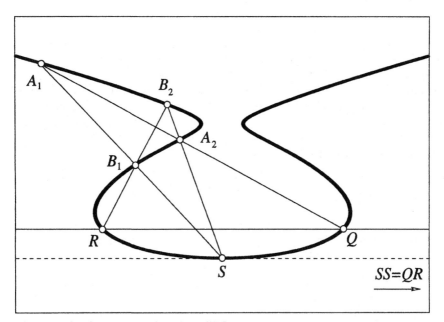

Fig. 3.4. Configuration II for Thm. GEO-1

5	$B_1 \cdot B_2 = R$	
6	$A_1 \cdot B_1 = S$	
7	$A_2 \cdot B_2 = S$	
13	$(x \cdot y) \cdot x = y$	$[3 \to 3]$
18	$A_2 \cdot Q = A_1$	$[4 \to 3]$
25	$B_2 \cdot R = B_1$	$[5 \to 3]$
32	$B_1 \cdot S = A_1$	$[6 \to 3]$
40	$B_2 \cdot S = A_2$	$[7 \to 3]$
134	$(B_2 \cdot R) \cdot S = A_1$	$[25 \to 32]$
171	$(B_2 \cdot S) \cdot Q = A_1$	$[40 \to 18]$
1849	$(x \cdot S) \cdot Q = (x \cdot R) \cdot S$	$[134 \to 171 :(\text{gL})]$
2161	$(Q \cdot R) \cdot S = S$	$[13 \to 1849, \text{flip}]$
6976	$S \cdot S = Q \cdot R$	$[2161 \to 3]$
6977	$\square$	$[6976,1]$

A second example is that D_{10}, the 10-point Desargues configuration, cannot be embedded in a nonsingular cubic curve. As shown in the proof of Thm. GEO-2 below (see Fig. 3.5), if D_{10} can be embedded in such a cubic, then $x \cdot x = O \cdot O$ for all the 10 points x in the configuration. In other words, the tangents at all these points are concurrent, and they meet at $O \cdot O$ on the curve. However, it is well known that cubic curves are of class 6; that is, at most six tangents can be drawn from a given point on the curve. Hence we have a contradiction. See [47] for more details and for an actual representation of the Desargues configuration on a *singular* cubic curve over a finite field.

Theorem GEO-2. Desargues configuration on a cubic curve.

$$\left\{ \begin{array}{l} x \cdot y = y \cdot x \\ x \cdot (y \cdot x) = y \end{array} \right\} \Rightarrow (gL) \Rightarrow \left\{ \left[\begin{array}{l} A \cdot P = O \\ B \cdot Q = O \\ C \cdot R = O \\ P \cdot Q = W \\ P \cdot R = V \\ Q \cdot R = U \\ A \cdot B = W \\ A \cdot C = V \\ B \cdot C = U \\ U \cdot V = W \end{array} \right] \to \left[\begin{array}{l} A \cdot A = O \cdot O \\ B \cdot B = O \cdot O \\ C \cdot C = O \cdot O \\ P \cdot P = O \cdot O \\ Q \cdot Q = O \cdot O \\ R \cdot R = O \cdot O \\ U \cdot U = O \cdot O \\ V \cdot V = O \cdot O \\ W \cdot W = O \cdot O \end{array} \right] \right\}.$$

Proof (found by Otter 3.0.4 on gyro at 103.17 seconds).

1        $x = x$
2        $O \cdot O = A \cdot A,\ O \cdot O = B \cdot B,\ O \cdot O = C \cdot C,\ P \cdot P = O \cdot O,$
         $Q \cdot Q = O \cdot O,\ R \cdot R = O \cdot O,\ U \cdot U = O \cdot O,$
         $V \cdot V = O \cdot O,\ W \cdot W = O \cdot O\ \to\ \square$
3        $x \cdot y = y \cdot x$

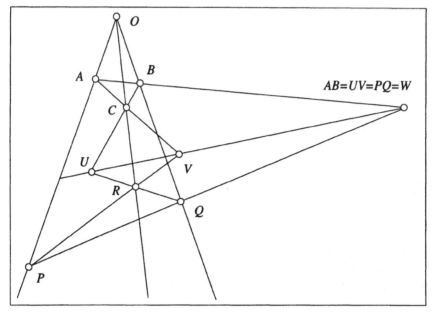

$AB=UV=PQ=W$

Fig. 3.5. Desargues Configuration, Thm. GEO-2

4	$x \cdot (y \cdot x) = y$	
7,6	$A \cdot P = O$	
9,8	$B \cdot Q = O$	
11,10	$C \cdot R = O$	
12	$P \cdot Q = W$	
14	$P \cdot R = V$	
16	$Q \cdot R = U$	
18	$A \cdot B = W$	
21,20	$A \cdot C = V$	
23,22	$B \cdot C = U$	
24	$U \cdot V = W$	
26	$(x \cdot y) \cdot z = z \cdot (y \cdot x)$	$[3 \rightarrow 3]$
27	$x \cdot (y \cdot z) = (z \cdot y) \cdot x$	$[3 \rightarrow 3]$
28	$P \cdot A = O$	$[3 \rightarrow 6]$
33,32	$R \cdot C = O$	$[3 \rightarrow 10]$
38	$Q \cdot W = P$	$[12 \rightarrow 4]$
43,42	$Q \cdot O = B$	$[8 \rightarrow 4]$
44	$P \cdot O = A$	$[6 \rightarrow 4]$
46	$(x \cdot y) \cdot x = y$	$[4 \rightarrow 4]$
48	$x \cdot (x \cdot y) = y$	$[3 \rightarrow 4]$
50	$(x \cdot y) \cdot y = x$	$[3 \rightarrow 4]$
54	$R \cdot V = P$	$[14 \rightarrow 4]$

58	$R \cdot U = Q$	$[16 \to 4]$
60	$B \cdot A = W$	$[3 \to 18]$
62	$B \cdot W = A$	$[18 \to 4]$
64	$C \cdot A = V$	$[3 \to 20]$
69,68	$C \cdot B = U$	$[3 \to 22]$
71,70	$C \cdot U = B$	$[22 \to 4]$
73,72	$V \cdot U = W$	$[3 \to 24]$
75,74	$V \cdot W = U$	$[24 \to 4]$
77,76	$A \cdot O = P$	$[28 \to 4]$
81,80	$C \cdot O = R$	$[32 \to 4]$
86	$(x \cdot y) \cdot (z \cdot u) = (u \cdot z) \cdot (y \cdot x)$	$[3 \to 26]$
88	$(x \cdot (y \cdot z)) \cdot (z \cdot y) = x$	$[4 \to 26, \text{flip}]$
102	$W \cdot Q = P$	$[3 \to 38]$
123,122	$V \cdot R = P$	$[3 \to 54]$
125,124	$V \cdot P = R$	$[54 \to 4]$
134	$U \cdot R = Q$	$[3 \to 58]$
139,138	$A \cdot W = B$	$[60 \to 4]$
140	$W \cdot B = A$	$[3 \to 62]$
150	$V \cdot C = A$	$[64 \to 46]$
161,160	$B \cdot U = C$	$[68 \to 4]$
163,162	$W \cdot V = U$	$[72 \to 46]$
169,168	$((x \cdot y) \cdot z) \cdot ((z \cdot (y \cdot x)) \cdot u) = u$	$[27 \to 48]$
194	$(x \cdot y) \cdot (z \cdot u) = (x \cdot z) \cdot (y \cdot u)$	$[(\text{gL}) \ 86]$
196	$(V \cdot x) \cdot (C \cdot y) = A \cdot (x \cdot y)$	$[150 \to 194, \text{flip}]$
640,639	$(V \cdot x) \cdot y = A \cdot (x \cdot (C \cdot y))$	$[48 \to 196]$
646,645	$U \cdot U = A \cdot A$	$[140 \to 196 : 75,69]$
647	$W \cdot O = A \cdot Q$	$[134 \to 196 : 73,11]$
651	$U \cdot (C \cdot Q) = O$	$[102 \to 196 : 75,7]$
653	$P \cdot B = A \cdot Q$	$[58 \to 196 : 123,71]$
658,657	$R \cdot R = A \cdot A$	$[44 \to 196 : 125,81]$
659	$P \cdot (C \cdot C) = P$	$[32 \to 196 : 123,77]$
666	$O \cdot O = A \cdot A, \ O \cdot O = B \cdot B, \ O \cdot O = C \cdot C, \ P \cdot P = O \cdot O,$	
	$Q \cdot Q = O \cdot O, \ V \cdot V = O \cdot O, \ W \cdot W = O \cdot O \ \to \ \square$	
		$[2 : 658,646]$
676	$U \cdot (A \cdot A) = U$	$[645 \to 48]$
690	$W \cdot (A \cdot Q) = O$	$[647 \to 48]$
710	$O \cdot (Q \cdot C) = U$	$[651 \to 88]$
715	$O \cdot (C \cdot Q) = U$	$[651 \to 50]$
733	$(A \cdot Q) \cdot B = P$	$[653 \to 50]$
736,735	$P \cdot (A \cdot Q) = B$	$[653 \to 48]$
738,737	$(A \cdot Q) \cdot P = B$	$[653 \to 46]$
763,762	$P \cdot P = C \cdot C$	$[659 \to 48]$
764	$O \cdot O = A \cdot A, \ O \cdot O = B \cdot B, \ O \cdot O = C \cdot C, \ Q \cdot Q = O \cdot O,$	
	$V \cdot V = O \cdot O, \ W \cdot W = O \cdot O \ \to \ \square$	$[666 : 763]$

801	$W \cdot (A \cdot x) = U \cdot (C \cdot x)$	$[651 \to 690 :(gL)]$
811	$U \cdot (C \cdot C) = U$	$[160 \to 801 :161,21,163, \text{flip}]$
816,815	$W \cdot x = U \cdot (C \cdot (A \cdot x))$	$[48 \to 801]$
827	$O \cdot O = A \cdot A,\ O \cdot O = B \cdot B,\ O \cdot O = C \cdot C,\ Q \cdot Q = O \cdot O,$	
¿	$V \cdot V = O \cdot O \to \square$	$[764 :816,139,69,646]$
951	$B \cdot (Q \cdot A) = P$	$[27 \to 733]$
1034	$U \cdot (x \cdot C) = O \cdot (x \cdot Q)$	$[715 \to 811 :(gL)]$
1036,1035	$U \cdot (C \cdot x) = O \cdot (Q \cdot x)$	$[710 \to 811 :(gL)]$
1038,1037	$x \cdot (C \cdot C) = x \cdot (A \cdot A)$	$[676 \to 811 :(gL)]$
1087	$O \cdot (Q \cdot Q) = O$	$[651 :1036]$
1099	$P \cdot (A \cdot A) = P$	$[659 :1038]$
1112,1111	$C \cdot C = A \cdot A$	$[762 \to 168 :763,1038,169, \text{flip}]$
1121	$O \cdot O = A \cdot A,\ O \cdot O = B \cdot B,\ Q \cdot Q = O \cdot O,$	
	$V \cdot V = O \cdot O \to \square$	$[827 :1112]$
1123,1122	$P \cdot P = A \cdot A$	$[762 :1112]$
1131,1130	$O \cdot O = A \cdot A$	$[737 \to 1034 :738,23,646,9, \text{flip}]$
1158	$B \cdot B = A \cdot A,\ Q \cdot Q = A \cdot A,\ V \cdot V = A \cdot A \to \square$	
		$[1121 :1131,1131,1131,1131 :1, \text{flip}]$
1234,1233	$Q \cdot Q = A \cdot A$	$[1087 \to 48 :1131, \text{flip}]$
1252	$B \cdot B = A \cdot A,\ V \cdot V = A \cdot A \to \square$	$[1158 :1234 :1]$
1253	$B \cdot (Q \cdot x) = P \cdot (A \cdot x)$	$[951 \to 1099 :(gL), \text{flip}]$
1266,1265	$B \cdot B = A \cdot A$	$[32 \to 1253 :33,43,77,1123]$
1282	$V \cdot V = A \cdot A \to \square$	$[1252 :1266 :1]$
1441	$(A \cdot (V \cdot (C \cdot x))) \cdot x = A \cdot A \to \square$	$[50 \to 1282 :640]$
1449	$A \cdot A = A \cdot A \to \square$	$[735 \to 1441 :736,69,73,139,1266]$
1450	$\square$	$[1449,3]$

3.2 The Median Law

We now return to the median law. The following three theorems, all first proved by Otter, show that the median law can be (gL)-derived from weaker hypotheses than those in Thm. MED-2.

Theorem MED-3. Median law for chord-tangent construction (2).

$$\{x(yx) = y\} \ =\!(gL)\!\Rightarrow\ \{(xy)(zu) = (xz)(yu)\}.$$

Proof (found by Otter 3.0.4 on gyro at 24.07 seconds).

2	$x \cdot (y \cdot x) = y$	
3	$(A \cdot C) \cdot (B \cdot D) \neq (A \cdot B) \cdot (C \cdot D)$	
4	$(x \cdot (y \cdot x)) \cdot (z \cdot y) = z$	$[2 \to 2]$
9	$(x \cdot y) \cdot x = y$	$[2 \to 2]$
38	$(x \cdot (y \cdot x)) \cdot (z \cdot y) = (u \cdot z) \cdot u$	$[9 \to 4]$
369	$(x \cdot (y \cdot z)) \cdot (u \cdot y) = (x \cdot u) \cdot z$	$[(gL)\ 38]$

| 371 | $(x \cdot y) \cdot (z \cdot u) = (x \cdot z) \cdot (y \cdot u)$ | $[2 \to 369]$ |
| 372 | □ | $[371,3]$ |

Theorem MED-4. Median law for chord-tangent construction (3).
$$\{x(ex) = e\} \;=\!(gL)\!\Rightarrow\; \{(xy)(zu) = (xz)(yu)\}.$$

Proof (found by Otter 3.0.4 on gyro at 1.30 seconds).

2	$x \cdot (e \cdot x) = e$	
4	$(A \cdot C) \cdot (B \cdot D) \neq (A \cdot B) \cdot (C \cdot D)$	
7	$x \cdot (y \cdot x) = z \cdot (y \cdot z)$	$[2 \to 2 :(gL)]$
8	$(x \cdot y) \cdot (z \cdot (x \cdot z)) = u \cdot (y \cdot u)$	$[7 \to 7]$
67	$(x \cdot y) \cdot (z \cdot u) = (x \cdot z) \cdot (y \cdot u)$	$[(gL) 8]$
68	□	$[67,4]$

The hypothesis of Thm. MED-4 has the model $-x - y$ in Abelian groups. The hypotheses of the following theorem have the four models $x + y$, $x - y$, $-x + y$, and $-x - y$ in Abelian groups, and all four operations are medial. We conjectured (correctly) that they (gL)-imply the median law.

Theorem MED-5. Median law for four group operations.
$$\left\{\begin{array}{l} e(ex) = x \\ ee = e \\ e(xe) = (ex)e \end{array}\right\} \;=\!(gL)\!\Rightarrow\; \{(xy)(zu) = (xz)(yu)\}.$$

Proof (found by Otter 3.0.4 on gyro at 16.26 seconds).

3,2	$e \cdot (e \cdot x) = x$	
4	$e \cdot e = e$	
6	$e \cdot (x \cdot e) = (e \cdot x) \cdot e$	
7	$(e \cdot x) \cdot e = e \cdot (x \cdot e)$	[flip 6]
9	$(A \cdot C) \cdot (B \cdot D) \neq (A \cdot B) \cdot (C \cdot D)$	
10	$(e \cdot x) \cdot (e \cdot y) = e \cdot (x \cdot y)$	$[4 \to 7 :(gL)]$
12	$(x \cdot y) \cdot e = (x \cdot e) \cdot (y \cdot e)$	$[4 \to 7 :(gL)]$
23	$x \cdot (y \cdot e) = e \cdot (y \cdot (x \cdot e))$	$[7 \to 10 :(gL) :3]$
25	$(x \cdot e) \cdot (y \cdot e) = (x \cdot y) \cdot e$	$[12 \to 2 :3]$
28	$x \cdot ((y \cdot e) \cdot z) = e \cdot ((y \cdot x) \cdot z)$	$[12 \to 23 :(gL)]$
29	$e \cdot (x \cdot (y \cdot e)) = y \cdot (x \cdot e)$	$[23 \to 2 :3]$
49	$x \cdot ((y \cdot z) \cdot e) = e \cdot ((y \cdot x) \cdot (z \cdot e))$	$[25 \to 28]$
66	$(x \cdot y) \cdot (z \cdot e) = (x \cdot z) \cdot (y \cdot e)$	$[49 \to 29 :3]$
67	$(x \cdot y) \cdot (z \cdot u) = (x \cdot z) \cdot (y \cdot u)$	$[2 \to 66 :(gL) :3]$
68	□	$[67,9]$

Cancellative median algebras with an idempotent can be embedded in (gL)-algebras by Thm. MED-1. In the following we remove cancellation and use the rule (gL) to derive the same equation. In this sense, (gL) can be viewed as a paracancellation law.

Theorem MED-6. Median (gL)-algebras.

$$\left\{ \begin{array}{l} (xy)(zu) = (xz)(yu) \\ ee = e \end{array} \right\} =\!\!(gL)\!\!\Rightarrow \{(x(yz))((uv)w) = (x(uz))((yv)w)\}.$$

Proof (found by Otter 3.0.4 on gyro at 2.27 seconds).

2	$(x \cdot y) \cdot (z \cdot u) = (x \cdot z) \cdot (y \cdot u)$	
4,3	$e \cdot e = e$	
5	$(A \cdot (D \cdot C)) \cdot ((B \cdot E) \cdot F) \neq (A \cdot (B \cdot C)) \cdot ((D \cdot E) \cdot F)$	
6	$(A \cdot (B \cdot C)) \cdot ((D \cdot E) \cdot F) \neq (A \cdot (D \cdot C)) \cdot ((B \cdot E) \cdot F)$	[flip 5]
7	$((x \cdot y) \cdot (z \cdot u)) \cdot (v \cdot w) = ((x \cdot z) \cdot v) \cdot ((y \cdot u) \cdot w)$	[2 → 2]
11	$(x \cdot e) \cdot (y \cdot e) = (x \cdot y) \cdot e$	[3 → 2]
17	$(e \cdot x) \cdot e = e \cdot (x \cdot e)$	[3 → 11 :4,4, flip]
54,53	$((e \cdot x) \cdot y) \cdot (e \cdot z) = (e \cdot (x \cdot e)) \cdot (y \cdot z)$	[17 → 2]
128	$(x \cdot (y \cdot z)) \cdot ((u \cdot v) \cdot w) = (x \cdot (u \cdot z)) \cdot ((y \cdot v) \cdot w)$	
		[17 → 7 :54 :(gL) :(gL)]
129	□	[128,6]

3.3 Abelian Groups

We start with the elementary theorem that groups satisfying (gL) are commutative.

Theorem ABGT-1. (gL)-groups are Abelian.

$$\left\{ \begin{array}{l} ex = x \\ xe = x \\ x'x = e \\ xx' = e \\ (xy)z = x(yz) \end{array} \right\} =\!\!(gL)\!\!\Rightarrow \{xy = yx\}.$$

Proof (found by Otter 3.0.4 on gyro at 4.36 seconds).

2	$e \cdot x = x$	
3	$x \cdot e = x$	
4	$x' \cdot x = e$	
5	$x \cdot x' = e$	
6	$(x \cdot y) \cdot z = x \cdot (y \cdot z)$	
7	$B \cdot A \neq A \cdot B$	
21	$e \cdot x = x \cdot e$	[3 → 2]
26	$(x \cdot e)' \cdot x = e$	[3 → 4]
30	$(e \cdot x)' \cdot x = e$	[2 → 4]
41,40	$e' = e$	[3 → 4]
45	$x' \cdot x = e \cdot e$	[3 → 4]

48,47	$(e \cdot e)' = e$	$[3 \to 40]$
56,55	$x \cdot (x \cdot e)' = e$	$[3 \to 5]$
85,84	$x \cdot (e \cdot y) = x \cdot y$	$[3 \to 6, \text{flip}]$
102,101	$(x \cdot y) \cdot z = e \cdot (x \cdot (y \cdot z))$	$[2 \to 6]$
103	$(x \cdot y)' \cdot x = (z \cdot y)' \cdot z$	$[26 \to 26 :(\text{gL})]$
104	$(x \cdot y)' \cdot y = (x \cdot z)' \cdot z$	$[30 \to 30 :(\text{gL})]$
106,105	$(x \cdot y)' \cdot x = e \cdot y'$	$[45 \to 103 :48, \text{flip}]$
112,111	$(x \cdot e)' \cdot e = e \cdot x'$	$[21 \to 103 :106]$
116,115	$(e \cdot x)' = e \cdot x'$	$[3 \to 103 :106]$
119,118	$(x \cdot y)' \cdot y = e \cdot x'$	$[55 \to 104 :56,112, \text{flip}]$
120	$e \cdot x'' = e \cdot x$	$[45 \to 104 :116,41,102,85,85,119, \text{flip}]$
137,136	$x'' = e \cdot x$	$[2 \to 120]$
140	$e \cdot (x \cdot y) = e \cdot (y \cdot x)$	$[105 \to 118 :116,137,85,102,85,137,85]$
144	$x \cdot y = e \cdot (y \cdot x)$	$[2 \to 140]$
155	$x \cdot y = y \cdot x$	$[2 \to 144]$
156	$\square$	$[155,7]$

In fact, we can prove a much stronger theorem, that a binary operation with a left and right identity element must be commutative and associative under (gL).

Theorem ABGT-2. Identity with (gL) is a commutative monoid.

$$\left\{ \begin{array}{l} ex = x \\ xe = x \end{array} \right\} =\!(gL)\!\Rightarrow \left\{ \begin{array}{l} xy = yx \\ (xy)z = x(yz) \end{array} \right\}.$$

Proof (found by Otter 3.0.4 on gyro at 51.46 seconds).

2	$B \cdot A = A \cdot B, (A \cdot B) \cdot C = A \cdot (B \cdot C) \to \square$	
3	$e \cdot x = x$	
4	$x \cdot e = x$	
5	$(e \cdot e) \cdot x = x$	$[3 \to 3]$
7	$(x \cdot e) \cdot e = x$	$[4 \to 4]$
8	$(e \cdot x) \cdot e = x$	$[3 \to 4]$
83	$(x \cdot y) \cdot e = (e \cdot y) \cdot x$	$[5 \to 7 :(\text{gL})]$
153	$(e \cdot x) \cdot y = x \cdot y$	$[4 \to 8 :(\text{gL})]$
201	$(x \cdot (y \cdot z)) \cdot e = (x \cdot y) \cdot z$	$[8 \to 153 :(\text{gL}), \text{flip}]$
384	$x \cdot y = (e \cdot y) \cdot x$	$[4 \to 83]$
884	$(x \cdot y) \cdot z = x \cdot (y \cdot z)$	$[4 \to 201, \text{flip}]$
1078	$x \cdot y = y \cdot x$	$[3 \to 384]$
1112	$\square$	$[2,1078,884]$

(Note that the proof of Thm. ABGT-1 is longer and uses more concepts than the proof of ABGT-2; but it is found much more quickly, because the additional richness of the language allows the use of rewriting with derived equalities, which reduces redundancy in the search. See Sec. 2.2.4).

The following theorem shows that given two operations, $+$ with a two-sided identity e, and $\cdot$ commutative and satisfying the Steiner quasigroup law, every element has an inverse with respect to $+$ and e.

Theorem ABGT-3. Existence of inverses under (gL).

The type is $\langle 2, 2, 0 \rangle$ with operations $+$, $\cdot$, and e.

$$\left\{ \begin{array}{l} x + e = x \\ e + x = x \\ x \cdot (y \cdot x) = y \\ x \cdot y = y \cdot x \end{array} \right\} \; =\!(gL)\!\Rightarrow \; \{\forall y \exists x (x + y = e)\}.$$

Proof (found by Otter 3.0.4 on gyro at 11.68 seconds).

2	$x + e = x$	
3	$e + x = x$	
4	$x \cdot (y \cdot x) = y$	
5	$x \cdot y = y \cdot x$	
6	$x + A = e \;\rightarrow\; \$Answer(x)$	
7	$x + e = x$	
12	$e + (x + A) = e \;\rightarrow\; \$Answer(x)$	$[3 \rightarrow 6]$
14	$(e + x) \cdot (y \cdot x) = y$	$[3 \rightarrow 4]$
20	$x \cdot (e + (y \cdot x)) = y$	$[3 \rightarrow 4]$
28	$(x \cdot y) \cdot y = x$	$[4 \rightarrow 5, \text{flip}]$
31,30	$e + (x \cdot y) = y \cdot x$	$[3 \rightarrow 5]$
33,32	$x \cdot (x \cdot y) = y$	$[20 :31]$
47	$(x + y) \cdot (z \cdot y) = (x + u) \cdot (z \cdot u)$	$[14 \rightarrow 14 :(gL)]$
56	$x \cdot (y \cdot e) = (x + z) \cdot (y \cdot z)$	$[2 \rightarrow 47]$
75	$(x + y) \cdot ((x + z) \cdot y) = z$	$[47 \rightarrow 32]$
242,241	$(x + y) \cdot (z \cdot y) = x \cdot (e \cdot z)$	$[5 \rightarrow 56, \text{flip}]$
243	$x \cdot (e \cdot (x + y)) = y$	$[75 \rightarrow 56 :7,242, \text{flip}]$
294	$e \cdot (x + y) = x \cdot y$	$[243 \rightarrow 32, \text{flip}]$
322,321	$x + y = e \cdot (x \cdot y)$	$[294 \rightarrow 32, \text{flip}]$
355	$e \cdot (x \cdot A) = e \;\rightarrow\; \$Answer(x)$	$[12 :322,322,33]$
356	$e \cdot x = e \;\rightarrow\; \$Answer(x \cdot A)$	$[28 \rightarrow 355]$
358	$\$Answer((e \cdot e) \cdot A)$	$[356,32]$

Note that we use an *answer literal* to record the instantiations of the existentially quantified variable in the goal. The last step of the proof, $\$Answer((e \cdot e) \cdot A)$, gives us an inverse operation, that is, $x' = (e \cdot e) \cdot x$. (We have presented the second proof found by Otter; the first proof gave us the (equal) operation $x' = (e \cdot ((z \cdot e) \cdot (e \cdot (e \cdot z)))) \cdot x$.) Also, step 322 of the proof, $x + y = e \cdot (x \cdot y)$, gives us the group operation in terms of $\cdot$ and e. This is the standard way of defining a group operation on a nonsingular cubic curve; that is, the identity e uniquely determines the group operation. See Fig. 3.6 and [68, p. 148].

Putting together Thms. ABGT-2 and ABGT-3, we have the following.

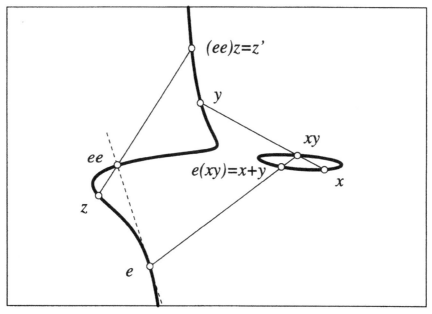

Fig. 3.6. A Group $\langle G; +, ', e \rangle$ on a Cubic Curve

Corollary ABGT-3a. Mumford-Ramanujam theorem for elliptic curves.

If $m : C \times C \longrightarrow C$ is a binary morphism of a nonsingular cubic curve C defined over the complex projective plane, and if for some e in C, $m(x, e) = m(e, x) = x$ for all x, then the structure $\langle C; m, e \rangle$ is an Abelian group.

This is a special case of a result of Mumford and Ramanujam that any complete variety with a binary morphism admitting a two-sided identity must be an Abelian group variety [49, appendix to Sec. 4].

3.4 Uniqueness of Group Laws

Theorem UAL-1. Uniqueness of inversive groupoids under (gL).

(Note that the two group operations have the same identity element.)

$$
\left\{
\begin{array}{ll}
f(e, x) = x, & F(e, x) = x \\
f(x, e) = x, & F(x, e) = x \\
f(g(x), x) = e, & F(G(x), x) = e \\
f(x, g(x)) = e, & F(x, G(x)) = e
\end{array}
\right\}
\xRightarrow{(gL)}
\left\{
\begin{array}{l}
f(x, y) = F(x, y) \\
g(x) = G(x)
\end{array}
\right\}.
$$

Proof (found by Otter 3.0.4 on gyro at 13.58 seconds).

2          $f(e, x) = x$

3	$f(x,e) = x$	
4	$f(g(x),x) = e$	
5	$f(x,g(x)) = e$	
6	$F(e,x) = x$	
7	$F(x,e) = x$	
8	$F(G(x),x) = e$	
9	$F(x,G(x)) = e$	
10	$f(A,B) = F(A,B),\ g(A) = G(A)\ \rightarrow\ \square$	
28,27	$f(x,e) = F(e,x)$	$[3 \rightarrow 6,\ \text{flip}]$
32,31	$f(e,x) = F(e,x)$	$[2 \rightarrow 6,\ \text{flip}]$
40	$F(e,x) = F(x,e)$	$[7 \rightarrow 6]$
41	$f(g(F(x,e)),x) = e$	$[7 \rightarrow 4]$
43	$f(g(x),F(x,e)) = e$	$[7 \rightarrow 4]$
82	$f(x,F(e,g(x))) = e$	$[6 \rightarrow 5]$
94,93	$F(x,f(y,g(y))) = x$	$[5 \rightarrow 7]$
112,111	$G(e) = e$	$[5 \rightarrow 8 :94]$
139,138	$G(F(G(x),x)) = e$	$[8 \rightarrow 111]$
144	$F(e,g(e)) = e$	$[111 \rightarrow 5 :112,32]$
176	$F(G(x),x) = F(y,G(y))$	$[9 \rightarrow 8]$
184,183	$F(F(x,G(x)),y) = y$	$[9 \rightarrow 6]$
186,185	$g(e) = e$	$[9 \rightarrow 144 :184]$
214,213	$f(F(G(x),x),y) = F(e,y)$	$[8 \rightarrow 31]$
231	$f(g(F(x,y)),x) = f(g(F(z,y)),z)$	$[41 \rightarrow 41 :(\text{gL})]$
236,235	$f(g(F(x,y)),x) = F(e,g(y))$	$[9 \rightarrow 231 :184,28,\ \text{flip}]$
243	$F(e,g(F(x,e))) = F(e,g(x))$	$[40 \rightarrow 231 :28,236]$
248,247	$F(e,g(x)) = F(e,G(x))$	$[8 \rightarrow 231 :186,32,236,\ \text{flip}]$
268,267	$F(x,g(F(y,e))) = F(x,g(y))$	$[243 :(\text{gL})]$
277	$f(x,F(e,G(x))) = e$	$[82 :248]$
289	$f(g(x),F(x,y)) = f(g(z),F(z,y))$	$[43 \rightarrow 43 :(\text{gL})]$
291,290	$g(x) = F(e,G(x))$	$[9 \rightarrow 267 :268,184,248]$
292	$f(F(e,G(x)),F(x,y)) = f(F(e,G(z)),F(z,y))$	$[289 :291,291]$
311	$f(A,B) = F(A,B)\ \rightarrow\ \square$	$[10 :291 :6]$
312	$f(F(A,e),B) = F(A,B)\ \rightarrow\ \square$	$[7 \rightarrow 311]$
350,349	$f(x,G(x)) = e$	$[9 \rightarrow 277 :184]$
409	$f(F(x,G(y)),F(y,z)) = f(F(x,G(u)),F(u,z))$	$[292 :(\text{gL})]$
443,442	$f(F(x,G(y)),F(y,z)) = f(F(x,e),F(e,z))$	
		$[349 \rightarrow 409 :350,112,\ \text{flip}]$
446,445	$f(F(x,e),F(e,y)) = f(F(x,e),y)$	$[176 \rightarrow 409 :139,184,443,\ \text{flip}]$
449,448	$f(F(x,e),y) = F(e,F(x,y))$	$[176 \rightarrow 409 :214,443,446,\ \text{flip}]$
478	$F(e,F(A,B)) = F(A,B)\ \rightarrow\ \square$	$[312 :449]$
479	$\square$	$[478,6]$

Corollary UAL-2. Uniqueness of group laws under (gL).

By Thm. UAL-1, there do not exist two distinct (gL)-groups with the same identity element.

Theorem MCV-1. Associativity of Mal'cev polynomial under (gL).

$$\left\{ \begin{array}{l} m(x,y,y) = x \\ m(x,x,y) = y \end{array} \right\} =\!(gL)\!\Rrightarrow \{m(x,y,m(z,u,v)) = m(v,u,m(x,y,z))\}.$$

Proof (found by Otter 3.0.4 on gyro at 17.08 seconds).

2	$m(x,y,y) = x$	
3	$m(x,x,y) = y$	
4	$m(A,B,m(C,D,E)) \neq m(E,D,m(A,B,C))$	
25	$m(x,x,y) = m(y,z,z)$	$[2 \rightarrow 3]$
70	$m(x,y,z) = m(z,y,x)$	$[(gL)\ 25]$
98	$m(x,x,m(y,z,u)) = m(u,z,y)$	$[3 \rightarrow 70]$
99	$m(m(x,y,z),u,u) = m(z,y,x)$	$[2 \rightarrow 70]$
135	$m(m(C,D,E),B,A) \neq m(E,D,m(A,B,C))$	$[70 \rightarrow 4]$
183	$m(x,y,m(z,u,y)) = m(x,u,z)$	$[(gL)\ 98]$
291	$m(m(x,y,z),x,u) = m(z,y,u)$	$[(gL)\ 99]$
304	$m(m(x,y,z),u,v) = m(z,y,m(v,u,x))$	$[183 \rightarrow 291]$
305	$\square$	$[304,135]$

Theorem MCV-2. Mal'cev polynomial under (gL).

$$\left\{ \begin{array}{l} m(x,y,y) = x \\ m(x,x,y) = y \\ x \cdot (y \cdot x) = y \\ x \cdot (x \cdot y) = y \end{array} \right\} =\!(gL)\!\Rrightarrow \{m(x,y,z) = y \cdot (x \cdot z)\}.$$

Proof (found by Otter 3.0.4 on gyro at 35.72 seconds).

2	$x \cdot (y \cdot x) = y$	
3	$x \cdot (x \cdot y) = y$	
4	$m(x,y,y) = x$	
5	$m(x,x,y) = y$	
6	$B \cdot (A \cdot C) \neq m(A,B,C)$	
19	$x \cdot (y \cdot m(x,z,z)) = y$	$[4 \rightarrow 2]$
34	$x \cdot (y \cdot m(z,z,x)) = y$	$[5 \rightarrow 2]$
36	$x \cdot (x \cdot m(y,y,z)) = z$	$[5 \rightarrow 3]$
48	$x \cdot (y \cdot (y \cdot (x \cdot z))) = z$	$[3 \rightarrow 3]$
69	$x \cdot m(y,y,z \cdot m(u,u,x)) = z$	$[5 \rightarrow 34]$
73	$x \cdot m(y,y,x \cdot m(z,z,u)) = u$	$[5 \rightarrow 36]$
140	$x \cdot m(y,y,z) = z \cdot m(u,u,x)$	$[69 \rightarrow 73]$
173	$x \cdot m(y,z,u) = u \cdot m(y,z,x)$	$[(gL)\ 140]$
222	$x \cdot (y \cdot m(x,y,z)) = z$	$[173 \rightarrow 19]$

| 239 | $x \cdot (y \cdot z) = m(y, x, z)$ | $[222 \to 48]$ |
| 241 | $\square$ | $[239, 6]$ |

Corollary UAL-3. Uniqueness of Mal'cev laws under (gL).

By Thm. MCV-2, there do not exist two distinct Mal'cev laws under (gL).

3.5 Uniqueness of n-ary Steiner Laws

Some of the material in this section is presented also in [58], where we give an equational characterization of the conic construction on a cubic curve. The key result in that paper, which we also present here, is the uniqueness of the 5-ary Steiner law under (gL). From there we have the well-known theorem in geometry that given five points of intersection between a cubic and a conic, the sixth point can be found by a simple ruler construction. (See the end of this section.) We start with the uniqueness of the binary Steiner law, then consider higher arities.

Theorem UAL-4. Uniqueness of binary Steiner law under (gL).

The type of theory is $\langle 2, 2, 0 \rangle$ with corresponding operations f, g, and e.

$$\left\{ \begin{array}{ll} f(x,y) = f(y,x), & g(x,y) = g(y,x) \\ f(x, f(y,x)) = y, & g(x, g(y,x)) = y \\ f(e,e) = e), & g(e,e) = e \end{array} \right\} =\!(gL)\!\Rrightarrow \{f(x,y) = g(x,y)\}.$$

Proof (found by Otter 3.0.4 on gyro at 161.14 seconds).

2	$f(x,y) = f(y,x)$	
3	$f(x, f(y,x)) = y$	
4	$f(e,e) = e$	
5	$g(x,y) = g(y,x)$	
6	$g(x, g(y,x)) = y$	
7	$g(e,e) = e$	
8	$g(A, B) \neq f(A, B)$	
9	$f(e,e) = e$	
10	$f(x, f(y,x)) = y$	
11	$f(x, g(e,e)) = f(e,x)$	$[7 \to 2]$
15	$f(x, f(x,y)) = y$	$[2 \to 3]$
22	$g(x, g(x,y)) = y$	$[5 \to 6]$
48	$g(x,y) = f(z, f(z, g(y,x)))$	$[15 \to 5]$
106	$f(x, g(e,e)) = f(e, g(y, g(y,x)))$	$[22 \to 11]$
493	$f(e, g(x, g(e,y))) = f(y, g(x,e))$	$[4 \to 106 :(gL) :9, \text{flip}]$
500	$f(x, g(x,e)) = e$	$[6 \to 493 :9, \text{flip}]$
536	$f(x, g(x,y)) = f(z, g(z,y))$	$[500 \to 500 :(gL)]$

551	$g(e,x) = f(x,e)$	$[500 \to 48]$
860	$f(x, g(x,y)) = y$	$[551 \to 536 :10]$
983	$g(x,y) = f(x,y)$	$[22 \to 860, \text{flip}]$
984	$\square$	$[983,8]$

Note in the preceding theorem that equations 9 and 10 are identical to equations 4 and 3. Those two equations are repeated because they are user-specified demodulators. The demodulation strategy is to use that pair of equations, and no others, as demodulators.

The next three theorems, STN-1, STN-2, and STN-3, show that for arities 3, 4, and 5, the universal Steiner law can be replaced with a weaker form in the presence of idempotence and permutation equations. In [60] we prove the corresponding general theorem for all arities.

Problem STN-1. (gL)-basis for ternary Steiner law.

The type is $\langle 3, 0 \rangle$.

$$\left\{ \begin{array}{l} f(w,x,y) = f(w,y,x) \\ f(e,e,e) = e \\ f(e,x,f(e,x,y)) = y \end{array} \right\} \overset{}{=}(gL)\Rightarrow \{f(w,x,f(w,x,y)) = y\}.$$

Proof (found by Otter 3.0.4 on gyro at 2.59 seconds).

1	$f(A,B,f(A,B,C)) \neq C$	
3	$f(x,y,z) = f(x,z,y)$	
4	$f(e,e,e) = e$	
7,6	$f(e,x,f(e,x,y)) = y$	
9,8	$f(e,x,f(e,y,x)) = y$	$[3 \to 6]$
12	$f(x,e,e) = f(x,y,f(e,y,e))$	$[6 \to 4 :(gL)]$
20	$f(e,x,f(y,f(z,e,e),x)) = f(z,u,f(y,u,e))$	$[8 \to 12 :(gL)]$
282	$f(e,e,x) = f(y,f(y,e,e),x)$	$[(gL)\ 20]$
347	$f(x,y,f(x,y,e)) = e$	$[282 \to 20 :9, \text{flip}]$
422	$f(x,y,f(x,y,z)) = z$	$[6 \to 347 :(gL) :7]$
424	$\square$	$[422,1]$

Theorem STN-2. (gL)-basis for 4-ary Steiner law.

$$\left\{ \begin{array}{l} f(v,w,x,y) = f(v,w,y,x) \\ f(e,e,e,e) = e \\ f(e,e,x,f(e,e,x,y)) = y \end{array} \right\} \overset{}{=}(gL)\Rightarrow \{f(v,w,x,f(v,w,x,y)) = y\}.$$

Proof (found by Otter 3.0.4 on gyro at 3.40 seconds).

1	$f(D,A,B,f(D,A,B,C)) \neq C$
3	$f(x,y,z,u) = f(x,y,u,z)$
4	$f(e,e,e,e) = e$
6	$f(e,e,x,f(e,e,x,y)) = y$

11,10	$f(e,e,f(e,e,x,y),x) = y$	$[3 \to 6]$
12	$f(x,y,e,e) = f(x,y,z,f(e,e,z,e))$	$[6 \to 4 :(gL) :(gL)]$
17	$f(x,y,z,f(u,v,z,w)) = f(x,y,v_6,f(u,v,v_6,w))$	
		$[12 \to 12 :(gL) :(gL) :(gL)]$
34	$f(x,y,f(z,u,v,w),v) = f(x,y,v_6,f(z,u,v_6,w))$	$[3 \to 17]$
120	$f(x,y,z,f(u,v,f(u,v,w,v_6),v_6)) = f(x,y,z,w)$	$[(gL)\ 34,\ \text{flip}]$
134	$f(x,y,f(x,y,z,u),u) = z$	$[120 \to 10 :11,\ \text{flip}]$
167	$f(x,y,z,f(x,y,z,u)) = u$	$[34 \to 134]$
169	$\square$	$[167,1]$

Theorem STN-3. (gL)-basis for 5-ary Steiner law.

$$\left\{ \begin{array}{l} f(u,v,w,x,y) = f(u,v,w,y,x) \\ f(e,e,e,e,e) = e \\ f(e,e,e,x,f(e,e,e,x,y)) = y \end{array} \right\} =\!(gL)\!\Rightarrow$$

$$\{f(u,v,w,x,f(u,v,w,x,y)) = y\}.$$

Proof (found by Otter 3.0.4 on gyro at 3.87 seconds).

1	$f(E,D,A,B,f(E,D,A,B,C)) \neq C$	
3	$f(x,y,z,u,v) = f(x,y,z,v,u)$	
4	$f(e,e,e,e,e) = e$	
7,6	$f(e,e,e,x,f(e,e,e,x,y)) = y$	
9,8	$f(e,e,e,x,f(e,e,e,y,x)) = y$	$[3 \to 6]$
12	$f(x,y,z,e,e) = f(x,y,z,u,f(e,e,e,u,e))$	$[6 \to 4 :(gL) :(gL) :(gL)]$
20	$f(e,e,e,x,f(y,z,u,f(v,w,v_6,e,e),x)) = f(v,w,v_6,v_7,f(y,z,u,v_7,e))$	
		$[8 \to 12 :(gL) :(gL) :(gL)]$
223	$f(e,e,e,e,x) = f(y,z,u,f(y,z,u,e,e),x)$	$[(gL)\ 20]$
277	$f(x,y,z,u,f(x,y,z,u,e)) = e$	$[223 \to 20 :9,\ \text{flip}]$
340	$f(x,y,z,u,f(x,y,z,u,v)) = v$	$[6 \to 277 :(gL) :7]$
342	$\square$	$[340,1]$

The following lemma leads to the uniqueness of the 5-ary Steiner law, the main result of this section.

Lemma UAL-5. Identity for two 5-ary Steiner laws.

The type is $\langle 5,5,0 \rangle$.

$$\left\{ \begin{array}{l} f(u,v,w,x,y) = f(u,v,w,y,x) \\ f(e,e,e,e,e) = e \\ f(u,v,w,x,f(u,v,w,x,y)) = y \\ g(u,v,w,x,y) = g(u,v,w,y,x) \\ g(e,e,e,e,e) = e \\ g(u,v,w,x,g(u,v,w,x,y)) = y \end{array} \right\} =\!(gL)\!\Rightarrow$$

$$\{f(x_1,x_2,x_3,y,g(x_4,x_5,x_6,y,x_7)) = f(x_1,x_2,x_3,z,g(x_4,x_5,x_6,z,x_7))\}.$$

Proof (found by Otter 3.0.4 on gyro at 22.00 seconds).

2    $f(u,v,w,x,y) = f(u,v,w,y,x)$

5    $g(u,v,w,x,y) = g(u,v,w,y,x)$

6    $g(e,e,e,e,e) = e$

7    $g(u,v,w,x,g(u,v,w,x,y)) = y$

8    $f(A_1,A_2,A_3,B,g(A_4,A_5,A_6,B,A_7)) \neq$
     $\quad f(A_1,A_2,A_3,C,g(A_4,A_5,A_6,C,A_7))$

10   $f(x,y,z,u,g(e,e,e,e,e)) = f(x,y,z,e,u)$ $[6 \to 2]$

14   $f(x,y,z,e,g(u,v,w,v_6,g(u,v,w,v_6,v_7))) = f(x,y,z,v_7,g(e,e,e,e,e))$
                                                                                 $[7 \to 10, \text{flip}]$

20   $f(x,y,z,e,g(u,v,w,v_6,g(u,v,w,v_7,v_6))) = f(x,y,z,v_7,g(e,e,e,e,e))$
                                                                                 $[5 \to 14]$

24   $f(x,y,z,e,g(u,v,w,v_6,g(u,v,w,e,v_7))) = f(x,y,z,v_7,g(e,e,e,v_6,e))$
                                                                                 $[(gL)\ 14]$

45   $f(x,y,z,e,g(u,v,w,e,v_6)) = f(x,y,z,v_7,g(u,v,w,v_7,v_6))$
                                                     $[20 \to 24 :(gL) :(gL) :(gL) :(gL)]$

52   $f(x,y,z,u,g(v,w,v_6,u,v_7)) = f(x,y,z,v_8,g(v,w,v_6,v_8,v_7))$ $[45 \to 45]$

53   $\square$ $[52,8]$

Theorem UAL-6. Uniqueness of 5-ary Steiner law under (gL).
The type is $\langle 5,5,0 \rangle$.

$$\left\{ \begin{array}{l} \text{full symmetry of } f \\ f(e,e,e,e,e) = e \\ f(e,e,e,x,f(e,e,e,x,y)) = y \\ \text{full symmetry of } g \\ g(e,e,e,e,e) = e \\ g(e,e,e,x,g(e,e,e,x,y)) = y \end{array} \right\} \relbar\joinrel(gL)\joinrel\rightarrow \{g(x,y,z,u,v) = f(x,y,z,u,v)\}.$$

By Thm. STN-3, we may also assume the universal 5-ary Steiner laws for f
and for g:
$$f(u,v,w,x,f(u,v,w,x,y)) = y,$$
$$g(u,v,w,x,g(u,v,w,x,y)) = y.$$

By Lem. UAL-5, we include the equation
$$f(x_1,x_2,x_3,y,g(x_4,x_5,x_6,y,x_7)) = f(x_1,x_2,x_3,z,g(x_4,x_5,x_6,z,x_7)).$$

Proof (found by Otter 3.0.4 on gyro at 3.68 seconds).

2    $g(x,y,z,u,v) = f(x,y,z,u,v) \;\to\; g(y,z,u,v,x) = f(y,z,u,v,x)$

3    $f(e,e,e,e,e) = e$

4    $f(u,v,w,x,f(u,v,w,x,y)) = y$

5    $g(e,e,e,e,e) = e$

6    $g(u,v,w,x,g(u,v,w,x,y)) = y$

7    $f(x_1, x_2, x_3, y, g(x_4, x_5, x_6, y, x_7)) = f(x_1, x_2, x_3, z, g(x_4, x_5, x_6, z, x_7))$

8    $g(A, B, C, D, E) = f(A, B, C, D, E)\ \rightarrow\ \square$

10	$f(e, e, e, e, g(e, e, e, e, e)) = e$	[5 → 3]
11	$f(e, e, e, x, g(e, e, e, x, e)) = e$	[7 → 10]
12	$g(e, e, e, x, e) = f(e, e, e, x, e)$	[11 → 4, flip]
13	$g(e, e, x, e, e) = f(e, e, x, e, e)$	[2,12]
15	$f(e, e, x, e, g(e, e, x, e, e)) = e$	[13 → 4]
18	$f(e, e, x, y, g(e, e, x, y, e)) = e$	[7 → 15]
19	$g(e, e, x, y, e) = f(e, e, x, y, e)$	[18 → 4, flip]
20	$g(e, x, y, e, e) = f(e, x, y, e, e)$	[2,19]
22	$f(e, x, y, e, g(e, x, y, e, e)) = e$	[20 → 4]
25	$f(e, x, y, z, g(e, x, y, z, e)) = e$	[7 → 22]
26	$g(e, x, y, z, e) = f(e, x, y, z, e)$	[25 → 4, flip]
27	$g(x, y, z, e, e) = f(x, y, z, e, e)$	[2,26]
29	$f(x, y, z, e, g(x, y, z, e, e)) = e$	[27 → 4]
32	$f(x, y, z, u, g(x, y, z, u, e)) = e$	[7 → 29]
33	$g(x, y, z, u, e) = f(x, y, z, u, e)$	[32 → 4, flip]
34	$g(x, y, z, e, u) = f(x, y, z, e, u)$	[2,33]
36	$f(x, y, z, e, g(x, y, z, e, u)) = u$	[6 → 34, flip]
39	$f(x, y, z, u, g(x, y, z, u, v)) = v$	[7 → 36]
40	$g(x, y, z, u, v) = f(x, y, z, u, v)$	[6 → 39, flip]
41	$\square$	[40,8]

The above Otter proofs of Thms. UAL-4, UAL-5, and UAL-6 were found with specialized search strategies. The first Otter proof of UAL-4, which was found with a straightforward strategy, required about 38 hours. Because it is an important theorem, and because we planned to attack problems of a similar nature (e.g., higher arities), we devised some strategies that would lead Otter quickly to a proof. The strategy that led to the above proof of UAL-4 (in a few minutes) discards clauses containing left-associated terms and prefers equations with variables.

Commutativity of a binary operation usually causes a minor explosion in the search space, and full permutability of a higher-arity operation is nearly always disastrous, especially when used with Otter's prolific and redundant strategies for (gL). So when we first considered searching for a proof of the 5-ary Steiner conjecture (Thm. UAL-6) with Otter, our feeling was "not a chance". Eight months later, we returned to the conjecture, and we looked to the binary case (Thm. UAL-4) for clues; we reduced the search time from 38 hours to several minutes by studying the form of the 38-hour proof. Although the cases with nice geometric interpretations are arities $3m + 2$ $(m \geq 0)$, we conjectured that the analogous theorems hold for all arities, and tried the ternary case next. After the search strategy was adjusted, a proof was found. We noticed that the universal Steiner law is a natural lemma for the theorem, and we proved the corresponding lemmas (Thms. STN-2 and STN-3) separately for the 4-ary and 5-ary cases.

The key strategies were to discard clauses that are not right associated, and to severely limit the use of symmetry. For the 4-ary and 5-ary cases, we directed the searches toward the form of proof we expected, and proofs were found. The first proof of the 5-ary universal Steiner law (Thm. STN-3) took half an hour (length 13), and the first 5-ary uniqueness proof (Thm. UAL-6) took 23 hours (length 46).

The strategies were further refined as we studied the proofs and worked out a proof (by hand) that the analogous proofs hold for all arities. Lemma UAL-5 was separated from the main proof at that time. The proofs above are the product of the specialized strategies. Independently, we found (entirely by hand) a higher-order proof for all arities. (The n-ary proofs will appear elsewhere [60].)

Corollary UAL-7. A ruler construction for cubic and conic.

The type is $\langle 5, 2, 0 \rangle$.

$$\left\{ \begin{array}{l} \text{full symmetry of } f \\ f(e,e,e,e,e) = e \\ f(e,e,e,x,f(e,e,e,x,y)) = y \\ h(x,y) = h(y,x) \\ h(e,e) = e \\ h(x,h(x,y)) = y \end{array} \right\} = \langle gL \rangle \Rightarrow$$

$$\{ h(h(h(x,y), h(z,u)), v) = f(x,y,z,u,v) \}.$$

Proof. This follows from Thm. UAL-5. Define

$$g(x,y,z,u,v) = h(h(h(x,y), h(z,u)), v).$$

Then g satisfies the hypotheses of Thm. UAL-6.

Figure 3.7 is a geometric interpretation of Thm. UAL-6. Let Γ be a nonsingular cubic curve, and let x, y, z, t, u be five points on the curve. Let Q be the unique conic determined by the 5 points. By the Bezout theorem of classical geometry, we have $|\Gamma \cap Q| = 6$, counting multiplicities. Now let $F(x,y,z,t,u)$ be the 5-ary morphism on Γ defined by the complete intersection cycle $\Gamma \cap Q$ $\{x, y, z, t, u, F(x,y,z,t,u)\}$. By Cor. UAL-7, we know that the unique sixth point $F(x,y,z,t,u)$ can be found by a simple ruler construction as shown. (A proof using the rigidity lemma was given by N. S. Mendelsohn, Padmanabhan, and B. Wolk in [46]).

3.6 Group Laws on a Quartic Curve

One of the intrinsic beauties of the classical group law on nonsingular cubics is the very connection between the algebra and the geometry. If three points P, Q, and R are collinear, then

$$P + Q + R = 0$$

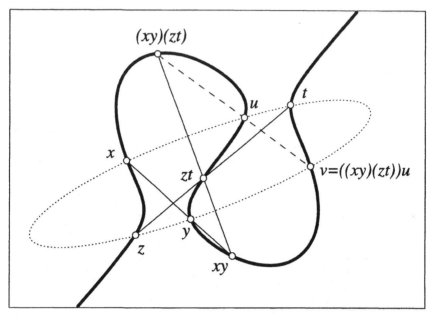

Fig. 3.7. The Conic Construction on a Cubic Curve

under the group law (with the 0 being an inflection point). Such a rule does not easily extend to higher projective curves unless their Riemann surfaces have just enough holes in them to bring down the genus. If, however, a complex projective curve Γ of degree n has enough singularities so that its genus becomes 1, then it is an elliptic curve (i.e., Abelian variety of dimension one), and hence there exists a birational transformation of the projective plane onto itself carrying this curve into a nonsingular cubic curve that does possess a synthetically constructed group law by means of the classical chord-tangent construction. But, of course, the interesting (and, relatively, down-to-earth) question is whether one can actually exhibit a synthetic group law on the n-th degree curve Γ such that if $P_1, P_2, P_3, \cdots, P_n$ on Γ are collinear, then

$$P_1 + P_2 + P_3 + \cdots + P_n = 0.$$

Here we give one such explicit construction for a quartic curve with two double points. The synthetic construction we employ here is taken from a suggestion given by Barry Mazur in a recent survey article [31].

Mazur Laws. Let A and B be the two nodes of the quartic Γ. For three points P, Q, and R on Γ, define a ternary law $m(P, Q, R)$ by the complete intersection cycle

$$\Gamma \cap \mathrm{K} = \{A, B, P, Q, R, m(P, Q, R)\}$$

for some conic K, counting proper multiplicities. Since every conic through the two nodes A and B has intersection multiplicity 2 at these two points, the final point (P, Q, R) is the unique eighth point common to the conic and the quartic. For example, if P, Q, and R happen to be collinear, then K is the reducible conic made up of the pair of straight lines PQ and the chord AB. Hence, in this case, $m(P, Q, R)$ is the unique fourth point where the line PQR meets the quartic again, quite a pleasing situation. Clearly the function m defines a ternary morphism from $\Gamma \times \Gamma \times \Gamma$ to Γ. In the sequel, we call this ternary m a Mazur law.

Theorem QUART-1 proves the validity of the identity

$$m(x, y, m(x, z, u)) = m(u, y, m(u, z, u)),$$

and from this we construct a desired ternary Mal'cev polynomial in Thm. QUART-2. This paves our way for constructing the synthetic group law on such quartics.

Theorem QUART-1. Mazur lemma.

$$\left\{ \begin{array}{l} m(x, y, m(x, y, z)) = z \\ m(x, y, z) = m(x, z, y) \end{array} \right\} =(gL)\!\Rightarrow$$
$$\{m(x, y, m(x, z, u)) = m(u, y, m(u, z, u))\}.$$

Our first version of this problem had a different conclusion:

$$m(x, y, m(x, z, w)) = m(u, y, m(u, z, w)).$$

After Otter failed to prove the first version, we found the following counterexample. Define $m(x, y, z)$ as $ax + by + cz$ in an Abelian group. Such an algebra is always a (gL)-algebra. The first equation $m(x, y, m(x, y, z)) = z$ demands that $a + ac = 0$, $b + bc = 0$, and $c^2 = 1$. Thus $c = -1$ would satisfy all the conditions without the number a being 0. Now the second equation $m(x, y, z) = m(x, z, y)$ demands that $b = c$. Hence choose $b = c = -1$ and $a = 2$, say. Now, $m(x, y, m(z, x, w)) = 2x - y - (2z - x - w) = x - y - 2z - w$, which is not independent of x in the Abelian group of integers! Thus our original goal is not a consequence of the assumptions. However, the following equation is true in this model: $m(x, y, m(x, z, w)) = m(u, y, m(u, z, w))$ because the left-hand side reduces to $2x - y - (2x - z - w) = z - y + w$, which is independent of x. Thus, the corrected formulation of Thm. QUART-1 is given above.

Proof (found by Otter 3.0.4 on gyro at 2.13 seconds).

2	$m(x, y, m(x, y, z)) = z$	
4	$m(A, B, m(A, C, D)) \neq m(E, B, m(E, C, D))$	
10	$m(x, y, m(x, y, z)) = m(u, v, m(u, v, z))$	$[2 \to 2]$
163	$m(x, y, m(x, z, u)) = m(v, y, m(v, z, u))$	$[(gL)\ 10]$
164	$\square$	$[163, 4]$

Guided by the intuition gained through the model described for the preceding problem, we suspected that $p(z, y, v) = m(x, y, m(x, z, v))$ must behave like the ternary Mal'cev polynomial $z - y + v$ in Abelian groups. Theorems QUART-1 and QUART-2, taken together, construct a synthetic group law $+$ on a quartic curve Γ such that if four points P, Q, R, S on Γ are collinear, then $P + Q + R + S = 0$ under the group law.

Theorem QUART-2. Conic construction on a quartic curve.

$$\left\{ \begin{array}{l} m(x, y, m(x, y, z)) = z \\ m(x, y, z) = m(x, z, y) \\ p(z, y, v) = m(x, y, m(x, z, v)) \end{array} \right\} =\!(gL)\!\Rightarrow$$

$$\{ p(x, y, p(z, v, u)) = p(u, v, p(x, y, z)) \}.$$

Proof (found by Otter 3.0.4 on gyro at 1622.09 seconds).

2	$m(x, y, m(x, y, z)) = z$	
3	$m(x, y, z) = m(x, z, y)$	
4	$p(z, y, v) = m(x, y, m(x, z, v))$	
5	$p(A, B, p(C, D, E)) \neq p(E, D, p(A, B, C))$	
6	$m(x, y, m(x, z, y)) = z$	$[3 \rightarrow 2]$
7	$p(x, y, y) = x$	$[6 \rightarrow 4]$
8	$p(x, x, y) = y$	$[2 \rightarrow 4]$
14	$p(x, x, y) = p(y, z, z)$	$[7 \rightarrow 8]$
78	$p(x, y, z) = p(z, y, x)$	$[(gL)\ 14]$
88	$p(x, y, p(z, u, u)) = p(z, y, x)$	$[7 \rightarrow 78]$
90	$p(x, x, p(y, z, u)) = p(u, z, y)$	$[8 \rightarrow 78]$
272	$p(x, y, p(z, u, y)) = p(x, u, z)$	$[(gL)\ 90]$
286	$p(x, y, p(z, u, y)) = p(z, u, p(x, v, v))$	$[88 \rightarrow 272]$
538	$p(x, y, p(z, u, v)) = p(z, u, p(x, y, v))$	$[(gL)\ 286]$
541	$p(x, y, p(z, u, v)) = p(v, u, p(x, y, z))$	$[78 \rightarrow 538]$
542	$\square$	$[541, 5]$

4. Other (gL)-Algebras

The rigidity property generalizes the local validity of an equation (say, around a special point) to its global validity. From the point of view of first-order logic with equality, the rigidity lemma captures the equational essence of the n-ary morphisms of nonsingular cubic curves. Various uniqueness theorems we proved in the preceding chapter and the equational proofs of incidence theorems in the projective geometry of cubic curves demonstrate this point of view.

In this chapter, we investigate the consistency of the rigidity lemma (and of (gL)) in other algebraic systems. For example, while we have exhibited many n-ary quasigroup morphisms on algebraic curves, a semilattice morphism cannot exist on a cubic curve. Why not? At a more intuitive level, take the Euclidean plane $R \times R$. While it is easy to construct a group addition geometrically (e.g., the parallelogram law of addition), no one has ever constructed, say, a lattice structure $\langle R \times R; \vee, \wedge \rangle$ on the plane such that both $\vee$ and $\wedge$ are ring polynomials.

It is this kind of question we consider in this chapter. Let Σ be a given equational theory of some given type. We say that rigidity *likes* Σ if there exists a nontrivial (gL)-algebra satisfying Σ.

If rigidity happens to like Σ, then our experience shows that it takes the algebra all the way to an Abelian group structure and produces a Σ-consistent group interpretation for various terms occurring in Σ. On the contrary, if rigidity does not like Σ, then

$$\Sigma \ =\!(gL)\!\Rightarrow\ x = y,$$

that is, Σ is strictly inconsistent with the rule (gL). In other words, if (gL) is applied as a rule of derivation on a lattice or a semilattice, it crushes that algebra to a singleton. Thus, once we establish that rigidity does not like Σ, it is clear that no nonsingular cubic curve can admit n-ary morphisms satisfying Σ.

Those who work in universal algebra know this phenomenon well (see, e.g., [44, Ch. 4, Ex. 4, 5, 7, and 8]). See the proofs in Sec. 4.2 for equational theories strictly inconsistent with cubic curves. This connection with algebraic geometry was first brought out by Padmanabhan in [55].

One can fruitfully use this idea to create nice mathematical problems for Otter. Start with a favorite group term, say, right division $x/y = x \cdot y'$. Now let $\Sigma = \{x/(x/y) = y, x/x = e\}$ be of type $\langle 2, 0\rangle$ with one binary and one nullary operation. Then the above scheme leads to the following conjecture:

$$\Sigma \ =\!(gL)\!\Rightarrow\ \{\text{all the equations true for } x/y \text{ in Abelian groups}\}.$$

This is, in fact, true and is proved in the following theorem.

4.1 Equations Consistent with (gL)

4.1.1 Abelian Groups

Theorem ABGT-4. A (gL)-basis for right division in Abelian groups (1).

$$\left\{ \begin{array}{l} x/(x/y) = y \\ x/x = e \end{array} \right\} \ =\!(gL)\!\Rightarrow\ \{(x/((x/y)/z))/y = z\}.$$

Here, x/y is right division, and the conclusion is a single axiom for Abelian groups in terms of right division.

Proof (found by Otter 3.0.4 on gyro at 13.29 seconds).

2	$x/x = e$	
3	$x/(x/y) = y$	
8	$(A/((A/B)/C))/B \neq C$	
22	$x/e = x$	$[2 \rightarrow 3]$
26	$(x/e)/e = x$	$[22 \rightarrow 22]$
27	$(x/(x/y))/e = y$	$[3 \rightarrow 22]$
167	$(x/e)/y = x/y$	$[22 \rightarrow 26 :(\text{gL})]$
239	$((x/y)/z)/e = (x/z)/y$	$[26 \rightarrow 167 :(\text{gL}), \text{flip}]$
3177	$(x/((x/y)/z))/y = z$	$[239 \rightarrow 27]$
3178	$\square$	$[3177,8]$

Theorem ABGT-5. A (gL)-basis for right division in Abelian groups (2).

$$\left\{ \begin{array}{l} x/e = x \\ x/x = e \end{array} \right\} \ =\!(gL)\!\Rightarrow\ \{x/((y/z)/(y/x)) = z\}.$$

The conclusion is another single axiom for Abelian groups in terms of right division. (Note: the first axioms we tried were $\{x/e = x, e/(e/x) = x\}$. Otter failed to find a proof. Then we found a counterexample and added axiom $x/x = e$ to kill the unwanted model $x + 3y$ (mod 8). Otter found a proof, but the proof did not use $e/(e/x) = x$. Hence the current basis.)

Proof (found by Otter 3.0.4 on gyro at 130.83 seconds).

2	$x/e = x$
3	$x/x = e$
5	$A/((B/C)/(B/A)) \neq C$

10	$(x/e)/e = x$	$[2 \to 2]$
15	$x/(x/e) = e$	$[2 \to 3]$
17	$x/(y/y) = x$	$[3 \to 2]$
152	$x/(x/y) = z/(z/y)$	$[15 \to 15 \text{ :(gL)}]$
634	$x/(x/y) = y$	$[17 \to 152, \text{ flip}]$
858	$(x/e)/(x/y) = y$	$[2 \to 634]$
884	$x/((x/y)/e) = y$	$[2 \to 634]$
9245	$(x/y)/(x/z) = (z/y)/e$	$[10 \to 858 \text{ :(gL)}]$
11150	$x/((y/z)/(y/x)) = z$	$[9245 \to 884]$
11151	$\square$	$[11150,5]$

Theorem ABGT-6. A (gL)-basis for left and right division in Abelian groups.

The type is $\langle 2, 0 \rangle$.

$$\left\{ \begin{array}{l} x \mid x = e \\ e \mid (e \mid x) = x \\ (x \mid e) \mid e = x \\ e \mid (x \mid e) = (e \mid x) \mid e \end{array} \right\} \; =\!(gL)\!\Rightarrow \; \{(x \mid y) \mid (((x \mid z) \mid (u \mid u)) \mid y) = z\}.$$

In view of [15], this gives us all Abelian group axioms satisfied by both left division \ and right division /.

Proof (found by Otter 3.0.4 on gyro at 2.97 seconds).

3,2	$x \mid x = e$	
5,4	$e \mid (e \mid x) = x$	
7,6	$(x \mid e) \mid e = x$	
8	$e \mid (x \mid e) = (e \mid x) \mid e$	
9	$(e \mid x) \mid e = e \mid (x \mid e)$	[flip 8]
11	$(A \mid B) \mid (((A \mid C) \mid (D \mid D)) \mid B) \neq C$	
12	$(A \mid B) \mid (((A \mid C) \mid e) \mid B) \neq C$	[copy,11 :3]
13	$(e \mid x) \mid (e \mid y) = e \mid (x \mid y)$	$[6 \to 9 \text{ :(gL) :3}]$
15	$(x \mid y) \mid e = (x \mid e) \mid (y \mid e)$	$[4 \to 9 \text{ :(gL) :3}]$
23,22	$(e \mid x) \mid y = e \mid (x \mid (e \mid y))$	$[4 \to 13]$
24	$(x \mid y) \mid (e \mid z) = (x \mid e) \mid (y \mid z)$	$[4 \to 13 \text{ :(gL) :3}]$
25	$x \mid (y \mid (e \mid z)) = e \mid (y \mid (x \mid z))$	$[13 \to 13 \text{ :(gL) :5}]$
26	$x \mid (y \mid e) = e \mid (y \mid (x \mid e))$	$[9 \to 13 \text{ :(gL) :5}]$
27	$((x \mid y) \mid z) \mid e = ((x \mid e) \mid z) \mid y$	$[15 \to 15 \text{ :(gL) :7}]$
39	$(A \mid B) \mid (((A \mid e) \mid (C \mid e)) \mid B) \neq C$	$[15 \to 12]$
53,52	$(x \mid y) \mid e = e \mid (y \mid x)$	$[15 \to 26 \text{ :7}]$
67,66	$((x \mid e) \mid y) \mid z = e \mid (y \mid (x \mid z))$	$[27 \text{ :53, flip}]$
71	$(A \mid B) \mid (e \mid ((C \mid e) \mid (A \mid B))) \neq C$	$[39 \text{ :67}]$
80	$(x \mid y) \mid (e \mid y) = x$	$[2 \to 24 \text{ :53,5}]$
113	$x \mid e = y \mid (e \mid (x \mid y))$	$[2 \to 25 \text{ :23,5}]$
137,136	$(x \mid (y \mid z)) \mid z = (x \mid e) \mid y$	$[80 \to 24 \text{ :5}]$

| 183 | $x \mid (e \mid ((y \mid e) \mid x)) = y$ | $[113 \rightarrow 113 : 53, 137, 3, 5, \text{flip}]$ |
| 185 | □ | $[183, 71]$ |

In each of the next two theorems, the right-hand side is a single axiom for Abelian groups in terms of the double inversion operation [35]. The context is Thm. 3 of [57]. Let e be an inflection point on the curve Γ. Then the binary morphism xy of chord-tangent construction on Γ is completely characterized by the single axiom. See Fig. 3.2 on page 41.

Theorem ABGT-7. A (gL)-basis for Abelian groups with double inversion (1).

$$\left\{ \begin{array}{l} x(yx) = y \\ ex = xe \\ ee = e \end{array} \right\} \begin{array}{c} =(gL)\!\!\Rightarrow \\ \Longleftarrow \end{array} \{(x((z(xy))(ey)))(ee) = z\}.$$

Proof $(\Rightarrow)$ found by Otter 3.0.4 on gyro at 7.48 seconds.

1	$x = x$	
3,2	$e \cdot e = e$	
4	$e \cdot x = x \cdot e$	
6,5	$x \cdot (y \cdot x) = y$	
7	$(A \cdot ((C \cdot (A \cdot B)) \cdot (e \cdot B))) \cdot (e \cdot e) \neq C$	
8	$(A \cdot ((C \cdot (A \cdot B)) \cdot (e \cdot B))) \cdot e \neq C$	$[\text{copy}, 7 : 3]$
13	$(e \cdot x) \cdot (y \cdot (x \cdot e)) = y$	$[4 \rightarrow 5]$
19,18	$(x \cdot y) \cdot x = y$	$[5 \rightarrow 5]$
21,20	$e \cdot (e \cdot x) = x$	$[4 \rightarrow 5]$
22	$x \cdot (x \cdot e) = e$	$[4 \rightarrow 5]$
35	$x \cdot (x \cdot y) = y$	$[20 \rightarrow 22 : (\text{gL}) : 21]$
41	$(x \cdot y) \cdot y = x$	$[18 \rightarrow 35]$
43	$x \cdot y = y \cdot x$	$[5 \rightarrow 35]$
44	$(x \cdot y) \cdot (z \cdot (y \cdot u)) = (x \cdot v) \cdot (z \cdot (v \cdot u))$	$[13 \rightarrow 13 : (\text{gL}) : (\text{gL})]$
46	$(e \cdot x) \cdot y = (x \cdot e) \cdot y$	$[18 \rightarrow 13]$
54	$(A \cdot (((A \cdot B) \cdot C) \cdot (e \cdot B))) \cdot e \neq C$	$[43 \rightarrow 8]$
65,64	$(x \cdot y) \cdot (z \cdot (y \cdot x)) = z$	$[43 \rightarrow 5]$
84	$(x \cdot y) \cdot z = (x \cdot u) \cdot (y \cdot (u \cdot z))$	$[35 \rightarrow 44]$
129,128	$((x \cdot y) \cdot e) \cdot z = (e \cdot (y \cdot x)) \cdot z$	$[43 \rightarrow 46, \text{flip}]$
473	$e \cdot ((((A \cdot B) \cdot C) \cdot (e \cdot B)) \cdot A) \neq C$	$[64 \rightarrow 54 : 129, 65]$
540,539	$((x \cdot y) \cdot z) \cdot u \doteq x \cdot (z \cdot (y \cdot u))$	$[41 \rightarrow 84]$
593	$C \neq C$	$[473 : 540, 6, 19, 6]$
594	□	$[593, 1]$

Proof $(\Leftarrow)$ found by Otter 3.0.4 on gyro at 0.56 seconds.

1	$x = x$	
2	$e \cdot e = e, \ e \cdot A = A \cdot e, \ A \cdot (B \cdot A) = B \ \rightarrow \ $ □	
4,3	$(x \cdot ((y \cdot (x \cdot z)) \cdot (e \cdot z))) \cdot (e \cdot e) = y$	

5	$(e \cdot (x \cdot (e \cdot e))) \cdot (e \cdot e) = y \cdot ((x \cdot (y \cdot z)) \cdot (e \cdot z))$	$[3 \to 3]$
7	$x \cdot ((y \cdot (x \cdot z)) \cdot (e \cdot z)) = u \cdot ((y \cdot (u \cdot v)) \cdot (e \cdot v))$	$[5 \to 5]$
8	$x \cdot (((y \cdot (e \cdot e)) \cdot (x \cdot z)) \cdot (e \cdot z)) = y$	$[3 \to 5, \text{flip}]$
11,10	$(e \cdot x) \cdot (e \cdot e) = e \cdot (x \cdot (e \cdot e))$	$[5 \to 3 :4]$
13,12	$e \cdot (((x \cdot (e \cdot y)) \cdot (e \cdot y)) \cdot (e \cdot e)) = x$	$[3 \to 10, \text{flip}]$
14	$(x \cdot (e \cdot ((x \cdot e) \cdot (e \cdot e)))) \cdot (e \cdot e) = e$	$[10 \to 3]$
16	$e \cdot (e \cdot (e \cdot (e \cdot (e \cdot (e \cdot (e \cdot e)))))) = e$	$[10 \to 8 :11,11,11,11,11]$
19,18	$e \cdot (((x \cdot y) \cdot y) \cdot (e \cdot e)) = x$	$[12 \to 12 :13]$
20	$e \cdot (((x \cdot (e \cdot e)) \cdot y) \cdot y) = x$	$[18 \to 8 :19]$
24	$((x \cdot e) \cdot x) \cdot (e \cdot e) = e$	$[18 \to 14]$
27,26	$e \cdot (e \cdot e) = e$	$[24 \to 24]$
29,28	$e \cdot e = e$	$[16 :27,27,27]$
30	$((x \cdot e) \cdot x) \cdot e = e$	$[24 :29]$
34	$e \cdot (((x \cdot e) \cdot y) \cdot y) = x$	$[20 :29]$
36	$e \cdot (((x \cdot y) \cdot y) \cdot e) = x$	$[18 :29]$
41,40	$(e \cdot x) \cdot e = e \cdot (x \cdot e)$	$[10 :29,29]$
44	$(x \cdot ((y \cdot (x \cdot z)) \cdot (e \cdot z))) \cdot e = y$	$[3 :29]$
46	$e \cdot A = A \cdot e, \ A \cdot (B \cdot A) = B \ \to \ \square$	$[2 :29 :1]$
47	$e \cdot ((e \cdot x) \cdot x) = (y \cdot e) \cdot y$	$[30 \to 34]$
49,48	$e \cdot ((e \cdot x) \cdot x) = e$	$[28 \to 34]$
50	$(x \cdot e) \cdot x = e$	$[\text{flip } 47 :49]$
53,52	$e \cdot (e \cdot x) = x$	$[50 \to 34]$
54	$((x \cdot e) \cdot y) \cdot y = e \cdot x$	$[34 \to 52, \text{flip}]$
62	$(e \cdot x) \cdot x = e$	$[48 \to 52 :29, \text{flip}]$
64	$x \cdot (e \cdot x) = e$	$[52 \to 62]$
67,66	$x \cdot ((y \cdot (x \cdot z)) \cdot (e \cdot z)) = e \cdot y$	$[64 \to 7 :53, \text{flip}]$
68	$x \cdot ((y \cdot e) \cdot x) = e \cdot y$	$[64 \to 7 :53,67]$
71,70	$e \cdot (x \cdot e) = x$	$[44 :67,41]$
75,74	$(x \cdot y) \cdot y = x$	$[36 :71]$
76	$x \cdot e = e \cdot x$	$[54 :75]$
77	$e \cdot x = x \cdot e$	$[\text{flip } 76]$
109,108	$x \cdot (y \cdot x) = y$	$[74 \to 68 :71]$
112	$\square$	$[46 :109 :77,1]$

Theorem ABGT-8. A (gL)-basis for Abelian groups with double inversion (2).

$$\left\{ \begin{array}{l} ee = e \\ (xe)e = x \\ x(xy) = y \end{array} \right\} \quad \overset{=(gL)}{\underset{\Longleftarrow}{\Longrightarrow}} \quad \{(x(((xy)z)(ye)))(ee) = z\}.$$

Proof ($\Rightarrow$) found by Otter 3.0.4 on gyro at 12.16 seconds.

1	$x = x$
3,2	$e \cdot e = e$
5,4	$(x \cdot e) \cdot e = x$

7,6	$x \cdot (y \cdot x) = y$	
8	$(A \cdot (((A \cdot B) \cdot C) \cdot (B \cdot e))) \cdot (e \cdot e) \neq C$	
9	$(A \cdot (((A \cdot B) \cdot C) \cdot (B \cdot e))) \cdot e \neq C$	[copy,8 :3]
11,10	$(x \cdot y) \cdot x = y$	$[6 \to 6]$
12	$e \cdot x = x \cdot e$	$[4 \to 6]$
14,13	$x \cdot (x \cdot e) = e$	$[4 \to 10]$
17,16	$e \cdot (e \cdot x) = x$	$[12 \to 12 :5]$
30	$x \cdot (x \cdot y) = z \cdot (z \cdot y)$	$[13 \to 13 :(gL)]$
34,33	$x \cdot (x \cdot y) = y$	$[13 \to 30 :14,17, \text{flip}]$
39	$(x \cdot y) \cdot y = x$	$[10 \to 30 :34]$
41	$x \cdot y = y \cdot x$	$[6 \to 30 :34]$
45	$(x \cdot y) \cdot z = z \cdot (y \cdot x)$	$[41 \to 41]$
121	$(x \cdot y) \cdot (z \cdot u) = (u \cdot z) \cdot (y \cdot x)$	$[41 \to 45]$
126	$e \cdot (((((A \cdot B) \cdot C) \cdot (B \cdot e)) \cdot A) \neq C$	$[45 \to 9]$
218	$(x \cdot y) \cdot (z \cdot u) = (x \cdot z) \cdot (y \cdot u)$	$[(gL) 121]$
222,221	$((x \cdot y) \cdot z) \cdot (y \cdot u) = x \cdot (z \cdot u)$	$[39 \to 218, \text{flip}]$
250	$C \neq C$	$[126 :222,11,7]$
251	$\square$	$[250,1]$

Proof $(\Leftarrow)$ found by Otter 3.0.4 on gyro at 0.40 seconds.

1	$x = x$	
2	$e \cdot e = e, \ (A \cdot e) \cdot e = A, \ A \cdot (B \cdot A) = B \ \to \ \square$	
3	$(x \cdot (((x \cdot y) \cdot z) \cdot (y \cdot e))) \cdot (e \cdot e) = z$	
8,7	$(((x \cdot e) \cdot y) \cdot z) \cdot (y \cdot e) = (x \cdot z) \cdot (e \cdot e)$	$[3 \to 3, \text{flip}]$
10	$x \cdot (((y \cdot x) \cdot (e \cdot e)) \cdot e) = (y \cdot (e \cdot e)) \cdot (e \cdot e)$	$[3 \to 7 :8]$
14,13	$((x \cdot e) \cdot ((x \cdot y) \cdot (e \cdot e))) \cdot (e \cdot e) = y$	$[7 \to 3]$
15	$(e \cdot e) \cdot ((x \cdot (e \cdot e)) \cdot e) = x \cdot (e \cdot e)$	$[13 \to 10 :14]$
18,17	$(e \cdot e) \cdot (x \cdot e) = x$	$[13 \to 15 :14]$
23	$((e \cdot e) \cdot e) \cdot (e \cdot e) = e$	$[17 \to 13]$
29,28	$(e \cdot (x \cdot (e \cdot e))) \cdot (e \cdot e) = x \cdot e$	$[17 \to 3]$
35	$((e \cdot e) \cdot e) \cdot e = e$	$[23 \to 28 :18, \text{flip}]$
40,39	$(e \cdot e) \cdot e = e$	$[17 \to 28 :18, \text{flip}]$
41	$(((x \cdot e) \cdot e) \cdot y) \cdot e = (x \cdot y) \cdot e$	$[7 \to 28 :29, \text{flip}]$
46,45	$e \cdot e = e$	$[35 :40]$
54,53	$e \cdot (x \cdot e) = x$	$[17 :46]$
58	$x \cdot (((y \cdot x) \cdot e) \cdot e) = (y \cdot e) \cdot e$	$[10 :46,46,46]$
65	$(A \cdot e) \cdot e = A, \ A \cdot (B \cdot A) = B \ \to \ \square$	$[2 :46 :1]$
68	$((x \cdot e) \cdot e) \cdot y = x \cdot y$	$[41 \to 53 :54, \text{flip}]$
87,86	$(x \cdot e) \cdot e = x$	$[68 \to 53 :54, \text{flip}]$
90	$A \cdot (B \cdot A) = B \ \to \ \square$	$[65 :87 :1]$
91	$x \cdot (y \cdot x) = y$	$[58 :87,87]$
93	$\square$	$[91,90]$

4.1.2 Quasigroups

Given a group, one can define at least three other binary operations $x \cdot y$ that are quasigroups; they are $x - y$, $-x + y$, and $-x - y$. In this section we present a set of one-variable laws true for all these three operations that (gL)-implies all the equations valid in Abelian groups common to those interpretations. The first conjecture we gave to Otter is not a theorem: the basis is what we intended, but the conclusion is too strong. However, Otter derived two equations that turned out to be sufficient for our purpose.

Consider type $\langle 2, 0 \rangle$ with constant e, and let

$$\text{A1} = \left\{ \begin{array}{l} (xe)e = x \\ e(ex) = x \\ ee = e \\ e(xe) = (ex)e \end{array} \right\}, \qquad \text{A2} = \text{A1} \cup \{((xx)e)(xx) = e\}.$$

We first asked Otter to prove

$$\text{A2} \ =\!\!(gL)\!\Rightarrow \ \left\{ \begin{array}{l} (xy)(zu) = (xz)(yu) \\ ((xx)y)(xx) = y \end{array} \right\}.$$

Otter failed, but we noticed several interesting (gL)-consequences of axioms A2, including

$$(x((xy)x))x = y, \qquad (5)$$
$$x((x(yx))x) = y. \qquad (6)$$

Electronic mail (slightly edited) from Padmanabhan to McCune, November 4, 1994:

Here what we want is already obtained by Otter, namely, the two equations (5) and (6). The only purpose of this problem is to prove that any (gL)-algebra satisfying the axioms A2 must also satisfy all the equations true for the three interpretations

$$a \cdot b = a - b, \ a \cdot b = -a + b, \ \text{or } a \cdot b = -a - b$$

in Abelian groups. What you have obtained is a basis for the equational theory of algebras of type $\langle 2, 0 \rangle$ satisfying all the equations common to the above three operations in Abelian groups. This is because the two equations (5) and (6) imply the two cancellation laws and make the algebra a quasigroup at the same time. David Kelly and I already have such a basis ([25, last line of Table 2]) but our set involved four variables. The above result shows that, modulo (gL), we can define the same equational theory by a set of *one*-variable sentences, i.e., the axioms A2. That is a *new* result and that was the intent of the problem.

Now coming to the actual problem posed as above, is $((xx)y)(xx) = y$ a (gL)-consequence? Unfortunately, no. Interpret $x \cdot y$ as $x - y$ and e as 0, the zero in an Abelian group. This is a (gL)-algebra satisfying all the axioms (A2). This is also medial. However, $((x \cdot x) \cdot y) \cdot (x \cdot x) = (0 \cdot y) \cdot 0 = (-y) \cdot 0 = -y \neq y$ as claimed in the conclusion. No wonder Otter could not derive this. I will write a new proof for the following.

Theorem. *A (gL)-algebra of type $\langle 2, 0 \rangle$ satisfies all the equations common to the three operations*

$$a \cdot b = a - b, \ a \cdot b = -a + b, \ or \ a \cdot b = -a - b$$

if and only if it satisfies the one-variable equations

$$\text{A2} = \left\{ \begin{array}{l} (xe)e = x \\ e(ex) = x \\ ee = e \\ e(xe) = (ex)e \\ ((xx)e)(xx) = e \end{array} \right\}.$$

The crucial step in the proof will be the two equations (5) and (6).

From the above example, we can make the conjecture [Thm. QGT-1 below] that

$$\text{A2} =\!\langle gL \rangle\!\!\Rightarrow \{((xx)y)(xx) = (ey)e\},$$

because this equation is valid for all the three models. Thus this forms a test case. This was the original intention of the problem.

Let us call a binary operation *generalized division* if it satisfies all equations common to $x - y$, $-x + y$, and $-x - y$ in Abelian groups.

Theorem QGT-1. Test case for generalized division.

$$\text{A2} =\!\langle gL \rangle\!\!\Rightarrow \{((xx)y)(xx) = (ey)e\}.$$

Proof (found by Otter 3.0.4 on gyro at 0.33 seconds).

2	$((A \cdot A) \cdot B) \cdot (A \cdot A) \neq (e \cdot B) \cdot e$	
3	$(x \cdot e) \cdot e = x$	
5	$e \cdot (e \cdot x) = x$	
8,7	$e \cdot e = e$	
9	$e \cdot (x \cdot e) = (e \cdot x) \cdot e$	
11,10	$(e \cdot x) \cdot e = e \cdot (x \cdot e)$	[flip 9]
12	$((x \cdot x) \cdot e) \cdot (x \cdot x) = e$	
14	$((A \cdot A) \cdot B) \cdot (A \cdot A) \neq e \cdot (B \cdot e)$	[2 :11]
15	$(e \cdot x) \cdot (e \cdot y) = e \cdot (x \cdot y)$	[7 → 10 :(gL)]
25,24	$(e \cdot x) \cdot y = e \cdot (x \cdot (e \cdot y))$	[5 → 15]
31	$((x \cdot x) \cdot y) \cdot (x \cdot x) = e \cdot (y \cdot e)$	[3 → 12 :(gL) :25,8]
33	□	[31,14]

Theorem QGT-2. A (gL)-basis for generalized division in Abelian groups.

$$\text{A2} \;=\!(gL)\!\Rightarrow\; \left\{ \begin{array}{l} (x((xy)x))x = y \\ x((x(yx))x) = y \end{array} \right\}.$$

Proof (found by Otter 3.0.4 on gyro at 85.32 seconds).

1	$x = x$	
2	$(A \cdot ((A \cdot B) \cdot A)) \cdot A = B, \; A \cdot ((A \cdot (B \cdot A)) \cdot A) = B \;\to\; \square$	
4,3	$(x \cdot e) \cdot e = x$	
6,5	$e \cdot (e \cdot x) = x$	
8,7	$e \cdot e = e$	
9	$e \cdot (x \cdot e) = (e \cdot x) \cdot e$	
10	$(e \cdot x) \cdot e = e \cdot (x \cdot e)$	[flip 9]
12	$((x \cdot x) \cdot e) \cdot (x \cdot x) = e$	
14	$(e \cdot x) \cdot (e \cdot y) = e \cdot (x \cdot y)$	$[7 \to 10 :(\text{gL})]$
16	$(x \cdot y) \cdot e = (x \cdot e) \cdot (y \cdot e)$	$[7 \to 10 :(\text{gL})]$
24,23	$(e \cdot x) \cdot y = e \cdot (x \cdot (e \cdot y))$	$[5 \to 14]$
25	$(x \cdot y) \cdot (e \cdot z) = (x \cdot e) \cdot (y \cdot z)$	$[7 \to 14 :(\text{gL})]$
27	$x \cdot (y \cdot e) = e \cdot (y \cdot (x \cdot e))$	$[10 \to 14 :(\text{gL}) :6]$
30	$((x \cdot x) \cdot y) \cdot (x \cdot x) = e \cdot (y \cdot e)$	$[3 \to 12 :(\text{gL}) :24,8]$
63	$x \cdot y = e \cdot ((y \cdot e) \cdot (x \cdot e))$	$[3 \to 27]$
66,65	$(x \cdot y) \cdot e = e \cdot (y \cdot x)$	$[16 \to 27 :4]$
74	$(x \cdot e) \cdot (y \cdot e) = e \cdot (y \cdot x)$	$[3 \to 27]$
108,107	$((x \cdot y) \cdot z) \cdot (e \cdot u) = e \cdot (z \cdot (e \cdot ((y \cdot x) \cdot u)))$	$[65 \to 25 :(\text{gL}) :24]$
111	$x \cdot ((y \cdot z) \cdot u) = z \cdot ((y \cdot x) \cdot u)$	$[25 \to 25 :108,24,6,6,66,6]$
160	$(x \cdot x) \cdot (y \cdot (x \cdot x)) = e \cdot (y \cdot e)$	$[30 \to 30 :(\text{gL}) :66,6]$
175	$(x \cdot y) \cdot (z \cdot e) = e \cdot (y \cdot (e \cdot (z \cdot x)))$	$[7 \to 74 :(\text{gL}) :24]$
294,293	$(x \cdot x) \cdot ((x \cdot x) \cdot y) = y$	$[65 \to 160 :(\text{gL}) :6]$
411	$(x \cdot (x \cdot x)) \cdot (x \cdot e) = e$	$[160 \to 175 :8,8,8]$
445	$(x \cdot (x \cdot x)) \cdot (x \cdot y) = y$	$[293 \to 411 :(\text{gL}) :294]$
467	$(x \cdot y) \cdot ((y \cdot y) \cdot y) = x$	$[63 \to 445 :66,66,24,6,6]$
645	$(x \cdot ((y \cdot y) \cdot y)) \cdot y = x$	$[467 \to 467 :294]$
1087,1086	$x \cdot ((x \cdot (y \cdot x)) \cdot x) = y$	$[467 \to 111, \text{flip}]$
1102	$(A \cdot ((A \cdot B) \cdot A)) \cdot A = B \;\to\; \square$	$[2 :1087 :1]$
1106	$(x \cdot ((x \cdot y) \cdot x)) \cdot x = y$	$[111 \to 645]$
1108	$\square$	$[1106,1102]$

Theorem STN-4. Stein quasigroups under (gL).

$$\left\{ \begin{array}{l} x(y(yx)) = yx \\ \text{cancellation} \end{array} \right\} \;=\!(gL)\!\Rightarrow\; \{x(xy) = yx\}.$$

Proof (found by Otter 3.0.4 on gyro at 0.44 seconds).

1	$A \cdot (A \cdot B) = B \cdot A \;\to\; \Box$	
4	$x \cdot y = z,\; u \cdot y = z \;\to\; x = u$	
6,5	$x \cdot (y \cdot (y \cdot x)) = y \cdot x$	
9	$(x \cdot y) \cdot (z \cdot (z \cdot u)) = z \cdot (y \cdot (x \cdot u))$	$[5 \to 5 :(\text{gL}) :6]$
10	$(x \cdot (x \cdot y)) \cdot (y \cdot z) = y \cdot (x \cdot z)$	$[5 \to 5 :(\text{gL})]$
33	$x \cdot (x \cdot y) = y \cdot x$	$[4,10,9]$
34	$\Box$	$[33,1]$

In correspondence with Padmanabhan, Ross Willard has shown that $x(y(yx)) = yx$ and the cancellation laws do not imply $x(xy) = yx$.

4.1.3 Boolean Groups

Theorem TBG-1. A (gL)-basis for ternary Boolean groups.

The pair of equations $\{p(x,x,x) = x, p(x,y,p(z,u,v)) = p(y,z,p(u,v,x))\}$ is a (gL)-basis for the variety of Boolean groups, where $p(x,y,z)$ is $x + (y + z)$. It is sufficient to derive a basis for Boolean groups as follows.

$$\left\{ \begin{array}{l} p(x,x,x) = x \\ p(x,y,p(z,u,v)) = p(y,z,p(u,v,x)) \\ p(x,y,z) = x + (y+z) \end{array} \right\} =\!(gL)\!\Rightarrow$$

$$\left\{ \begin{array}{l} (x+y) + z = x + (y+z) \\ x + x = y + y \\ x + (y+y) = x \end{array} \right\}.$$

Proof (found by Otter 3.0.4 on gyro at 10.57 seconds).

1	$x = x$	
2	$(A+B) + C = A + (B+C),\; B+B = A+A,$	
	$A + (B+B) = A \;\to\; \Box$	
3	$p(x,x,x) = x$	
4	$p(x,y,p(z,u,v)) = p(y,z,p(u,v,x))$	
5	$p(x,y,z) = x + (y+z)$	
7,6	$(x + (x+x)) + (x+x) = x$	$[3 \to 3 :5,5]$
13,12	$x + (x+x) = x$	$[3 \to 6 :5,7]$
14	$x + (y+z) = y + (z + (z + (z+x)))$	$[3 \to 4 :5,5,5]$
16,15	$x + (y + (z + (x+x))) = y + (z+x)$	$[3 \to 4 :5,5,5]$
17	$x + (y+z) = y + (x+z)$	$[(\text{gL})\ 4 :5,5]$
20,19	$x + (y + (x+x)) = y + x$	$[12 \to 17,\ \text{flip}]$
26,25	$(x+x) + (y+x) = y + x$	$[12 \to 14 :13]$
32	$x + (y + (y + (y+z))) = z + (x+y)$	$[14 \to 12 :13]$
35,34	$(x+x) + ((x+x) + y) = x + (x+y)$	$[19 \to 19 :(\text{gL})]$
36	$x + (x + (y+x)) = y + x$	$[17 \to 19]$

38	$x + (x + (x + (x + (x + y)))) = y + x$	$[14 \to 19]$
43,42	$(x + x) + x = x$	$[12 \to 19 : 26,13]$
56	$((x + x) + y) + (x + x) = (x + y) + x$	$[42 \to 12 : (gL) : 43]$
61,60	$((x + y) + (x + y)) + z = (y + y) + z$	$[25 \to 25 : (gL)]$
65,64	$(x + x) + (y + (z + x)) = y + (z + x)$	$[19 \to 25 : (gL) : 61,20]$
68,67	$(x + y) + ((x + y) + z) = y + (y + z)$	
		$[25 \to 34 : 26,26,26,26,61,61,35,26, \text{flip}]$
76,75	$x + ((y + x) + x) = y + x$	$[14 \to 36 : 68,13]$
80	$(x + (y + y)) + (y + y) = (x + y) + y$	$[36 \to 56 : 65,20]$
84	$(x + (y + y)) + y = (x + y) + (y + y)$	$[12 \to 56 : (gL) : 26, \text{flip}]$
96,95	$(x + y) + (z + y) = x + (y + (z + y))$	$[12 \to 75 : (gL) : 13,16]$
99	$((x + y) + y) + (x + (x + y)) = x + y$	$[75 \to 75 : 76,96,76]$
110	$(x + (y + y)) + y = x + y$	$[84 : 96,13]$
125,124	$(x + (y + y)) + z = x + z$	$[42 \to 110 : (gL) : 43]$
143,142	$(x + y) + y = x + (y + y)$	$[80 : 125, \text{flip}]$
145,144	$x + (x + (x + y)) = x + y$	$[99 : 143,125]$
148	$x + y = y + x$	$[38 : 145,145]$
149	$x + (y + z) = z + (x + y)$	$[32 : 145]$
167	$x + (y + (z + y)) = x + z$	$[36 \to 148 : (gL)]$
170,169	$x + (y + (z + z)) = x + y$	$[19 \to 148 : (gL)]$
176,175	$(x + y) + z = x + (y + z)$	$[15 \to 148 : 170, \text{flip}]$
194	$x + (y + (x + z)) = y + z$	$[36 \to 148 : (gL) : 176,176]$
197,196	$x + (y + (y + z)) = x + z$	$[19 \to 148 : (gL) : 176,176]$
209	$B + B = A + A, \ A + (B + B) = A \ \to \ \square$	$[2 : 176 : 1]$
222,221	$x + (y + y) = x$	$[12 \to 167 : 176,197]$
224	$B + B = A + A \ \to \ \square$	$[209 : 222 : 1]$
231,230	$x + (y + x) = y$	$[149 \to 221]$
236	$x + x = A + A \ \to \ \square$	$[194 \to 224 : 231]$
237	$\square$	$[236,221]$

The preceding theorem arose by accident as we were trying to write axioms for semilattices in terms of a ternary operation p. Because semilattices are inconsistent with (gL) (Thm. LT-1), we expected Otter to derive $x = y$ from our equations. Otter failed to do so, but we noticed that p was shown to be a minority polynomial, that is, $p(x, x, y) = p(x, y, x) = p(y, x, x) = y$. We soon realized that what we had written was in fact a (gL)-basis for Boolean groups in terms of a ternary operation, a system we had not previously considered. Note also that the preceding Otter proof shows that a nonregular equation can be derived by (gL) from a set of regular equations; this situation cannot happen in ordinary equational logic. (An equation is *regular* if the same set of variables occurs on both sides of the equal sign).

4.1.4 Cancellative Semigroups

Theorem CS-GL-1. Cancellative (gL)-semigroups are commutative.

$$\left\{ \begin{array}{l} (xy)z = x(yz) \\ \text{cancellation} \end{array} \right\} \; =\!\!(gL)\!\!\Rightarrow \; \{xy = yx\}.$$

Proof (found by Otter 3.0.4 on gyro at 6.13 seconds).

2	$x \cdot y = u, \; x \cdot z = u \; \rightarrow \; y = z$	
3	$y \cdot x = u, \; z \cdot x = u \; \rightarrow \; y = z$	
4	$(x \cdot y) \cdot z = x \cdot (y \cdot z)$	
5	$B \cdot A = A \cdot B \; \rightarrow \; \square$	
6	$(((x \cdot y) \cdot z) \cdot u) \cdot v = (x \cdot (y \cdot z)) \cdot (u \cdot v)$	$[4 \rightarrow 4]$
11	$(x \cdot (y \cdot z)) \cdot u = (x \cdot y) \cdot (z \cdot u)$	$[4 \rightarrow 4]$
432	$(((x \cdot y) \cdot z) \cdot u) \cdot v = (x \cdot u) \cdot ((y \cdot z) \cdot v)$	$[(gL)\ 6]$
433	$((x \cdot y) \cdot z) \cdot u = x \cdot (u \cdot (y \cdot z))$	$[3,11,432,\ \text{flip}]$
1205	$x \cdot y = y \cdot x$	$[2,432,433]$
1206	$\square$	$[1205,5]$

Electronic mail (slightly edited) from Padmanabhan to McCune, May 11, 1993:

The proof of

$$\{CS\} \; =\!\!(gL)\!\!\Rightarrow \; \{xy = yx\}$$

is especially excellent. This means that not only a group law, but even a cancellative semigroup law on a projective curve must be commutative! This is a new result.

Now the commutative law has only two variables and (gL) is an inference rule. Believing in the cosmic orderliness of mathematics, one can ask the question whether the two cancellation laws along with the set of all two-variable consequences of associativity will already (gL)-imply $xy = yx$.

The following theorem answers the question, using just two of the three two-variable instances of associativity.

Theorem CS-GL-2. Diassociative cancellative (gL)-groupoids are commutative.

$$\left\{ \begin{array}{l} x(yx) = (xy)x \\ x(xy) = (xx)y \\ \text{cancellation} \end{array} \right\} \; =\!\!(gL)\!\!\Rightarrow \; \{xy = yx\}.$$

Proof (found by Otter 3.0.4 on gyro at 1.52 seconds).

1	$B \cdot A = A \cdot B \to \square$	
2	$x = x$	
3	$x \cdot y = z,\ x \cdot u = z \to y = u$	
4	$x \cdot y = z,\ u \cdot y = z \to x = u$	
5	$(x \cdot x) \cdot y = x \cdot (x \cdot y)$	
6	$(x \cdot y) \cdot x = x \cdot (y \cdot x)$	
12	$(x \cdot x) \cdot (y \cdot z) = x \cdot ((x \cdot y) \cdot z)$	$[6 \to 5 :(gL)]$
13	$x \cdot ((y \cdot z) \cdot u) = x \cdot (y \cdot (z \cdot u))$	$[5 \to 12 :(gL), \text{flip}]$
14	$(x \cdot x) \cdot (y \cdot z) = x \cdot (y \cdot (x \cdot z))$	$[5 \to 12 :(gL)]$
19	$(x \cdot y) \cdot z = x \cdot (y \cdot z)$	$[3,2,13, \text{flip}]$
22	$(x \cdot y) \cdot z = y \cdot (x \cdot z)$	$[3,12,14, \text{flip}]$
38	$x \cdot y = y \cdot x$	$[4,19,22]$
39	$\square$	$[38,1]$

The next few theorems generalize Thm. CS-GL-1, the commutativity of cancellative semigroups under (gL), to weaker forms of associativity.

Theorem CS-GL-3. Nearly (1) associative cancellative (gL)-groupoids are commutative.

$$\left\{ \begin{array}{l} \text{cancellation} \\ x(y(zu)) = ((xy)z)u \end{array} \right\} =\!(gL)\!\Rightarrow \{xy = yx\}.$$

Proof (found by Otter 3.0.4 on gyro at 18.78 seconds).

1	$B \cdot A = A \cdot B \to \square$	
2	$x = x$	
3	$x \cdot y = z,\ x \cdot u = z \to y = u$	
4	$x \cdot y = z,\ u \cdot y = z \to x = u$	
5	$x \cdot (y \cdot (z \cdot u)) = ((x \cdot y) \cdot z) \cdot u$	
6	$((x \cdot y) \cdot z) \cdot u = x \cdot (y \cdot (z \cdot u))$	$[\text{flip } 5]$
7	$((((x \cdot y) \cdot z) \cdot u) \cdot v) \cdot w = x \cdot ((y \cdot (z \cdot u)) \cdot (v \cdot w))$	$[6 \to 6]$
8	$((x \cdot (y \cdot (z \cdot u))) \cdot v) \cdot w = ((x \cdot y) \cdot z) \cdot (u \cdot (v \cdot w))$	$[6 \to 6]$
14	$((x \cdot y) \cdot z) \cdot (u \cdot v) = x \cdot (((y \cdot z) \cdot u) \cdot v)$	$[6 \to 6]$
38	$((x \cdot y) \cdot z) \cdot (u \cdot v) = x \cdot ((y \cdot (z \cdot u)) \cdot v)$	$[6 \to 7 :(gL)]$
42	$x \cdot (((y \cdot z) \cdot u) \cdot v) = x \cdot ((y \cdot (z \cdot u)) \cdot v)$	$[14 \to 38]$
53	$((x \cdot y) \cdot z) \cdot u = (x \cdot (y \cdot z)) \cdot u$	$[3,2,42, \text{flip}]$
63	$(x \cdot y) \cdot z = x \cdot (y \cdot z)$	$[4,2,53, \text{flip}]$
111	$((x \cdot (y \cdot (z \cdot u))) \cdot v) \cdot w = ((x \cdot y) \cdot v) \cdot (u \cdot (z \cdot w))$	$[(gL)\ 8]$
125	$(x \cdot (y \cdot z)) \cdot u = (x \cdot y) \cdot (z \cdot u)$	$[63 \to 63]$
273	$(x \cdot y) \cdot (z \cdot (u \cdot v)) = (x \cdot u) \cdot (z \cdot (y \cdot v))$	$[8 \to 111 :(gL)]$
453	$(x \cdot y) \cdot z = y \cdot (x \cdot z)$	$[3,273,125, \text{flip}]$
540	$x \cdot y = y \cdot x$	$[4,63,453]$
541	$\square$	$[540,1]$

Theorem CS-GL-4. Nearly (2) associative cancellative (gL)-groupoids are commutative.

$$\left\{ \begin{array}{l} \text{cancellation} \\ x(y(z(uv))) = (((xy)z)u)v \end{array} \right\} =\!(gL)\!\Rightarrow \{xy = yx\}.$$

Proof (found by Otter 3.0.4 on gyro at 16.28 seconds).

1	$B \cdot A = A \cdot B \rightarrow \square$	
3	$x \cdot y = z, \ x \cdot u = z \rightarrow y = u$	
4	$x \cdot y = z, \ u \cdot y = z \rightarrow x = u$	
5	$x \cdot (y \cdot (z \cdot (u \cdot v))) = (((x \cdot y) \cdot z) \cdot u) \cdot v$	
6	$(((x \cdot y) \cdot z) \cdot u) \cdot v = x \cdot (y \cdot (z \cdot (u \cdot v)))$	[flip 5]
7	$((x \cdot (y \cdot (z \cdot (u \cdot v)))) \cdot w) \cdot v_6 = ((x \cdot y) \cdot z) \cdot (u \cdot (v \cdot (w \cdot v_6)))$	
		[6 → 6]
8	$(x \cdot (y \cdot (z \cdot (u \cdot v)))) \cdot w = (x \cdot y) \cdot (z \cdot (u \cdot (v \cdot w)))$	[6 → 6]
25	$((x \cdot (y \cdot (z \cdot (u \cdot v)))) \cdot w) \cdot v_6 = ((x \cdot y) \cdot w) \cdot (u \cdot (v \cdot (z \cdot v_6)))$	
		[(gL) 7]
32	$(x \cdot (y \cdot (z \cdot (u \cdot v)))) \cdot w = (x \cdot y) \cdot (w \cdot (u \cdot (v \cdot z)))$	[4,8,25, flip]
33	$(x \cdot (y \cdot (z \cdot u))) \cdot v = (x \cdot (y \cdot (v \cdot u))) \cdot z$	[4,7,25 :(gL)]
37	$(x \cdot y) \cdot (z \cdot (u \cdot (v \cdot w))) = (x \cdot v) \cdot (z \cdot (u \cdot (y \cdot w)))$	[7 → 25 :(gL)]
56	$x \cdot (y \cdot (z \cdot (u \cdot v))) = x \cdot (v \cdot (z \cdot (u \cdot y)))$	[8 → 32 :(gL)]
61	$x \cdot (y \cdot (z \cdot u)) = u \cdot (y \cdot (z \cdot x))$	[3,33,56]
208	$x \cdot y = y \cdot x$	[4,61,37]
209	$\square$	[208,1]

Theorem CS-GL-5. Nearly (3) associative cancellative (gL)-groupoids are commutative.

$$\left\{ \begin{array}{l} \text{cancellation} \\ x(y(zu)) = (xy)(zu) \end{array} \right\} =\!(gL)\!\Rightarrow \{xy = yx\}.$$

Proof (found by Otter 3.0.4 on gyro at 0.48 seconds).

1	$B \cdot A = A \cdot B \rightarrow \square$	
3	$x \cdot y = z, \ x \cdot u = z \rightarrow y = u$	
4	$x \cdot y = z, \ u \cdot y = z \rightarrow x = u$	
5	$x \cdot (y \cdot (z \cdot u)) = (x \cdot y) \cdot (z \cdot u)$	
6	$(x \cdot y) \cdot (z \cdot u) = x \cdot (y \cdot (z \cdot u))$	[flip 5]
13	$(x \cdot y) \cdot (z \cdot u) = x \cdot ((y \cdot z) \cdot u)$	[6 → 6 :(gL)]
15	$(x \cdot y) \cdot (z \cdot u) = x \cdot (z \cdot (y \cdot u))$	[(gL) 6]
42	$(x \cdot y) \cdot z = y \cdot (x \cdot z)$	[3,13,15, flip]
91	$x \cdot y = y \cdot x$	[4,15,42]
92	$\square$	[91,1]

Example CS-GL-6. Nearly (4) associative cancellative (gL)-groupoids are not necessarily commutative.

$$\left\{ \begin{array}{l} \text{cancellation} \\ x(y(zu)) = (x(yz))u \end{array} \right\} = \!\!\!\!(gL)\!\!\!\!\nRightarrow \{xy = yx\}.$$

Proof. The hypotheses have the noncommutative (gL)-model $x \cdot y = x + 2y \pmod{3}$.

Theorem CS-GL-7. Nearly (5) associative cancellative (gL)-groupoids are commutative.

$$\left\{ \begin{array}{l} \text{cancellation} \\ (x(yz))u = (xy)(zu) \end{array} \right\} = \!\!\!\!(gL)\!\!\!\!\Rightarrow \{xy = yx\}.$$

Proof (found by Otter 3.0.4 on gyro at 2.32 seconds).

1	$B \cdot A = A \cdot B \ \rightarrow \ \square$	
3	$x \cdot y = z, \ x \cdot u = z \ \rightarrow \ y = u$	
4	$x \cdot y = z, \ u \cdot y = z \ \rightarrow \ x = u$	
5	$(x \cdot (y \cdot z)) \cdot u = (x \cdot y) \cdot (z \cdot u)$	
8	$((x \cdot (y \cdot z)) \cdot u) \cdot v = ((x \cdot y) \cdot z) \cdot (u \cdot v)$	$[5 \rightarrow 5]$
24	$((x \cdot (y \cdot z)) \cdot u) \cdot v = ((x \cdot y) \cdot u) \cdot (z \cdot v)$	$[(gL)\ 8]$
29	$(x \cdot (y \cdot z)) \cdot u = (x \cdot z) \cdot (y \cdot u)$	$[5 \rightarrow 24 :(gL)]$
32	$(x \cdot y) \cdot (z \cdot u) = (x \cdot z) \cdot (y \cdot u)$	$[8 \rightarrow 24 :(gL)]$
67	$x \cdot (y \cdot z) = x \cdot (z \cdot y)$	$[4,5,29]$
183	$x \cdot y = y \cdot x$	$[3,32,67]$
184	$\square$	$[183,1]$

4.2 Theories Strictly Inconsistent with (gL)

The philosophy behind Thms. PIX-1, MAJ-1, LT-1, and SD-1 (all previously known) is that there are no polynomially defined semilattice or lattice laws on, say, the reals, rationals, or complex numbers. Hence, modulo the process (gL), a semilattice will be crushed into a singleton. The validity of the implication made in Thm. PIX-1 was proved by Padmanabhan, actually via Thm. MAJ-1.

The following theorem shows that any theory containing a term satisfying the so-called Pixley properties [61], which characterize the permutability and distributivity of congruences of an equational class of algebras, is inconsistent with (gL). An example is Boolean algebra, with $p(x, y, z) = xy' + xz + y'z$.

Theorem PIX-1. Inconsistency of a Pixley polynomial with (gL).

$$\{p(y, y, x) = p(x, z, z) = p(x, u, x) = x\} = \!\!\!\!(gL)\!\!\!\!\Rightarrow \{x = y\}.$$

Proof (found by Otter 3.0.4 on gyro at 0.11 seconds).

1	$B \neq A$	
4,3	$p(x, x, y) = y$	
6,5	$p(x, y, y) = x$	
7	$p(x, y, x) = x$	
11,10	$p(x, y, z) = x$	[5 → 7 :(gL) :6,6]
12	$x = y$	[3 → 7 :(gL) :4,11,4]
13	□	[12,1]

The next theorem says that no theory containing a ternary majority polynomial [53] will satisfy the rule (gL). In particular, lattice theory does not satisfy this rule.

Theorem MAJ-1. Inconsistency of a majority polynomial with (gL).

$$\{p(x, x, y) = p(x, y, x) = p(y, x, x) = x\} \;=\!(gL)\!\Rightarrow\; \{x = y\}.$$

Proof (found by Otter 3.0.4 on gyro at 0.31 seconds).

1	$B \neq A$	
3	$p(x, x, y) = x$	
4	$p(x, y, x) = x$	
5	$p(x, y, y) = y$	
11	$p(x, y, z) = p(x, y, u)$	[3 → 3 :(gL)]
41	$p(x, y, z) = x$	[4 → 11, flip]
53	$p(x, y, z) = p(u, y, z)$	[5 → 5 :(gL)]
65	$p(x, y, z) = u$	[53 → 41]
66	$x = y$	[65 → 65]
67	□	[66,1]

The following theorem shows directly that any theory containing a semilattice operation is inconsistent with (gL).

Theorem LT-1. Inconsistency of a semilattice operation with (gL).

$$\left\{ \begin{array}{l} xx = x \\ x(yz) = y(zx) \end{array} \right\} \;=\!(gL)\!\Rightarrow\; \{x = y\}.$$

Proof (found by Otter 3.0.4 on gyro at 0.10 seconds).

1	$B \neq A$	
4,3	$x \cdot x = x$	
5	$x \cdot (y \cdot z) = y \cdot (z \cdot x)$	
10	$x \cdot (y \cdot (z \cdot y)) = x \cdot y$	[3 → 5 :(gL), flip]
12	$(x \cdot y) \cdot (z \cdot u) = z \cdot (y \cdot (u \cdot x))$	[5 → 5]
15,14	$x \cdot (y \cdot x) = y \cdot x$	[3 → 5]
21,20	$x \cdot (y \cdot z) = x \cdot z$	[10 :15]

25,24	$x \cdot y = y$	[14 :21,4, flip]
26	$x = y$	[12 :25,25,25,25,25,25]
27	$\square$	[26,1]

The next theorem shows that a fragment of Boolean algebra satisfying some set difference properties (studied by J. Kalman [21]) is inconsistent with (gL).

Theorem SD-1. Inconsistency of set difference with (gL).

$$\left\{ \begin{array}{l} x - (y - x) = x \\ x - (x - y) = y - (y - x) \end{array} \right\} =\!(gL)\!\Rightarrow \{x = y\}.$$

Proof (found by Otter 3.0.4 on gyro at 0.29 seconds).

1	$A \neq B$	
4,3	$x - (y - x) = x$	
5	$x - (x - y) = y - (y - x)$	
8,7	$(x - y) - z = x - z$	[3 → 3 :(gL)]
10,9	$x - (y - (z - u)) = x - u$	[3 → 7 :(gL), flip]
16,15	$x - x = x$	[5 → 5 :10,8,8,4]
18,17	$x - y = y$	[3 → 5 :16,8,16, flip]
19	$x = y$	[7 → 5 :18,18,18,18,18]
20	$\square$	[19,1]

4.3 Quasigroups and the Overlay Principle

The equational theory of morphisms in Abelian varieties is full of formal principles that enable one to derive complicated equations from seemingly simpler ones. We have already seen that the rule (gL) is one such principle. The overlay principle is another. First let us give an illustrative example.

Suppose $f(x, y)$ is a binary operation defined over a nonsingular complex cubic curve, and let $f(x, y)$ admit a two-sided identity e: $f(x, e) = f(e, x) = x$ for all x. Then an elegant theorem of Mumford-Ramanujam (see page 51) says that f is, indeed, an Abelian group law on the curve. Let us derive, for example, the associativity of the binary law f by means of a formal equational technique and later prove that technique is valid for cubic curves. Consider the two derived ternary terms $g(x, y, z)$ and $h(x, y, z)$ defined by $g(x, y, z) = f(x, f(y, z))$ and $h(x, y, z) = f(f(x, y), z)$. We have

$$g(x, e, e) = h(x, e, e),$$
$$g(e, y, e) = h(e, y, e),$$
$$g(e, e, z) = h(e, e, z).$$

Imagine writing these three equations on three different transparencies where we leave a blank space for e. Now if we overlay the three transparencies one on top of another, we get the desired associative law

$$g(x, y, z) = h(x, y, z).$$

For this intuitive reason, we call this technique the ternary overlay principle. Similarly, one can establish the commutativity of f by using a binary overlay principle.

Now to prove that nonsingular cubic curves enjoy these overlay principles, it is enough if we show that

{presence of quasigroup operations} $=(gL)\Rightarrow$ {the overlay principle},

since nonsingular cubic curves do admit such quasigroup morphisms (e.g., the binary Steiner law). This is precisely what is done in the following two theorems.

Recall that we can define quasigroups equationally as algebras of type $\langle 2, 2, 2 \rangle$ satisfying the four equations

$$x \cdot (x \backslash y) = y, \quad (x/y) \cdot y = x,$$
$$x \backslash (x \cdot y) = y, \quad (x \cdot y)/y = x,$$

where $\cdot$ is the quasigroup operation, $/$ is right division, and $\backslash$ is left division.

Theorem OC-1. Validity of binary overlay for quasigroups.

$$\left\{ \begin{array}{l} \cdot \text{ is a quasigroup} \\ g(x, e) = f(x, e) \\ g(e, x) = f(e, x) \end{array} \right\} \quad =(gL)\Rightarrow \quad \{g(x, y) = f(x, y)\}.$$

Proof (found by Otter 3.0.4 on gyro at 57.71 seconds).

1	$g(A, B) \neq f(A, B)$	
6,5	$x \backslash (x \cdot y) = y$	
7	$(x/y) \cdot y = x$	
9	$(x \cdot y)/y = x$	
11	$g(x, e) = f(x, e)$	
12	$g(e, x) = f(e, x)$	
13	$(x/y) \backslash x = y$	$[7 \rightarrow 5]$
17	$(x \cdot g(y, e))/f(y, e) = x$	$[11 \rightarrow 9]$
33	$(x \cdot g(e, y))/f(e, y) = x$	$[12 \rightarrow 9]$
65	$(x \cdot g(y, z))/f(y, u) = (x \cdot g(v, z))/f(v, u)$	$[17 \rightarrow 17 :(\text{gL}) :(\text{gL})]$
124	$(x \cdot g(y, z))/f(y, z) = x$	$[65 \rightarrow 33]$
592	$g(x, y) = f(x, y)$	$[124 \rightarrow 13 :6]$
593	$\square$	$[592,1]$

Corollary OC-2. Binary overlay with Steiner law.

$$\left\{ \begin{array}{l} x \cdot (y \cdot x) = y \\ g(x, e) = f(x, e) \\ g(e, x) = f(e, x) \end{array} \right\} \quad =(gL)\Rightarrow \quad \{g(x, y) = f(x, y)\}.$$

This follows immediately from Thm. OC-1, because the Steiner law implies that $\cdot$ is a quasigroup.

Theorem OC-3. Validity of ternary overlay for quasigroups.

$$\left\{ \begin{array}{l} \cdot \text{ is a quasigroup} \\ g(x,e,e) = f(x,e,e) \\ g(e,x,e) = f(e,x,e) \\ g(e,e,x) = f(e,e,x) \end{array} \right\} \;=\!(gL)\!\Rightarrow\; \{g(x,y,z) = f(x,y,z)\}.$$

Proof (found by Otter 3.0.4 on gyro at 723.61 seconds).

1	$g(A,B,C) \neq f(A,B,C)$	
6,5	$x\backslash(x \cdot y) = y$	
7	$(x/y) \cdot y = x$	
9	$(x \cdot y)/y = x$	
11	$g(x,e,e) = f(x,e,e)$	
12	$g(e,x,e) = f(e,x,e)$	
13	$g(e,e,x) = f(e,e,x)$	
14	$(x/y)\backslash x = y$	$[7 \to 5]$
18	$(x \cdot g(y,e,e))/f(y,e,e) = x$	$[11 \to 9]$
42	$(x \cdot g(e,y,e))/f(e,y,e) = x$	$[12 \to 9]$
66	$(x \cdot g(e,e,y))/f(e,e,y) = x$	$[13 \to 9]$
90	$(x \cdot g(y,z,u))/f(y,v,w) = (x \cdot g(v_6,z,u))/f(v_6,v,w)$	
		$[18 \to 18 \;:(gL) \;:(gL) \;:(gL) \;:(gL)]$
107	$((x \cdot g(y,z,u))/f(y,v,w))\backslash(x \cdot g(v_6,z,u)) = f(v_6,v,w)$	$[90 \to 14]$
224	$(x \cdot g(y,z,e))/f(y,z,e) = x$	$[90 \to 42]$
406	$(x \cdot g(y,z,u))/f(v,w,u) = (x \cdot g(y,z,v_6))/f(v,w,v_6)$	
		$[66 \to 66 \;:(gL) \;:(gL) \;:(gL) \;:(gL)]$
1602	$(x \cdot g(y,z,u))/f(y,z,u) = x$	$[224 \to 406, \text{flip}]$
2470	$g(x,y,z) = f(x,y,z)$	$[1602 \to 107 \;:6]$
2471	$\square$	$[2470,1]$

Corollary OC-4. Ternary overlay with Steiner law.

$$\left\{ \begin{array}{l} x \cdot (y \cdot x) = y \\ g(x,e,e) = f(x,e,e) \\ g(e,x,e) = f(e,x,e) \\ g(e,e,x) = f(e,e,x) \end{array} \right\} \;=\!(gL)\!\Rightarrow\; \{g(x,y,z) = f(x,y,z)\}.$$

This follows immediately from Thm. OC-3, because the Steiner law implies that $\cdot$ is a quasigroup.

4.4 Closure Conditions and (gL)

Closure conditions occur in the so-called web geometry associated with quasigroups. These are usually algebraic reformulations of the closure of certain geometric configurations in the associated web geometry. See [10, Sec. X.6] for complete details. For example, associativity on a web is equivalent to the following closure condition:

$$\left\{ \begin{array}{l} z_1 \cdot x = z_2 \cdot y \\ z_3 \cdot x = z_4 \cdot y \\ z_1 \cdot u = z_2 \cdot v \end{array} \right\} \Rightarrow \{z_3 \cdot u = z_4 \cdot v\}. \tag{RC}$$

This is the famous Reidemeister condition (RC). From the universal algebraic point of view, this implication may be taken to be the algebraic essence of a binary operation being "closest" to a group operation. Of course, one cannot derive the associativity of a quasigroup from (RC) because, for example, the nonassociative operation of right division in groups is a quasigroup and it will also satisfy the condition (RC). However, such a quasigroup will be isotopic to a group. This is what we mean by being closest to a group law. In fact, the condition (RC) happens to be the same as Ore's left quotient condition (Sec. 5.3) because that is what one needs to embed a cancellative semigroup into a group. Similarly, the Thomsen condition (TC),

$$\left\{ \begin{array}{l} z_1 \cdot x = z_2 \cdot y \\ z_1 \cdot z = z_3 \cdot y \end{array} \right\} \Rightarrow \{z_2 \cdot z = z_3 \cdot x\}, \tag{TC}$$

when imposed on a quasigroup will make it isotopic to an Abelian group. In other words, (TC) is the algebraic essence of a binary operation being "closest" to an Abelian group.

For our purpose, these closure conditions are implications in the language of one binary operation. We explore the interrelations among these closure conditions and the rule (gL) in the context of quasigroups, semigroups, etc. As we demonstrate in the following, (RC), (TC), and (gL) have a lot in common and behave in a similar fashion. In selecting our problems, we are guided by [10, Tab. 6.1, p. 292]. Let us mention a few examples, most proved below, so that the reader can get a feeling for the close affinity among these conditions.

$\{(RC), xe = x, ex = x\} \Rightarrow \{\text{associativity}\}$
$\{(TC), xe = x, ex = x\} \Rightarrow \{\text{associativity, commutativity}\}$
$\{(TC)\} \Rightarrow \{\text{associativity of the Mal'cev operation on a quasigroup}\}$
$\{(TC), \text{Steiner law}\} \Rightarrow \{\text{medial law}\}$
$\{(RC), xe = x, x(xy) = y)\} \Rightarrow \{xy \text{ is right division in Abelian groups}\}$
$\{(TC) \text{ or } (RC), \text{semilattice}\} \Rightarrow \{x = y\}$

The reader may recall that all the above statements are true with (TC) or (RC) being replaced by (gL). Apart from (TC) and (RC) there are many more closure conditions associated with the web geometry (e.g., hexagonal condition, Bol condition; see [10]), and these provide good sources of problems for Otter. We invite readers to modify the Otter input files[1] for the following

---

[1] See the Preface for information on obtaining the input files.

theorems, replacing (TC) or (RC) by the hexagonal or Bol conditions, and try the conjecture with Otter.

Preliminary experiments with some of the following theorems showed that the following forms of (RC) and (TC) are usually more effective than the forms given above.

$$\left.\begin{array}{l} x \cdot y = z \\ u \cdot v = z \\ x \cdot w = v_6 \\ v_7 \cdot v = v_6 \end{array}\right\} \Rightarrow \{u \cdot w = v_7 \cdot y\} \qquad \text{(TC')}$$

$$\left.\begin{array}{l} x \cdot y = z \\ u \cdot v = z \\ w \cdot y = v_6 \\ v_7 \cdot v = v_6 \\ x \cdot v_8 = v_9 \\ u \cdot v_{10} = v_9 \end{array}\right\} \Rightarrow \{w \cdot v_8 = v_7 \cdot v_{10}\} \qquad \text{(RC')}$$

These forms are analogous to the more effective forms of cancellation we ordinarily use (p. 19). However, when our goal is (RC) or (TC), we deny the original forms.

Theorem TC-1. TC quasigroups satisfy RC.

$$\left\{\begin{array}{l} \text{TC} \\ \text{quasigroup} \end{array}\right\} \Rightarrow \text{RC}.$$

Proof (found by Otter 3.0.4 on gyro at 5.22 seconds).

1	$x = x$	
2	$x \cdot y = z,\ u \cdot v = z,\ x \cdot w = v_6,\ v_7 \cdot v = v_6 \to u \cdot w = v_7 \cdot y$	
3	$x \cdot (x \backslash y) = y$	
6,5	$x \backslash (x \cdot y) = y$	
7	$(x/y) \cdot y = x$	
11	$C_4 \cdot A = C_3 \cdot B$	
13	$C_2 \cdot A = C_1 \cdot B$	
15	$C_4 \cdot F = C_3 \cdot E$	
17	$C_2 \cdot F = C_1 \cdot E \to \square$	
31	$C_4 \cdot (C_3 \backslash x) = (x/A) \cdot B$	[2,3,7,1,11, flip]
85	$C_2 \cdot (C_1 \backslash x) = (x/A) \cdot B$	[2,3,7,1,13, flip]
1078	$C_2 \cdot F = C_1 \cdot E$	[2,15,3,31,85 :6, flip]
1080	$\square$	[1078,17]

Theorem TC-2. TC with an identity element satisfies RC.

$$\left\{\begin{array}{l} \text{TC} \\ e \cdot x = x \\ x \cdot e = x \end{array}\right\} \Rightarrow \text{RC}.$$

Proof (found by Otter 3.0.4 on gyro at 2.23 seconds).

1	$x = x$	
2	$x \cdot y = z,\ u \cdot v = z,\ x \cdot w = v_6,\ v_7 \cdot v = v_6\ \rightarrow\ u \cdot w = v_7 \cdot y$	
3	$e \cdot x = x$	
5	$x \cdot e = x$	
7	$C_4 \cdot A = C_3 \cdot B$	
9	$C_2 \cdot A = C_1 \cdot B$	
11	$C_4 \cdot F = C_3 \cdot E$	
13	$C_2 \cdot F = C_1 \cdot E\ \rightarrow\ \square$	
15	$x \cdot y = y \cdot x$	[2,3,5,3,5]
40	$C_4 \cdot C_1 = C_2 \cdot C_3$	[2,15,9,15,7, flip]
77	$E \cdot C_1 = C_2 \cdot F$	[2,11,15,40,1]
212	$C_2 \cdot F = C_1 \cdot E$	[15 → 77, flip]
214	$\square$	[212,13]

Example TC-3. TC with a right identity does not necessarily satisfy RC.

$$\left\{ \begin{array}{c} \text{TC} \\ x \cdot e = x \end{array} \right\} \not\Rightarrow \text{RC}.$$

Model. The clauses

$$z_1 \cdot x = u,\ z_2 \cdot y = u,\ z_1 \cdot z = w,\ z_3 \cdot y = w\ \rightarrow\ z_2 \cdot z = z_3 \cdot x$$
$$x \cdot e = x$$
$$C_4 \cdot A = C_3 \cdot B$$
$$C_2 \cdot A = C_1 \cdot B$$
$$C_4 \cdot F = C_3 \cdot E$$
$$C_2 \cdot F \neq C_1 \cdot E$$

have the following model (found by MACE 1.2.0 on gyro at 0.89 seconds).

e: 0	C3: 0	.	0 1 2
A: 0	C4: 0	--+-------	
B: 0	F: 0	0	0 0 0	
C1: 1	E: 2	1	1 1 2	
C2: 1		2	2 2 1	

Theorem TC-4. TC with a weak identity element satisfies RC (1).

$$\left\{ \begin{array}{c} \text{TC} \\ (x \cdot e) \cdot e = x \\ e \cdot (e \cdot x) = x \end{array} \right\} \Rightarrow \text{RC}.$$

Proof (found by Otter 3.0.4 on gyro at 55.76 seconds).

1	$x = x$
2	$x \cdot y = z,\ u \cdot v = z,\ x \cdot w = v_6,\ v_7 \cdot v = v_6\ \rightarrow\ u \cdot w = v_7 \cdot y$
4,3	$(x \cdot e) \cdot e = x$

6,5	$e \cdot (e \cdot x) = x$	
7	$C_4 \cdot A = C_3 \cdot B$	
9	$C_2 \cdot A = C_1 \cdot B$	
11	$C_4 \cdot F = C_3 \cdot E$	
13	$C_2 \cdot F = C_1 \cdot E \;\to\; \square$	
20	$((e \cdot x) \cdot e) \cdot (e \cdot (y \cdot e)) = y \cdot x$	[2,1,3,5,1]
22	$x \cdot y = ((e \cdot y) \cdot e) \cdot (e \cdot (x \cdot e))$	[2,5,1,1,3]
39	$((e \cdot E) \cdot e) \cdot A = ((e \cdot B) \cdot e) \cdot F$	[2,11,20,7,20]
60	$((e \cdot E) \cdot e) \cdot (e \cdot (C_1 \cdot e)) = C_2 \cdot F$	[2,20,9,1,39, flip]
70	$C_2 \cdot F = C_1 \cdot E$	[22 $\to$ 60 :6,4,4,6, flip]
72	$\square$	[70,13]

Theorem TC-5. TC with a weak identity element satisfies RC (2).

$$\left\{ \begin{array}{l} \text{TC} \\ ((x \cdot e) \cdot e) \cdot e = x \\ e \cdot (e \cdot (e \cdot x)) = x \end{array} \right\} \Rightarrow \text{RC.}$$

Proof (found by Otter 3.0.4 on gyro at 74.62 seconds).

1	$x = x$	
2	$x \cdot y = z,\; u \cdot v = z,\; x \cdot w = v_6,\; v_7 \cdot v = v_6 \;\to\; u \cdot w = v_7 \cdot y$	
4,3	$((x \cdot e) \cdot e) \cdot e = x$	
6,5	$e \cdot (e \cdot (e \cdot x)) = x$	
7	$C_4 \cdot A = C_3 \cdot B$	
9	$C_2 \cdot A = C_1 \cdot B$	
11	$C_4 \cdot F = C_3 \cdot E$	
13	$C_2 \cdot F = C_1 \cdot E \;\to\; \square$	
14	$(((x \cdot y) \cdot e) \cdot e) \cdot z = (((x \cdot z) \cdot e) \cdot e) \cdot y$	[2,1,3,1,3]
18,17	$(((C_2 \cdot x) \cdot e) \cdot e) \cdot A = (((C_1 \cdot B) \cdot e) \cdot e) \cdot x$	[2,1,3,9,3]
33	$x \cdot (e \cdot (e \cdot y)) = ((y \cdot e) \cdot e) \cdot (e \cdot (e \cdot (x \cdot e)))$	[2,5,1,5,3]
36	$(((e \cdot x) \cdot e) \cdot e) \cdot (e \cdot (e \cdot (y \cdot e))) = y \cdot x$	[2,1,3,5,1]
293	$(((e \cdot E) \cdot e) \cdot e) \cdot A = (((e \cdot B) \cdot e) \cdot e) \cdot F$	[2,11,36,7,36]
418,417	$(((e \cdot E) \cdot e) \cdot e) \cdot (e \cdot (e \cdot (C_1 \cdot B))) = (((C_2 \cdot F) \cdot e) \cdot e) \cdot B$	
		[2,36,17,36,293 :4,6,4, flip]
423	$(((C_2 \cdot F) \cdot e) \cdot e) \cdot B = (((C_1 \cdot B) \cdot e) \cdot e) \cdot E$	
		[2,33,17,293,293 :4,6,18,418, flip]
515	$C_2 \cdot F = C_1 \cdot E$	[2,3,14,3,423 :4,4, flip]
517	$\square$	[515,13]

Theorem TC-6. Inconsistency of TC with semilattices.

$$\left\{ \begin{array}{l} \text{TC} \\ x \cdot x = x \\ x \cdot y = y \cdot x \\ (x \cdot y) \cdot z = x \cdot (y \cdot z) \end{array} \right\} \Rightarrow \{x = y\}.$$

Proof (found by Otter 3.0.4 on gyro at 0.41 seconds).

1	$x = x$	
2	$x \cdot y = z, \ u \cdot v = z, \ x \cdot w = v_6, \ v_7 \cdot v = v_6 \ \rightarrow \ u \cdot w = v_7 \cdot y$	
4,3	$x \cdot x = x$	
5	$x \cdot y = y \cdot x$	
7,6	$(x \cdot y) \cdot z = x \cdot (y \cdot z)$	
8	$B = A \ \rightarrow \ \square$	
24	$x \cdot (x \cdot y) = x \cdot y$	[2,6,3,6,1 :7,4]
27,26	$x \cdot (y \cdot x) = y \cdot x$	[2,5,3,6,1 :7,4]
43,42	$x \cdot y = y$	[2,5,3,5,24 :7,27,4]
44	$x = y$	[2,5,1,3,24 :43,43,43]
45	$\square$	[44,8]

Theorem RC-1. Inconsistency of RC with semilattices.

$$\left\{ \begin{array}{l} \text{RC} \\ x \cdot x = x \\ x \cdot y = y \cdot x \\ (x \cdot y) \cdot z = x \cdot (y \cdot z) \end{array} \right\} \Rightarrow \{x = y\}.$$

Proof (found by Otter 3.0.4 on gyro at 0.55 seconds).

1	$x = x$	
2	$x \cdot y = z, \ u \cdot v = z, \ w \cdot y = v_6, \ v_7 \cdot v = v_6, \ x \cdot v_8 = v_9,$	
	$\quad u \cdot v_{10} = v_9 \ \rightarrow \ w \cdot v_8 = v_7 \cdot v_{10}$	
4,3	$x \cdot x = x$	
5	$x \cdot y = y \cdot x$	
7,6	$(x \cdot y) \cdot z = x \cdot (y \cdot z)$	
8	$B = A \ \rightarrow \ \square$	
9	$x \cdot (y \cdot x) = y \cdot x$	[2,5,3,5,3,1,5 :7,4]
16,15	$x \cdot y = y$	[2,9,9,9,3,1,9 :4,4,7,4, flip]
17	$x = y$	[2,9,9,5,5,5,9 :16,16,16]
18	$\square$	[17,8]

Theorem TC-7. Associativity of Mal'cev polynomial under TC.

$$\left\{ \begin{array}{l} \text{TC} \\ m(x,y,z) = ((y \cdot y) \cdot x) \cdot (z \cdot y) \\ x \cdot (y \cdot x) = y \end{array} \right\} \Rightarrow$$

$$\{m(x,y,m(z,u,v)) = m(m(x,y,z),u,v)\}.$$

Proof (found by Otter 3.0.4 on gyro at 15.38 seconds).

1	$x = x$	
2	$x \cdot y = z, \ u \cdot v = z, \ x \cdot w = v_6, \ v_7 \cdot v = v_6 \ \rightarrow \ u \cdot w = v_7 \cdot y$	

4,3	$m(x, y, z) = ((y \cdot y) \cdot x) \cdot (z \cdot y)$	
5	$x \cdot (y \cdot x) = y$	
7	$m(A, B, m(C, D, E)) = m(m(A, B, C), D, E) \; \to \; \square$	
8	$((B \cdot B) \cdot A) \cdot ((((D \cdot D) \cdot C) \cdot (E \cdot D)) \cdot B) =$	
	$\quad ((D \cdot D) \cdot (((B \cdot B) \cdot A) \cdot (C \cdot B))) \cdot (E \cdot D) \; \to \; \square$	
		[copy,7 :4,4,4,4]
14	$(x \cdot y) \cdot x = y$	$[5 \to 5]$
17	$(x \cdot ((y \cdot z) \cdot u)) \cdot y = (x \cdot z) \cdot u$	[2,14,14,1,14, flip]
19	$x \cdot (y \cdot (z \cdot u)) = ((u \cdot x) \cdot y) \cdot z$	[2,14,5,5,14]
784	$x \cdot (((y \cdot z) \cdot u) \cdot v) = (y \cdot ((((v \cdot w) \cdot x) \cdot z) \cdot u)) \cdot w$	[2,1,5,5,17]
1734,1733	$(x \cdot ((((y \cdot z) \cdot u) \cdot v) \cdot w)) \cdot z = (x \cdot (u \cdot (v \cdot y))) \cdot w$	$[19 \to 17]$
1752	$x \cdot (((y \cdot z) \cdot u) \cdot v) = (y \cdot (x \cdot (z \cdot v))) \cdot u$	[784 :1734]
1753	$\square$	[1752,8]

Theorem TC-8. TC Steiner quasigroups are medial.

(Cf. Thm. MED-3, p. 46, for (gL).)

$$\left\{ \begin{array}{l} \text{TC} \\ x \cdot (y \cdot x) = y \end{array} \right\} \Rightarrow \{(x \cdot y) \cdot (z \cdot u) = (x \cdot z) \cdot (y \cdot u)\}.$$

Proof (found by Otter 3.0.4 on gyro at 0.09 seconds).

2	$x \cdot y = z, \; u \cdot v = z, \; x \cdot w = v_6, \; v_7 \cdot v = v_6 \; \to \; u \cdot w = v_7 \cdot y$	
3	$x \cdot (y \cdot x) = y$	
5	$(A \cdot C) \cdot (B \cdot D) = (A \cdot B) \cdot (C \cdot D) \; \to \; \square$	
11	$(x \cdot y) \cdot x = y$	$[3 \to 3]$
19	$(x \cdot y) \cdot (z \cdot u) = (x \cdot z) \cdot (y \cdot u)$	[2,3,11,3,11]
20	$\square$	[19,5]

Theorem RC-2. Commutative RC Steiner quasigroups are medial.

(Cf. Thm. MED-3, p. 46, for (gL).)

$$\left\{ \begin{array}{l} \text{RC} \\ x \cdot y = y \cdot x \\ x \cdot (y \cdot x) = y \end{array} \right\} \Rightarrow \{(x \cdot y) \cdot (z \cdot u) = (x \cdot z) \cdot (y \cdot u)\}.$$

Proof (found by Otter 3.0.4 on gyro at 10.48 seconds).

2	$x \cdot y = z, \; u \cdot v = z, \; w \cdot y = v_6, \; v_7 \cdot v = v_6, \; x \cdot v_8 = v_9,$	
	$\quad u \cdot v_{10} = v_9 \; \to \; w \cdot v_8 = v_7 \cdot v_{10}$	
3	$x \cdot y = y \cdot x$	
4	$x \cdot (y \cdot x) = y$	
6	$(A \cdot C) \cdot (B \cdot D) = (A \cdot B) \cdot (C \cdot D) \; \to \; \square$	
68	$(x \cdot y) \cdot x = y$	$[4 \to 4]$

70	$x \cdot (x \cdot y) = y$	$[3 \to 4]$
1096	$(x \cdot y) \cdot (z \cdot u) = (x \cdot z) \cdot (y \cdot u)$	$[2,70,3,4,68,70,70]$
1097	$\square$	$[1096,6]$

Theorem RC-3. RC groupoids with identity are associative.

$$\left\{ \begin{array}{l} \text{RC} \\ x \cdot e = x \\ e \cdot x = x \end{array} \right\} \Rightarrow \{(x \cdot y) \cdot z = x \cdot (y \cdot z)\}.$$

Proof (found by Otter 3.0.4 on gyro at 0.14 seconds).

1	$x = x$	
2	$x \cdot y = z,\ u \cdot v = z,\ w \cdot y = v_6,\ v_7 \cdot v = v_6,\ x \cdot v_8 = v_9,$	
	$\quad u \cdot v_{10} = v_9 \to w \cdot v_8 = v_7 \cdot v_{10}$	
3	$x \cdot e = x$	
5	$e \cdot x = x$	
7	$(A \cdot B) \cdot C = A \cdot (B \cdot C) \to \square$	
8	$(x \cdot y) \cdot z = x \cdot (y \cdot z)$	$[2,3,5,3,1,1,5]$
10	$\square$	$[8,7]$

Theorem RC-4. RC basis for right division in Abelian groups.

$$\left\{ \begin{array}{l} \text{RC} \\ x \cdot e = x \\ x \cdot (x \cdot y) = y \end{array} \right\} \Rightarrow \{((e \cdot (((x \cdot y) \cdot z) \cdot x)) \cdot z) = y\}.$$

The conclusion is a single axiom for Abelian groups in terms of $x - y$.

Proof (found by Otter 3.0.4 on gyro at 8.41 seconds).

1	$x = x$	
2	$x \cdot y = z,\ u \cdot v = z,\ w \cdot y = v_6,\ v_7 \cdot v = v_6,\ x \cdot v_8 = v_9,$	
	$\quad u \cdot v_{10} = v_9 \to w \cdot v_8 = v_7 \cdot v_{10}$	
4,3	$x \cdot e = x$	
6,5	$x \cdot (x \cdot y) = y$	
7	$(e \cdot (((A \cdot B) \cdot C) \cdot A)) \cdot C = B \to \square$	
13,12	$(x \cdot (y \cdot z)) \cdot u = x \cdot (y \cdot (z \cdot u))$	$[2,3,5,3,1,1,5]$
24	$e \cdot (((A \cdot B) \cdot C) \cdot (A \cdot C)) = B \to \square$	$[7 :13]$
25	$x \cdot x = e$	$[3 \to 5]$
29,28	$(x \cdot y) \cdot x = e \cdot y$	$[2,5,5,25,25,3,25 :4, \text{flip}]$
34	$x \cdot y = e \cdot (y \cdot x)$	$[2,1,5,25,25,25,25 :29]$
47,46	$(x \cdot y) \cdot z = e \cdot (y \cdot (x \cdot z))$	$[2,5,3,25,25,1,5]$
55	$C \cdot (A \cdot (B \cdot (e \cdot (A \cdot C)))) = B \to \square$	$[24 :47,47,47,47,6,6]$
130	$x \cdot (y \cdot (z \cdot (e \cdot (y \cdot x)))) = z$	$[2,25,25,5,34,3,5 :4, \text{flip}]$
132	$\square$	$[130,55]$

Theorem TC-9. TC basis for double inversion in Abelian groups (1).

$$\left\{\begin{array}{l} \text{TC} \\ e \cdot e = e \\ (x \cdot e) \cdot e = x \\ x \cdot (x \cdot y) = y \end{array}\right\} \Rightarrow \{(x \cdot (((x \cdot y) \cdot z) \cdot (y \cdot e))) \cdot (e \cdot e) = z\}.$$

The conclusion is a single axiom for Abelian groups in terms of $-x - y$.

Proof (found by Otter 3.0.4 on gyro at 1.04 seconds).

1	$x = x$	
2	$x \cdot y = z,\ u \cdot v = z,\ x \cdot w = v_6,\ v_7 \cdot v = v_6 \ \rightarrow\ u \cdot w = v_7 \cdot y$	
4,3	$e \cdot e = e$	
6,5	$(x \cdot e) \cdot e = x$	
7	$x \cdot (y \cdot x) = y$	
9	$(A \cdot (((A \cdot B) \cdot C) \cdot (B \cdot e))) \cdot (e \cdot e) = C \ \rightarrow\ \square$	
10	$(A \cdot (((A \cdot B) \cdot C) \cdot (B \cdot e))) \cdot e = C \ \rightarrow\ \square$	[copy,9 :4]
30	$e \cdot x = x \cdot e$	[2,5,3,7,1 :4,6]
39,38	$(x \cdot y) \cdot x = y$	[7 $\rightarrow$ 7]
50	$x \cdot y = y \cdot x$	[2,30,5,30,5 :6,6]
55,54	$(x \cdot y) \cdot y = x$	[2,38,38,5,38 :6, flip]
180,179	$(((x \cdot y) \cdot z) \cdot u) \cdot x = z \cdot (u \cdot y)$	[2,7,38,50,7]
195	$C = C \ \rightarrow\ \square$	[50 $\rightarrow$ 10 :180,39,55]
196	$\square$	[195,1]

Theorem TC-10. TC basis for double inversion in Abelian groups (2).

$$\left\{\begin{array}{l} \text{TC} \\ e \cdot e = e \\ e \cdot x = x \cdot e \\ x \cdot (x \cdot y) = y \end{array}\right\} \Rightarrow \{(x \cdot (((x \cdot y) \cdot z) \cdot (y \cdot e))) \cdot (e \cdot e) = z\}.$$

The conclusion is a single axiom for Abelian groups in terms of $-x - y$.

Proof (found by Otter 3.0.4 on gyro at 0.64 seconds).

1	$x = x$	
2	$x \cdot y = z,\ u \cdot v = z,\ x \cdot w = v_6,\ v_7 \cdot v = v_6 \ \rightarrow\ u \cdot w = v_7 \cdot y$	
4,3	$e \cdot e = e$	
5	$e \cdot x = x \cdot e$	
6	$x \cdot (y \cdot x) = y$	
8	$(A \cdot (((A \cdot B) \cdot C) \cdot (B \cdot e))) \cdot (e \cdot e) = C \ \rightarrow\ \square$	
9	$(A \cdot (((A \cdot B) \cdot C) \cdot (B \cdot e))) \cdot e = C \ \rightarrow\ \square$	[copy,8 :4]
11	$x \cdot y = y \cdot x$	[2,5,1,5,1]
19	$(x \cdot y) \cdot x = y$	[6 $\rightarrow$ 6]
36,35	$(x \cdot y) \cdot y = x$	[6 $\rightarrow$ 11, flip]
38,37	$x \cdot (x \cdot y) = y$	[11 $\rightarrow$ 6]
77,76	$((x \cdot y) \cdot z) \cdot (x \cdot u) = y \cdot (z \cdot u)$	[2,6,6,6,19, flip]

| 132 | $C = C \to \square$ | $[11 \to 9 : 77,38,36]$ |
| 133 | $\square$ | $[132,1]$ |

Theorem TC-11. TC groupoids with identity are commutative semigroups.

$$\left\{ \begin{array}{l} \text{TC} \\ e \cdot x = x \\ x \cdot e = x \end{array} \right\} \Rightarrow \left\{ \begin{array}{l} x \cdot y = y \cdot x \\ (x \cdot y) \cdot z = x \cdot (y \cdot z) \end{array} \right\}.$$

Proof (found by Otter 3.0.4 on gyro at 0.55 seconds).

1	$x = x$	
2	$x \cdot y = z, \; u \cdot v = z, \; x \cdot w = v_6, \; v_7 \cdot v = v_6 \to u \cdot w = v_7 \cdot y$	
3	$B \cdot A = A \cdot B, \; (A \cdot B) \cdot C = A \cdot (B \cdot C) \to \square$	
4	$e \cdot x = x$	
7,6	$x \cdot e = x$	
8	$x \cdot (y \cdot z) = y \cdot (x \cdot z)$	$[2,4,1,4,1]$
9	$x \cdot y = y \cdot x$	$[2,4,6,4,6]$
32,31	$(x \cdot y) \cdot z = y \cdot (x \cdot z)$	$[2,6,9,8,9 : 7, \text{flip}]$
45	$\square$	$[3 : 32 : 9,8]$

Theorem TC-12. TC Steiner quasigroups with $xe = ex$ are Abelian groups.

$$\left\{ \begin{array}{l} \text{TC} \\ f(x,y) = e \cdot (y \cdot x) \\ x \cdot (y \cdot x) = y \\ x \cdot e = e \cdot x \end{array} \right\} \Rightarrow \left\{ \begin{array}{l} f(x,y) = f(y,x) \\ f(f(x,y),z) = f(x,f(y,z)) \\ f(e,x) = x \\ \forall y \exists x, \; f(x,y) = e \end{array} \right\}.$$

Proof (found by Otter 3.0.4 on gyro at 2.15 seconds).

1	$x = x$	
2	$x \cdot y = z, \; u \cdot v = z, \; x \cdot w = v_6, \; v_7 \cdot v = v_6 \to u \cdot w = v_7 \cdot y$	
3	$f(B,A) = f(A,B), \; f(f(A,B),C) = f(A,f(B,C)),$	
	$\quad f(e,A) = A, \; f(x,A) = e \to \square$	
5,4	$x \cdot (y \cdot x) = y$	
6	$x \cdot e = e \cdot x$	
8,7	$f(x,y) = e \cdot (y \cdot x)$	
10	$e \cdot (B \cdot A) = e \cdot (A \cdot B), \; e \cdot ((e \cdot (C \cdot B)) \cdot A) = e \cdot (C \cdot (e \cdot (B \cdot A))),$	
	$\quad e \cdot (A \cdot x) = e \to \square$	$[3 : 8,8,8,8,8,8,8,5,8 : 1, \text{flip,flip}]$
17,16	$(x \cdot y) \cdot x = y$	$[4 \to 4]$
19	$x \cdot y = y \cdot x$	$[2,4,6,1,6 : 17]$
115	$x \cdot (x \cdot y) = y$	$[16 \to 19, \text{flip}]$
117	$(x \cdot y) \cdot y = x$	$[4 \to 19, \text{flip}]$
119	$x \cdot (y \cdot z) = x \cdot (z \cdot y)$	$[2,115,115,4,115]$
243,242	$(x \cdot (y \cdot z)) \cdot u = y \cdot (x \cdot (z \cdot u))$	$[2,117,115,4,117, \text{flip}]$
348	$e \cdot (A \cdot x) = e \to \square$	$[10 : 243 : 119,1]$
349	$e \cdot x = e \to \square$	$[115 \to 348]$
350	$\square$	$[349,115]$

Theorem RC-5. Quasigroups do not necessarily satisfy RC.

Model. The clauses

$$C_4 \cdot A = C_3 \cdot B$$
$$C_2 \cdot A = C_1 \cdot B$$
$$C_4 \cdot F = C_3 \cdot E$$
$$C_2 \cdot F \neq C_1 \cdot E$$

have the following quasigroup model (found by MACE 1.2.0 on gyro at 7.17 seconds).

```
A:   0           . | 0 1 2 3 4
B:   2           --+----------
C1:  2           0 | 0 1 2 3 4
C2:  1           1 | 1 4 0 2 3
C3:  1           2 | 2 3 1 4 0
C4:  0           3 | 3 0 4 1 2
F:   1           4 | 4 2 3 0 1
E:   0
```

Theorem TC-13. Cancellative medial algebras satisfy TC.

$$\left\{ \begin{array}{l} (x \cdot y) \cdot (z \cdot u) = (x \cdot z) \cdot (y \cdot u) \\ \text{cancellation} \end{array} \right\} \Rightarrow \text{TC}.$$

Proof (found by Otter 3.0.4 on gyro at 1.09 seconds).

2	$x \cdot y = z,\ x \cdot u = z \rightarrow y = u$	
4	$(x \cdot y) \cdot (z \cdot u) = (x \cdot z) \cdot (y \cdot u)$	
5	$C_3 \cdot A = C_2 \cdot B$	
7	$C_3 \cdot D = C_1 \cdot B$	
9	$C_2 \cdot D = C_1 \cdot A \rightarrow \square$	
14	$(x \cdot y) \cdot (C_2 \cdot B) = (x \cdot C_3) \cdot (y \cdot A)$	$[5 \rightarrow 4]$
17	$(x \cdot C_3) \cdot (y \cdot A) = (x \cdot y) \cdot (C_2 \cdot B)$	$[\text{flip } 14]$
18	$(x \cdot y) \cdot (C_1 \cdot B) = (x \cdot C_3) \cdot (y \cdot D)$	$[7 \rightarrow 4]$
101	$(x \cdot C_3) \cdot (y \cdot D) = (x \cdot C_1) \cdot (y \cdot B)$	$[4 \rightarrow 18, \text{flip}]$
235	$C_2 \cdot D = C_1 \cdot A$	$[2,17,101, \text{flip}]$
237	$\square$	$[235,9]$

The last two theorems of this section show that the two set subtraction equations shown to be inconsistent with (gL) in Thm. SD-1 (p. 79) are also inconsistent with RC and with TC.

Theorem RC-6. Set difference is inconsistent with RC.

$$\left\{ \begin{array}{l} \text{RC} \\ x - (y - x) = x \\ x - (x - y) = y - (y - x) \end{array} \right\} \Rightarrow \{x = y\}.$$

Proof (found by Otter 3.0.4 on gyro at 0.79 seconds).

1	$A = B \rightarrow \square$	
2	$x = x$	
3	$x - y = z, \ u - v = z, \ w - y = v_6, \ v_7 - v = v_6, \ x - v_8 = v_9,$	
	$\quad u - v_{10} = v_9 \ \rightarrow \ w - v_8 = v_7 - v_{10}$	
5,4	$x - (y - x) = x$	
6	$x - (x - y) = y - (y - x)$	
9,8	$(x - y) - y = x - y$	$[4 \rightarrow 4]$
15,14	$x - (y - (z - u)) = x - u$	$[3,8,8,2,8,4,8 :9]$
17,16	$(x - y) - z = x - z$	$[3,2,2,8,2,8,8]$
23,22	$x - y = x$	$[3,16,6,6,6,2,6 :17,5,15, \text{flip}]$
24	$x = y$	$[3,16,6,6,2,16,6 :23,23,23,23]$
25	$\square$	$[24,1]$

Theorem TC-14. Set difference is inconsistent with TC.
$$\left\{ \begin{array}{l} \text{TC} \\ x - (y - x) = x \\ x - (x - y) = y - (y - x) \end{array} \right\} \Rightarrow \{x = y\}.$$

Proof (found by Otter 3.0.4 on gyro at 0.75 seconds).

1	$A = B \rightarrow \square$	
2	$x = x$	
3	$x - y = z, \ u - v = z, \ x - w = v_6, \ v_7 - v = v_6 \ \rightarrow \ u - w = v_7 - y$	
5,4	$x - (y - x) = x$	
6	$x - (x - y) = y - (y - x)$	
9,8	$(x - y) - y = x - y$	$[4 \rightarrow 4]$
11,10	$x - (y - (x - z)) = x - z$	$[3,8,2,4,8 :9]$
14	$(x - y) - ((x - y) - z) = z - y$	$[3,8,4,2,6]$
41,40	$(x - y) - (z - ((u - y) - v)) = (u - y) - v$	$[3,10,10,14,14 :11, \text{flip}]$
44,43	$(x - y) - ((z - y) - x) = x - y$	$[3,10,2,8,14 :9,41, \text{flip}]$
49,48	$x - y = x$	$[3,14,10,14,4 :5,44, \text{flip}]$
50	$x = y$	$[3,14,8,14,8 :49,49,49,49,49,49,49,49]$
51	$\square$	$[50,1]$

4.5 A Discovery Rule

Let us conclude this chapter with a heuristic principle for discovering new theorems about (gL)-algebras and TC-algebras. Take, for example, the group term $p(x,y,z) = x - y - z$ in Abelian groups (ABGT). Write a set of two- or three-variable laws in the language of one ternary operation p that are valid for this interpretation. For example, p is symmetric in the last two variables,

$p(x, x, y)$ is independent of x, and $p(x, y, p(x, y, z))$ is independent of both x and y. Thus, so far, we have the following set:

$$S = \left\{ \begin{array}{l} p(x, y, z) = p(x, z, y) \\ p(x, x, z) = p(y, y, z) \\ p(x, y, p(x, y, z)) = z \end{array} \right\}.$$

Now solve for integers l, m, and n so that $p(x, y, z) = lx + my + nz$ satisfies the equations in the set S. If it turns out that $l = 1$ and $m = n = -1$, then we have the conjectures

$$S \xRightarrow{(gL)} \{\text{all the identities true for } x - y - z \text{ in ABGT}\},$$
$$S \cup \{(\text{TC})\} \Rightarrow \{\text{all the identities true for } x - y - z \text{ in ABGT}\},$$

where (TC) is the Thomsen closure condition. If, on the other hand, we find more than one set of solutions, our conjectures take the form

$$S \xRightarrow{(gL)} \{\text{identities common to all the interpretations in ABGT}\},$$
$$S \cup \{(\text{TC})\} \Rightarrow \{\text{identities common to all the interpretations in ABGT}\}.$$

Table 4.1 is a sample of such results we have proved.

Similar experiments may be carried out with other closure conditions. This is what we mean when we say that both (gL) and the Thomsen closure condition (TC) extract the essence of Abelian group theory, while the Reidemeister closure condition (RC) extracts only that of group theory. Thus the rules (RC), (TC), and (gL) may be viewed as group filters, which retain the group-like properties and take them all the way to group theory or Abelian group theory. In particular, they derive $x = y$ when supplied with identities with no group models, such as semilattices, difference algebras, or majority polynomials.

Table 4.1. Discovery Rule Examples

Set	Solutions	Theorem
$xe = x$ $ex = x$	$m = n = 1$	ABGT-2
$x\|x = e$ $e\|(e\|x) = x$ $e\|x = x\|e$	$m = 1, n = -1$ $m = -1, n = 1$	ABGT-6
$x \wedge x = x$ $x \wedge y = y \wedge x$ $x \wedge (y \wedge z) = (x \wedge y) \wedge z$	no solutions	LT-1 RC-1 TC-6
Pixley conditions	no solutions	PIX-1
$x \cdot (y \cdot x) = y$	$n = \omega, m = 1/n$	TC-12
$e \cdot e = e$ $e \cdot x = x \cdot e$ $x \cdot (y \cdot x) = x$	$n = m = -1$	ABGT-7 TC-10
A2 (p. 69)	$m = 1, n = -1$ $m = -1, n = 1$ $m = n = -1$	QGT-1 QGT-2

5. Semigroups

5.1 A Conjecture in Cancellative Semigroups

Electronic mail (slightly edited) from Padmanabhan to McCune, May 7, 1993:

Let me now state more problems for Otter all in one area: this time it is semigroups. These are based on an unproven conjecture of mine. So, any negative solution will disprove my conjecture once and for all. But every positive solution, apart from being a new theorem in mathematics, will generalize the corresponding result from group theory. It will thus strengthen the conjecture and also may pave the way for the general proof. So much for the preamble.

Groups are of type $\langle 2, 1, 0 \rangle$ with an associative multiplication, an inverse and an identity element. The semigroup of a group is, of course, cancellative: $ab = ac \Rightarrow a'ab = a'ac \Rightarrow b = c$. The cancellative law, as a first-order property, is within the language of type $\langle 2 \rangle$. This is the backdrop of my conjecture.

Conjecture. Let A be a nonempty set of equations of type $\langle 2 \rangle$, and let a be an equation of the same type. If every group satisfying A also satisfies a, then every cancellative semigroup satisfying A must satisfy a as well. To put it more formally, let GT be the axioms of group theory and CS be the associative law and the two cancellation laws. Then

$$\text{if } (A, \text{GT} \Rightarrow a) \text{ then } (A, \text{CS} \Rightarrow a).$$

From the computational point of view, this conjecture says that if one uses the luxury of the richer language of group theory {associativity, inverse, identity} to derive a from A, then one can do the same thing within the limited language of just one binary operation that is associative and cancellative. Over the past ten years, I have mentioned this problem to many famous group theorists (including S. I. Adjan, B. H. Neumann, Sevrin, Narain Gupta, and many others). No one believed it right away and all tried to give counterexamples. But none of the examples really worked, and the problem is still open.

For our purpose, all we need to do is to pick "good" A's in the language of one binary operation implying a nontrivial a in groups and try to prove them under the milder hypothesis of just CS. Over the years, I have collected some examples. Naturally, Otter can play a very useful role here.

5.2 Theorems Supporting the Conjecture

In the rest of this chapter, binary terms without parentheses should be interpreted as right associated.

Theorem CS-1. Support (1) for the CS conjecture.

$$\{CS, \; xyzyx = yxzxy\} \Rightarrow \{xyyx = yxxy\}.$$

Proof (found by Otter 3.0.4 on gyro at 7.14 seconds).

1	$A \cdot B \cdot B \cdot A = B \cdot A \cdot A \cdot B \; \rightarrow \; \square$	
2	$x = x$	
3	$x \cdot y = z, \; x \cdot u = z \; \rightarrow \; y = u$	
6,5	$(x \cdot y) \cdot z = x \cdot y \cdot z$	
7	$x \cdot y \cdot z \cdot y \cdot x = y \cdot x \cdot z \cdot x \cdot y$	
11	$x \cdot y \cdot z \cdot u \cdot y \cdot x = y \cdot x \cdot z \cdot u \cdot x \cdot y$	$[5 \rightarrow 7 :6]$
17	$x \cdot y \cdot z \cdot y \cdot x \cdot u = y \cdot x \cdot z \cdot x \cdot y \cdot u$	$[7 \rightarrow 5 :6,6,6,6,6,6]$
22	$x \cdot y \cdot z \cdot u \cdot v \cdot y \cdot x = y \cdot x \cdot z \cdot u \cdot v \cdot x \cdot y$	$[5 \rightarrow 11 :6]$
262	$x \cdot y \cdot x \cdot z \cdot z \cdot x = x \cdot y \cdot z \cdot x \cdot x \cdot z$	$[3,17,22]$
329	$x \cdot y \cdot z \cdot z \cdot y = x \cdot z \cdot y \cdot y \cdot z$	$[3,2,262]$
376	$x \cdot y \cdot y \cdot x = y \cdot x \cdot x \cdot y$	$[3,2,329]$
377	$\square$	$[376,1]$

To prove the preceding theorem for groups, simply substitute $z = e$ to derive the right-hand side. There is a second-order reason for Thm. CS-1: such semigroups can be embedded in groups satisfying the same equation and hence satisfy the conclusion. The embedding theorem was proved by A. I. Mal'cev, B. H. Neumann, and others around the 1950s. A first-order proof was given (unpublished) by Padmanabhan.

Theorem CS-2. Support (2) for the CS conjecture.

$$\{CS, \; xy^2 = y^2x\} \Rightarrow \{(xy)^4 = x^4y^4\}.$$

This was previously proved by Padmanabhan (unpublished). In fact, Otter proves a stronger theorem, $xyxyzyxu = zxxxyyyu$, without using the cancellation laws.

Proof (found by Otter 3.0.4 on gyro at 187.41 seconds).

1	$A \cdot B \cdot A \cdot B \cdot A \cdot B \cdot A \cdot B \neq A \cdot A \cdot A \cdot A \cdot B \cdot B \cdot B \cdot B$	
4,3	$(x \cdot y) \cdot z = x \cdot y \cdot z$	
5	$x \cdot y \cdot y = y \cdot y \cdot x$	
8	$x \cdot y \cdot z \cdot y \cdot z = y \cdot z \cdot y \cdot z \cdot x$	$[3 \rightarrow 5 :4,4]$
13	$x \cdot x \cdot y \cdot z = y \cdot x \cdot x \cdot z$	$[5 \rightarrow 3 :4,4,4]$
14	$x \cdot y \cdot y \cdot z = y \cdot y \cdot x \cdot z$	$[\text{flip } 13]$
48	$x \cdot y \cdot x \cdot y \cdot z \cdot u = z \cdot x \cdot y \cdot x \cdot y \cdot u$	$[8 \rightarrow 3 :4,4,4,4,4,4,4]$
50	$x \cdot y \cdot z \cdot y \cdot z \cdot u = y \cdot z \cdot y \cdot z \cdot x \cdot u$	$[\text{flip } 48]$
78	$x \cdot y \cdot z \cdot z \cdot y \cdot u = y \cdot y \cdot x \cdot z \cdot z \cdot u$	$[14 \rightarrow 14]$
3988	$x \cdot y \cdot y \cdot y \cdot z \cdot z \cdot z \cdot u = y \cdot z \cdot y \cdot z \cdot x \cdot z \cdot y \cdot u$	$[78 \rightarrow 50]$
4008	$x \cdot y \cdot x \cdot y \cdot z \cdot y \cdot x \cdot u = z \cdot x \cdot x \cdot x \cdot x \cdot y \cdot y \cdot y \cdot u$	$[\text{flip } 3988]$
4009	$\square$	$[4008,1]$

Theorem CS-3. Support (3) for the CS conjecture.

$$\{CS, \ xy^3 = y^3x\} \Rightarrow \{(xy)^9 = x^9y^9\}.$$

The analogous statement for group theory was first proved by Narain Gupta in 1969 [17]. In fact, we prove this for ordinary semigroups, without cancellation. (Note the nonstandard representation, and see the comments after the proof.)

Proof (found by Otter 3.0.4 on gyro at 203.10 seconds).

2	$(x \cdot y) \cdot z = x \cdot y \cdot z$	
3	$x \cdot y \cdot y \cdot y = y \cdot y \cdot y \cdot x$	
4	$p(A \cdot B \cdot A \cdot B \cdot A \cdot B \cdot A \cdot B \cdot A \cdot B \cdot A \cdot B \cdot A \cdot B \cdot A \cdot B \cdot A \cdot B)$	
5	$-p(A \cdot A \cdot A \cdot A \cdot A \cdot A \cdot A \cdot A \cdot A \cdot B \cdot B \cdot B \cdot B \cdot B \cdot B \cdot B \cdot B \cdot B)$	
6	$(x \cdot y) \cdot z = x \cdot y \cdot z$	
8	$x \cdot y \cdot z \cdot y \cdot z \cdot y \cdot z = y \cdot z \cdot y \cdot z \cdot y \cdot z \cdot x$	$[2 \rightarrow 3 :6,6,6,6,6,6]$
9	$x \cdot y \cdot z \cdot z \cdot z = z \cdot z \cdot z \cdot x \cdot y$	$[2 \rightarrow 3]$
12	$x \cdot x \cdot x \cdot y \cdot z = y \cdot x \cdot x \cdot x \cdot z$	$[3 \rightarrow 2 :6,6,6,6]$
15	$x \cdot x \cdot x \cdot y = y \cdot x \cdot x \cdot x$	$[3 \rightarrow 2 :6]$
16	$x \cdot y \cdot y \cdot y \cdot z = y \cdot y \cdot y \cdot x \cdot z$	$[3 \rightarrow 2 :6,6,6,6]$
20	$x \cdot y \cdot z \cdot u \cdot y \cdot z \cdot u \cdot y \cdot z \cdot u = y \cdot z \cdot u \cdot y \cdot z \cdot u \cdot y \cdot z \cdot u \cdot x$	
		$[2 \rightarrow 8 :6,6,6,6,6,6]$
22	$x \cdot y \cdot z \cdot u \cdot z \cdot u \cdot z \cdot u = z \cdot u \cdot z \cdot u \cdot z \cdot u \cdot x \cdot y$	$[2 \rightarrow 8]$
40	$p(A \cdot B \cdot A \cdot B \cdot A \cdot B \cdot A \cdot B \cdot A \cdot B \cdot A \cdot B \cdot B \cdot A \cdot B \cdot A \cdot B \cdot A)$	
		$[8 \rightarrow 4]$
42	$x \cdot y \cdot x \cdot y \cdot x \cdot y \cdot z = z \cdot x \cdot y \cdot x \cdot y \cdot x \cdot y$	$[8 \rightarrow 2 :6]$
65	$p(A \cdot B \cdot A \cdot B \cdot A \cdot B \cdot B \cdot A \cdot B \cdot A \cdot B \cdot A \cdot B \cdot A \cdot B \cdot A \cdot B \cdot A)$	
		$[8 \rightarrow 40 :6,6,6,6,6,6]$
69	$p(A \cdot B \cdot A \cdot B \cdot A \cdot B \cdot B \cdot A \cdot B \cdot A \cdot B \cdot B \cdot A \cdot B \cdot A \cdot B \cdot A \cdot A)$	
		$[8 \rightarrow 65]$

92    $p(B \cdot A \cdot B \cdot A \cdot B \cdot B \cdot A \cdot B \cdot A \cdot B \cdot A \cdot A \cdot A \cdot B \cdot A \cdot B \cdot A \cdot B)$
$$[8 \rightarrow 69 :6,6,6,6,6,6,6,6,6,6,6]$$

98    $x \cdot x \cdot x \cdot y \cdot z \cdot u \cdot u \cdot u = u \cdot x \cdot x \cdot x \cdot u \cdot u \cdot y \cdot z$ $[9 \rightarrow 12]$

144   $p(B \cdot A \cdot B \cdot A \cdot B \cdot B \cdot A \cdot B \cdot A \cdot A \cdot A \cdot A \cdot B \cdot B \cdot A \cdot B \cdot A \cdot B)$
$$[12 \rightarrow 92]$$

174   $p(B \cdot A \cdot B \cdot A \cdot B \cdot B \cdot A \cdot A \cdot A \cdot A \cdot B \cdot A \cdot B \cdot B \cdot A \cdot B \cdot A \cdot B)$
$$[12 \rightarrow 144]$$

176   $p(B \cdot A \cdot B \cdot A \cdot B \cdot A \cdot A \cdot A \cdot B \cdot A \cdot B \cdot A \cdot B \cdot B \cdot A \cdot B \cdot A \cdot B)$
$$[12 \rightarrow 174]$$

180   $p(A \cdot A \cdot B \cdot A \cdot B \cdot A \cdot B \cdot B \cdot A \cdot B \cdot A \cdot B \cdot B \cdot A \cdot B \cdot A \cdot B \cdot A)$
$$[8 \rightarrow 176 :6,6,6,6,6,6,6,6,6,6,6]$$

272   $p(A \cdot A \cdot B \cdot A \cdot B \cdot A \cdot B \cdot B \cdot A \cdot B \cdot B \cdot A \cdot B \cdot A \cdot B \cdot A \cdot A \cdot B)$
$$[22 \rightarrow 180]$$

296   $p(A \cdot A \cdot B \cdot A \cdot A \cdot B \cdot A \cdot A \cdot B \cdot B \cdot A \cdot B \cdot B \cdot A \cdot B \cdot B \cdot A \cdot B)$
$$[20 \rightarrow 272 :6,6,6,6]$$

351   $p(A \cdot A \cdot B \cdot A \cdot A \cdot B \cdot A \cdot A \cdot B \cdot B \cdot B \cdot A \cdot B \cdot B \cdot A \cdot B \cdot B \cdot A)$
$$[20 \rightarrow 296]$$

390   $p(A \cdot A \cdot B \cdot A \cdot A \cdot B \cdot A \cdot A \cdot A \cdot B \cdot B \cdot B \cdot B \cdot B \cdot A \cdot B \cdot B \cdot A)$
$$[16 \rightarrow 351]$$

449   $p(A \cdot A \cdot B \cdot A \cdot A \cdot B \cdot B \cdot B \cdot B \cdot B \cdot A \cdot B \cdot B \cdot A \cdot A \cdot A \cdot A)$
$$[15 \rightarrow 390 :6,6,6,6,6,6,6,6]$$

687   $p(A \cdot A \cdot B \cdot A \cdot A \cdot A \cdot B \cdot B \cdot A \cdot A \cdot A \cdot A \cdot B \cdot B \cdot B \cdot B \cdot B \cdot B)$
$$[42 \rightarrow 449 :6,6,6,6,6,6]$$

2438 $p(A \cdot A \cdot A \cdot A \cdot A \cdot A \cdot A \cdot A \cdot A \cdot B \cdot B \cdot B \cdot B \cdot B \cdot B \cdot B \cdot B \cdot B)$
$$[98 \rightarrow 687 :6,6,6,6,6,6,6,6]$$

2439 $\square$ $[2438,5]$

A digression. The preceding proof is difficult for Otter, and a nonstandard representation and search strategy were used to obtain it. Our first proof, which uses cancellation, took about seven hours (on a Sun SPARC 1+) with the following special strategy: (1) `max_weight=37`, (2) `pick_given_ratio=3`, (3) Skolem constants in the denial with weight 0, and (4) cancellation as demodulators only. The proof is bidirectional, with mostly forward steps, and with 23 uses of the left cancellation law. Because a proof of Thm. CS-2, $\{CS, xy^2 = y^2x\} \Rightarrow \{(xy)^4 = x^4y^4\}$, was obtained without use of cancellation, we tried for a similar proof of Thm. CS-3. After several failed attempts, a proof was found (without cancellation) with a similar strategy. That proof was then dissected and rewritten (by hand) into the following more readable form, in which one half of the conclusion is rewritten, using the hypothesis, into the other half. In each step, the rewritten subterm is listed on the right.

0. $xxxxxxxxyyyyyyyyy$
1. $xxxxxxyyyyyyyyyxxxy$ $[xxx(yyyyyyyy)]$
2. $xxxxxxyyyyyyyyxxxyy$ $[yxxx]$
3. $xxxxxxyyyyxxyyyyxyy$ $[yyy(xx)]$
4. $xxxxyyyxxyxxyyyxyy$ $[(xx)yyy]$
5. $xyyxxxyxxyxxyyyxyy$ $[xxx(yy)]$
6. $xyyxyyxyxxyxxyxxyy$ $[(xxy)(xxy)(xxy)(yyxy)]$
7. $xyyxyyxyxyxxyxxyxy$ $[(xyx)(xyx)(xyx)(xy)]$
8. $xyyxyxyxxyxyxyxyxy$ $[(yx)(yx)(yx)(xyxxyx)]$
9. $xyxyxyxyxyxyxyxyxy$ $[y(yx)(yx)(yx)]$

This can be viewed as a backward proof, with paramodulation from (both sides of) the hypothesis into the right-hand sides of negative equalities, with associativity built into unification. (Otter, without associative unification, must derive various instantiated and associated versions of the hypothesis to carry out the proof.) This humanized proof led us to the nonstandard representation

$$\{\text{S}, \ xy^3 = y^3x, P((AB)^9)\} \Rightarrow \{P(A^9B^9)\}$$

used in the above Otter proof. The strategy for that proof was (1) max_weight=39, (2) paramodulation from both sides of equations, (3) paramodulation into both sides of equations and positive P clauses, and (4) all equalities with weight 1, all P clauses with weight 0, and pick_given_ratio=5, so that five P clauses are used for each equality that is used. This strategy leads Otter to a proof in a few minutes. The motivation for such a finely tuned strategy was the next problem of the sequence,

$$\{\text{S}, \ xy^4 = y^4x\} \Rightarrow \{(xy)^{16} = x^{16}y^{16}\},$$

but we have not found a proof of this. (The problem is decidable, but we haven't fully investigated it.) *End of digression.*

Theorem CS-4. Support (4) for the CS conjecture.

$$\{\text{CS}, \ xy^4 = y^4x\} \Rightarrow \{(xy)^4 = (yx)^4\}.$$

Proof (found by Otter 3.0.4 on gyro at 4.67 seconds).

1	$B \cdot A \cdot B \cdot A \cdot B \cdot A \cdot B \cdot A = A \cdot B \cdot A \cdot B \cdot A \cdot B \cdot A \cdot B \ \rightarrow \ \square$	
2	$x = x$	
3	$x \cdot y = z, \ x \cdot u = z \ \rightarrow \ y = u$	
6,5	$(x \cdot y) \cdot z = x \cdot y \cdot z$	
7	$x \cdot y \cdot y \cdot y \cdot y = y \cdot y \cdot y \cdot y \cdot x$	
10	$x \cdot y \cdot z \cdot y \cdot z \cdot y \cdot z \cdot y \cdot z = y \cdot z \cdot y \cdot z \cdot y \cdot z \cdot y \cdot z \cdot x$	
		$[5 \rightarrow 7 : 6,6,6,6,6,6]$
211	$x \cdot y \cdot x \cdot y \cdot x \cdot y \cdot x \cdot y = y \cdot x \cdot y \cdot x \cdot y \cdot x \cdot y \cdot x$	$[3,2,10]$
212	$\square$	$[211,1]$

Theorem CS-5. Support (5) for the CS conjecture.

$$\left\{ \begin{array}{l} (xy)z = x(yz) \\ xyzu = yzux \end{array} \right\} \Rightarrow \left\{ \begin{array}{l} (xy)^3 = x^3y^3 \\ x^3y^3 = y^3x^3 \end{array} \right\}.$$

Proof (found by Otter 3.0.4 on gyro at 4.89 seconds).

4	$A \cdot B \cdot A \cdot B \cdot A \cdot B = A \cdot A \cdot A \cdot B \cdot B \cdot B,$	
	$B \cdot B \cdot B \cdot A \cdot A \cdot A = A \cdot A \cdot A \cdot B \cdot B \cdot B \rightarrow \square$	
6,5	$(x \cdot y) \cdot z = x \cdot y \cdot z$	
7	$x \cdot y \cdot z \cdot u = y \cdot z \cdot u \cdot x$	
8	$x \cdot y \cdot z \cdot u = u \cdot x \cdot y \cdot z$	[flip 7]
9	$x \cdot y \cdot z \cdot u \cdot v \cdot w = y \cdot w \cdot z \cdot u \cdot v \cdot x$	[7 → 7 :6,6]
22	$x \cdot y \cdot z \cdot u = z \cdot u \cdot x \cdot y$	[7 → 8]
111	$x \cdot y \cdot z \cdot u \cdot v \cdot w = v \cdot u \cdot w \cdot y \cdot z \cdot x$	[22 → 9 :6]
134	$x \cdot y \cdot z \cdot u \cdot v \cdot w = w \cdot u \cdot v \cdot y \cdot x \cdot z$	[flip 111]
373	$x \cdot y \cdot z \cdot u \cdot v \cdot w = z \cdot w \cdot y \cdot u \cdot x \cdot v$	[8 → 134 :6]
412	$x \cdot y \cdot z \cdot u \cdot v \cdot w = v \cdot z \cdot x \cdot u \cdot w \cdot y$	[flip 373]
462	$\square$	[4,412,134]

Theorem CS-6. Support (6) for the CS conjecture.

$$\left\{ \begin{array}{l} (xy)z = x(yz) \\ xyzuv = yzuvx \end{array} \right\} \Rightarrow \left\{ \begin{array}{l} (xy)^4 = x^4y^4 \\ x^4y^4 = y^4x^4 \end{array} \right\}.$$

Proof (found by Otter 3.0.4 on gyro at 269.12 seconds).

4	$A \cdot B \cdot A \cdot B \cdot A \cdot B \cdot A \cdot B = A \cdot A \cdot A \cdot A \cdot B \cdot B \cdot B \cdot B,$	
	$B \cdot B \cdot B \cdot B \cdot A \cdot A \cdot A \cdot A = A \cdot A \cdot A \cdot A \cdot B \cdot B \cdot B \cdot B \rightarrow \square$	
6,5	$(x \cdot y) \cdot z = x \cdot y \cdot z$	
7	$x \cdot y \cdot z \cdot u \cdot v = y \cdot z \cdot u \cdot v \cdot x$	
8	$x \cdot y \cdot z \cdot u \cdot v = v \cdot x \cdot y \cdot z \cdot u$	[flip 7]
9	$x \cdot y \cdot z \cdot u \cdot v \cdot w \cdot v_6 \cdot v_7 = y \cdot z \cdot v_7 \cdot u \cdot v \cdot w \cdot v_6 \cdot x$	[7 → 7 :6,6,6]
13	$x \cdot y \cdot z \cdot u \cdot v \cdot w = z \cdot u \cdot v \cdot w \cdot x \cdot y$	[5 → 7]
14	$x \cdot y \cdot z \cdot u \cdot v \cdot w \cdot v_6 \cdot v_7 = v_7 \cdot x \cdot y \cdot u \cdot v \cdot w \cdot v_6 \cdot z$	[flip 9]
34	$x \cdot y \cdot z \cdot u \cdot v \cdot w = z \cdot u \cdot v \cdot w \cdot y \cdot x$	[8 → 7 :6]
37	$x \cdot y \cdot z \cdot u \cdot v \cdot w = w \cdot v \cdot x \cdot y \cdot z \cdot u$	[flip 34]
2779	$x \cdot y \cdot z \cdot u \cdot v \cdot w \cdot v_6 \cdot v_7 = v \cdot w \cdot v_6 \cdot v_7 \cdot u \cdot z \cdot x \cdot y$	[37 → 13 :6,6]
2799	$x \cdot y \cdot z \cdot u \cdot v \cdot w \cdot v_6 \cdot v_7 = z \cdot x \cdot v \cdot v_6 \cdot v_7 \cdot u \cdot y \cdot w$	[37 → 14 :6]
3136	$\square$	[4,2799,2779]

The next two theorems arose during our work on self-dual equations in group theory (Sec. 7.1). We realized that if an equation in product, inverse, and variables holds for a group, then its dual (i.e., its mirror image) holds also. We then realized that if the cancellative semigroup conjecture is true, there must be a corresponding duality theorem for cancellative semigroups. This fact opens a new class of problems for testing the conjecture, and we present two examples here.

Theorem CS-7. Support (7) for the CS conjecture.

$$\{CS,\ xxxyy = yxxyx\} \Rightarrow \{zzwww = wzwwz\}.$$

Note that the conclusion is the dual (i.e., reverse) of the hypothesis. In group theory, substitute for each variable its inverse, take the inverse of both sides, then simplify. Note also that in the proof below, Otter proves that squares commute.

Proof (found by Otter 3.0.4 on gyro at 172.79 seconds).

1	$x = x$	
2	$x \cdot y = z,\ x \cdot u = z\ \rightarrow\ y = u$	
5,4	$(x \cdot y) \cdot z = x \cdot y \cdot z$	
6	$x \cdot x \cdot x \cdot y \cdot y = y \cdot x \cdot x \cdot y \cdot x$	
7	$B \cdot B \cdot A \cdot A \cdot A = A \cdot B \cdot A \cdot A \cdot B\ \rightarrow\ \square$	
8	$x \cdot y \cdot y \cdot x \cdot y = y \cdot y \cdot y \cdot x \cdot x$	[flip 6]
9	$x \cdot x \cdot x \cdot y \cdot z \cdot y \cdot z = y \cdot z \cdot x \cdot x \cdot y \cdot z \cdot x$	$[4 \rightarrow 6\ :5,5]$
11	$x \cdot y \cdot z \cdot z \cdot x \cdot y \cdot z = z \cdot z \cdot z \cdot x \cdot y \cdot x \cdot y$	[flip 9]
13	$x \cdot y \cdot y \cdot x \cdot y \cdot z = y \cdot y \cdot y \cdot x \cdot x \cdot z$	$[6 \rightarrow 4\ :5,5,5,5,5,5]$
14	$x \cdot x \cdot x \cdot y \cdot y \cdot z = y \cdot x \cdot x \cdot y \cdot x \cdot z$	[flip 13]
31	$x \cdot y \cdot y \cdot y \cdot x \cdot y = y \cdot y \cdot y \cdot x \cdot y \cdot x$	[2,1,9]
45	$x \cdot x \cdot y \cdot y \cdot y \cdot x \cdot x = y \cdot x \cdot x \cdot y \cdot x \cdot x \cdot y$	$[8 \rightarrow 14]$
53	$x \cdot y \cdot y \cdot x \cdot y \cdot y \cdot x = y \cdot y \cdot x \cdot x \cdot x \cdot y \cdot y$	[flip 45]
55	$x \cdot y \cdot z \cdot z \cdot z \cdot x \cdot y \cdot z = z \cdot z \cdot z \cdot x \cdot y \cdot z \cdot x \cdot y$	$[4 \rightarrow 31\ :5,5]$
57	$x \cdot x \cdot y \cdot y \cdot y \cdot x \cdot y \cdot x = y \cdot x \cdot x \cdot y \cdot x \cdot y \cdot x \cdot y$	$[31 \rightarrow 14]$
59	$x \cdot x \cdot x \cdot y \cdot x \cdot y \cdot z = x \cdot x \cdot x \cdot y \cdot x \cdot z$	$[31 \rightarrow 4\ :5,5,5,5,5,5,5,5,5]$
60	$x \cdot y \cdot y \cdot x \cdot y \cdot x \cdot y \cdot x = y \cdot y \cdot x \cdot x \cdot x \cdot y \cdot x \cdot y$	[flip 57]
62	$x \cdot y \cdot y \cdot y \cdot x \cdot y \cdot z = y \cdot y \cdot y \cdot x \cdot y \cdot x \cdot z$	[flip 59]
65	$x \cdot x \cdot y \cdot x \cdot y \cdot y = y \cdot x \cdot x \cdot y \cdot y \cdot x$	[2,13,11]
82	$x \cdot y \cdot y \cdot x \cdot x \cdot y \cdot z = y \cdot y \cdot x \cdot y \cdot x \cdot x \cdot z$	$[65 \rightarrow 4\ :5,5,5,5,5,5,5,5,5]$
135	$x \cdot x \cdot y \cdot y \cdot y \cdot x \cdot y \cdot z = y \cdot y \cdot y \cdot x \cdot y \cdot x \cdot x \cdot z$	$[59 \rightarrow 62]$
140	$x \cdot x \cdot x \cdot y \cdot x \cdot y \cdot y \cdot z = y \cdot y \cdot x \cdot x \cdot x \cdot y \cdot x \cdot z$	[flip 135]
171	$x \cdot y \cdot x \cdot x \cdot x \cdot y \cdot y = y \cdot y \cdot x \cdot y \cdot x \cdot x \cdot x$	$[8 \rightarrow 82]$
487	$x \cdot x \cdot y \cdot x \cdot x \cdot y \cdot y = y \cdot x \cdot x \cdot x \cdot y \cdot y \cdot x$	[2,13,55]
598	$x \cdot x \cdot y \cdot x \cdot y \cdot x \cdot y = y \cdot y \cdot x \cdot y \cdot x \cdot x \cdot x$	[2,60,140]
612	$x \cdot y \cdot y \cdot y \cdot x \cdot x = y \cdot x \cdot y \cdot x \cdot y \cdot x$	[2,171,598]
613	$x \cdot x \cdot y \cdot y \cdot x \cdot y = y \cdot x \cdot y \cdot x \cdot y \cdot x$	[2,82,598]
643	$x \cdot y \cdot x \cdot y \cdot x \cdot y \cdot z = y \cdot x \cdot x \cdot x \cdot y \cdot y \cdot z$	$[612 \rightarrow 4\ :5,5,5,5,5,5,5,5,5]$
714	$x \cdot y \cdot x \cdot x \cdot y \cdot y = y \cdot x \cdot y \cdot x \cdot y \cdot x$	[2,487,643]
777	$x \cdot y \cdot y \cdot x \cdot y = y \cdot x \cdot x \cdot y \cdot y$	[2,613,714]
778	$x \cdot x \cdot x \cdot y \cdot y = x \cdot y \cdot y \cdot x \cdot x$	[2,612,714]
841	$x \cdot y \cdot y \cdot x \cdot x \cdot z = y \cdot x \cdot x \cdot y \cdot x \cdot z$	$[777 \rightarrow 4\ :5,5,5,5,5,5]$
869	$x \cdot x \cdot y \cdot y = y \cdot y \cdot x \cdot x$	[2,1,778]
920	$x \cdot x \cdot y \cdot y \cdot z = y \cdot y \cdot x \cdot x \cdot z$	$[869 \rightarrow 4\ :5,5,5,5,5]$
997	$x \cdot x \cdot y \cdot x \cdot x \cdot y = y \cdot x \cdot x \cdot y \cdot x \cdot x$	[2,53,920]

1949	$x \cdot x \cdot y \cdot y \cdot y = y \cdot x \cdot y \cdot y \cdot x$	[2,841,997]
1950	□	[1949,7]

Theorem CS-8. Support (8) for the CS conjecture.

$$\{CS,\ xyyyxy = yyyyxx\} \Rightarrow \{zwzzzw = wwzzzz\}.$$

Note that the conclusion is the dual (i.e., reverse) of the hypothesis. In group theory, substitute for each variable its inverse, take the inverse of both sides, then simplify.

Proof (found by Otter 3.0.4 on gyro at 955.82 seconds).

1	$x = x$	
2	$x \cdot y = z,\ x \cdot u = z \ \to \ y = u$	
5,4	$(x \cdot y) \cdot z = x \cdot y \cdot z$	
6	$x \cdot y \cdot y \cdot y \cdot x \cdot y = y \cdot y \cdot y \cdot y \cdot x \cdot x$	
7	$B \cdot A \cdot B \cdot B \cdot B \cdot A = A \cdot A \cdot B \cdot B \cdot B \cdot B \ \to \ \square$	
8	$x \cdot x \cdot x \cdot x \cdot y \cdot y = y \cdot x \cdot x \cdot x \cdot y \cdot x$	[flip 6]
9	$x \cdot y \cdot z \cdot z \cdot z \cdot x \cdot y \cdot z = z \cdot z \cdot z \cdot z \cdot x \cdot y \cdot x \cdot y$	[4 → 6 :5,5]
13	$x \cdot x \cdot x \cdot x \cdot y \cdot y \cdot z = y \cdot x \cdot x \cdot x \cdot y \cdot x \cdot z$	[6 → 4 :5,5,5,5,5,5,5,5,5]
19	$x \cdot x \cdot x \cdot x \cdot y \cdot z \cdot z \cdot z \cdot y \cdot z = z \cdot x \cdot x \cdot x \cdot z \cdot x \cdot z \cdot z \cdot y \cdot y$	[8 → 13]
38	$x \cdot x \cdot x \cdot x \cdot y \cdot x \cdot y = y \cdot x \cdot x \cdot x \cdot x \cdot y \cdot x$	[2,1,9]
41	$x \cdot y \cdot z \cdot u \cdot u \cdot u \cdot x \cdot y \cdot z \cdot u = u \cdot u \cdot u \cdot u \cdot x \cdot y \cdot z \cdot x \cdot y \cdot z$	
		[4 → 9 :5,5]
66	$x \cdot y \cdot y \cdot y \cdot y \cdot x \cdot y \cdot z = y \cdot y \cdot y \cdot y \cdot x \cdot y \cdot x \cdot z$	
		[38 → 4 :5,5,5,5,5,5,5,5,5,5,5]
277	$x \cdot x \cdot x \cdot y \cdot y \cdot y \cdot y \cdot z \cdot z = x \cdot x \cdot x \cdot z \cdot y \cdot y \cdot y \cdot z \cdot y$	[2,13,19]
566	$x \cdot x \cdot x \cdot x \cdot y \cdot x \cdot y \cdot x \cdot y = x \cdot y \cdot x \cdot x \cdot x \cdot y \cdot x \cdot y \cdot x$	[2,66,41]
695	$x \cdot x \cdot y \cdot z \cdot z \cdot z \cdot y \cdot z = x \cdot x \cdot z \cdot z \cdot z \cdot z \cdot y \cdot y$	[2,1,277]
699	$x \cdot y \cdot y \cdot y \cdot y \cdot z \cdot z = x \cdot z \cdot y \cdot y \cdot y \cdot z \cdot y$	[2,1,695]
717	$x \cdot y \cdot z \cdot z \cdot z \cdot y \cdot z \cdot u = x \cdot z \cdot z \cdot z \cdot z \cdot y \cdot y \cdot u$	
		[699 → 4 :5,5,5,5,5,5,5,5,5,5]
718	$x \cdot y \cdot y \cdot y \cdot y \cdot z \cdot z \cdot u = x \cdot z \cdot y \cdot y \cdot y \cdot z \cdot y \cdot u$	[flip 717]
905	$x \cdot x \cdot x \cdot x \cdot y \cdot y \cdot y \cdot x = x \cdot x \cdot x \cdot y \cdot x \cdot y \cdot x \cdot y$	[2,718,566]
925	$x \cdot x \cdot y \cdot x \cdot y \cdot x \cdot y = x \cdot x \cdot x \cdot y \cdot y \cdot y \cdot x$	[2,1,905]
939	$x \cdot x \cdot y \cdot y \cdot y \cdot x = x \cdot y \cdot x \cdot y \cdot x \cdot y$	[2,1,925]
977	$x \cdot y \cdot x \cdot y \cdot x = y \cdot x \cdot x \cdot x \cdot y$	[2,1,939]
978	$x \cdot y \cdot y \cdot y \cdot x = y \cdot x \cdot y \cdot x \cdot y$	[2,939,1]
1017	$x \cdot y \cdot z \cdot x \cdot y \cdot z \cdot x = y \cdot z \cdot x \cdot x \cdot x \cdot y \cdot z$	[4 → 977 :5,5]
1054	$x \cdot y \cdot y \cdot y \cdot x \cdot z = y \cdot x \cdot y \cdot x \cdot y \cdot z$	[977 → 4 :5,5,5,5,5,5,5]
1088	$x \cdot y \cdot x \cdot y \cdot x \cdot z = y \cdot x \cdot x \cdot x \cdot y \cdot z$	[flip 1054]
1132	$B \cdot B \cdot A \cdot B \cdot A \cdot B = A \cdot A \cdot B \cdot B \cdot B \cdot B \ \to \ \square$	[978 → 7]
1505	$x \cdot x \cdot y \cdot y \cdot y \cdot x \cdot z = y \cdot x \cdot x \cdot x \cdot y \cdot y \cdot z$	[1088 → 1088]
1566	$x \cdot y \cdot y \cdot y \cdot x \cdot x \cdot z = y \cdot y \cdot x \cdot x \cdot x \cdot y \cdot z$	[flip 1505]
3129	$x \cdot x \cdot x \cdot y \cdot x \cdot y = y \cdot x \cdot x \cdot y \cdot x \cdot x$	[2,38,1017]
4718	$x \cdot x \cdot y \cdot x \cdot x \cdot y = x \cdot x \cdot x \cdot y \cdot y \cdot x$	[2,1017,1566]

4942	$x \cdot x \cdot y \cdot y \cdot x = x \cdot y \cdot x \cdot x \cdot y$	[2,1,4718]
5054	$x \cdot y \cdot y \cdot x = y \cdot x \cdot x \cdot y$	[2,1,4942]
5272	$x \cdot y \cdot y \cdot x \cdot z = y \cdot x \cdot x \cdot y \cdot z$	[5054 → 4 :5,5,5,5,5]
5383	$x \cdot x \cdot y \cdot x \cdot y = y \cdot y \cdot x \cdot x \cdot x$	[2,3129,5272]
5972	$x \cdot x \cdot y \cdot y \cdot y \cdot z = y \cdot y \cdot x \cdot y \cdot x \cdot z$	[5383 → 4 :5,5,5,5,5,5,5]
6049	$x \cdot x \cdot y \cdot x \cdot y \cdot z = y \cdot y \cdot x \cdot x \cdot x \cdot z$	[flip 5972]
6050	□	[6049,1132]

The last four theorems in this section are stated for arbitrary parameter n, so they are outside of Otter's scope. (Otter easily proved all of the particular cases we gave to it.) The proofs of the last three are left as exercises for the reader.

Theorem CS-9. Support (9) for the CS conjecture.
$$\{CS, \ xy^n = y^n x\} \Rightarrow \{(xy)^n = (yx)^n\}.$$

Otter easily proves the statement for $n = 2, 3, 4, 5$, and it is easy to construct by hand a proof of the general theorem by looking at the Otter proofs.

Proof.

$$
\begin{aligned}
xy^n \ &= y^n x &&\text{[hypothesis]} \\
x(xw)^n \ &= (xw)^n x &&\text{[set } y = xw\text{]} \\
&= x(wx)^n &&\text{[change power notation]} \\
(xw)^n \ &= (wx)^n &&\text{[cancel } x\text{]}
\end{aligned}
$$

Theorem CS-10. Support (10) for the CS conjecture.
$$\left\{ \begin{array}{l} (xy)z = x(yz) \\ x_1 x_2 \cdots x_n = x_2 \cdots x_n x_1 \end{array} \right\} \Rightarrow \left\{ \begin{array}{l} \forall m \geq \lceil n/2 \rceil, x^m y^m = y^m x^m \\ \forall m \geq n-1, (xy)^m = x^m y^m \end{array} \right\}.$$

Theorem CS-11. Support (11) for the CS conjecture.
$$\left\{ \begin{array}{l} CS \\ x_1 x_2 \cdots x_n = x_2 \cdots x_n x_1 \end{array} \right\} \Rightarrow \{xy = yx\}.$$

Theorem CS-12. Support (12) for the CS conjecture.
$$\left\{ \begin{array}{l} CS \\ x^n y = yz^n \end{array} \right\} \Rightarrow \{x^n y = y\}.$$

5.3 Meta-Abelian Cancellative Semigroups and the Quotient Condition

A classical result of Mal'cev [30] says that the following theorem in group theory does not hold for cancellative semigroups. Note that this theorem does not satisfy the conditions of the conjecture, because the equations added to group theory contain constants; that is, we are proving a universally quantified implication, $\forall x(\alpha(x) \Rightarrow \beta(x))$, rather than universally quantified equations implying a universally quantified equation, $(\forall x \alpha(x)) \Rightarrow (\forall x \beta(x))$.

Theorem GT-1. Groups satisfy the quotient condition.

Let $a, b, c, d, a_0, b_0, c_0, d_0$ be constants.

$$\left\{ \begin{array}{l} \text{GT} \\ aa_0 = bb_0 \\ ca_0 = db_0 \\ ac_0 = bd_0 \end{array} \right\} \Rightarrow \{cc_0 = dd_0\}.$$

Proof (found by Otter 3.0.4 on gyro at 0.35 seconds).

1	$d \cdot d_0 \neq c \cdot c_0$	
4,3	$(x \cdot y) \cdot z = x \cdot (y \cdot z)$	
15	$x' \cdot (x \cdot y) = y$	
18,17	$x \cdot (x' \cdot y) = y$	
23	$a \cdot a_0 = b \cdot b_0$	
24	$b \cdot b_0 = a \cdot a_0$	[flip 23]
26	$c \cdot a_0 = d \cdot b_0$	
27	$d \cdot b_0 = c \cdot a_0$	[flip 26]
29	$a \cdot c_0 = b \cdot d_0$	
30	$b \cdot d_0 = a \cdot c_0$	[flip 29]
32	$b \cdot (b_0 \cdot x) = a \cdot (a_0 \cdot x)$	[24 → 3 :4, flip]
34	$d \cdot (b_0 \cdot x) = c \cdot (a_0 \cdot x)$	[27 → 3 :4, flip]
44	$b' \cdot (a \cdot c_0) = d_0$	[30 → 15]
62	$b' \cdot (a \cdot (a_0 \cdot x)) = b_0 \cdot x$	[32 → 15]
97,96	$b' \cdot (a \cdot x) = b_0 \cdot (a_0' \cdot x)$	[17 → 62]
106	$b_0 \cdot (a_0' \cdot c_0) = d_0$	[44 :97]
114	$d \cdot d_0 = c \cdot c_0$	[106 → 34 :18]
116	□	[114,1]

However, if we include the equation $xyzyx = yxzxy$, we can prove the analogous theorem for cancellative semigroups.

Theorem CS-13. Nilpotent CS satisfy the quotient condition.

Let $a, b, c, d, a_0, b_0, c_0, d_0$ be constants.

$$\left\{ \begin{array}{l} \text{CS} \\ xyzyx = yxzxy \\ aa_0 = bb_0 \\ ca_0 = db_0 \\ ac_0 = bd_0 \end{array} \right\} \Rightarrow \{cc_0 = dd_0\}.$$

Proof (found by Otter 3.0.4 on gyro at 25288.37 seconds).

1	$d \cdot d_0 = c \cdot c_0 \;\rightarrow\; \square$
2	$x = x$

3	$x \cdot y = z, \ x \cdot u = z \ \rightarrow \ y = u$	
6,5	$(x \cdot y) \cdot z = x \cdot y \cdot z$	
7	$x \cdot y \cdot z \cdot y \cdot x = y \cdot x \cdot z \cdot x \cdot y$	
8	$b \cdot b_0 = a \cdot a_0$	
11,10	$d \cdot b_0 = c \cdot a_0$	
12	$b \cdot d_0 = a \cdot c_0$	
15,14	$b \cdot b_0 \cdot x = a \cdot a_0 \cdot x$	$[8 \rightarrow 5 :6, \text{flip}]$
17,16	$d \cdot b_0 \cdot x = c \cdot a_0 \cdot x$	$[10 \rightarrow 5 :6, \text{flip}]$
18	$b_0 \cdot d \cdot x \cdot c \cdot a_0 = c \cdot a_0 \cdot x \cdot b_0 \cdot d$	$[10 \rightarrow 7 :17]$
20	$b_0 \cdot b \cdot x \cdot a \cdot a_0 = a \cdot a_0 \cdot x \cdot b_0 \cdot b$	$[8 \rightarrow 7 :15]$
22	$x \cdot y \cdot z \cdot u \cdot y \cdot x = y \cdot x \cdot z \cdot u \cdot x \cdot y$	$[5 \rightarrow 7 :6]$
27	$x \cdot y \cdot z \cdot y \cdot x \cdot u = y \cdot x \cdot z \cdot x \cdot y \cdot u$	$[7 \rightarrow 5 :6,6,6,6,6,6,6]$
28	$d_0 \cdot b \cdot x \cdot a \cdot c_0 = b \cdot d_0 \cdot x \cdot d_0 \cdot b$	$[12 \rightarrow 7]$
30,29	$b \cdot d_0 \cdot x = a \cdot c_0 \cdot x$	$[12 \rightarrow 5 :6, \text{flip}]$
31	$a \cdot c_0 \cdot x \cdot d_0 \cdot b = d_0 \cdot b \cdot x \cdot a \cdot c_0$	$[\text{flip } 28 :30]$
33	$b \cdot x \cdot b_0 \cdot y \cdot b_0 \cdot x = a \cdot a_0 \cdot x \cdot y \cdot x \cdot b_0$	$[7 \rightarrow 14]$
53	$x \cdot d_0 \cdot a \cdot c_0 \cdot x = d_0 \cdot x \cdot b \cdot x \cdot d_0$	$[29 \rightarrow 7]$
66	$c \cdot a_0 \cdot x \cdot b_0 \cdot d \cdot y = b_0 \cdot d \cdot x \cdot c \cdot a_0 \cdot y$	$[18 \rightarrow 5 :6,6,6,6,6,6,6]$
81	$a \cdot a_0 \cdot x \cdot b_0 \cdot b \cdot y = b_0 \cdot b \cdot x \cdot a \cdot a_0 \cdot y$	$[20 \rightarrow 5 :6,6,6,6,6,6,6]$
95	$d_0 \cdot b \cdot x \cdot a \cdot c_0 \cdot y = a \cdot c_0 \cdot x \cdot d_0 \cdot b \cdot y$	$[31 \rightarrow 5 :6,6,6,6,6,6,6]$
96	$a \cdot c_0 \cdot x \cdot d_0 \cdot b \cdot y = d_0 \cdot b \cdot x \cdot a \cdot c_0 \cdot y$	$[\text{flip } 95]$
137	$x \cdot b_0 \cdot y \cdot a \cdot a_0 \cdot x = b_0 \cdot x \cdot y \cdot b \cdot x \cdot b_0$	$[14 \rightarrow 22]$
138	$x \cdot y \cdot z \cdot u \cdot v \cdot y \cdot x = y \cdot x \cdot z \cdot u \cdot v \cdot x \cdot y$	$[5 \rightarrow 22 :6]$
235	$d_0 \cdot x \cdot b \cdot x \cdot d_0 \cdot y = x \cdot d_0 \cdot a \cdot c_0 \cdot x \cdot y$	$[53 \rightarrow 5 :6,6,6,6,6,6,6]$
236	$x \cdot d_0 \cdot a \cdot c_0 \cdot x \cdot y = d_0 \cdot x \cdot b \cdot x \cdot d_0 \cdot y$	$[\text{flip } 235]$
309	$a \cdot a_0 \cdot d_0 \cdot x \cdot d_0 \cdot b_0 = a \cdot c_0 \cdot b_0 \cdot x \cdot b_0 \cdot d_0$	$[29 \rightarrow 33, \text{flip}]$
430,429	$b_0 \cdot d \cdot b \cdot c \cdot a_0 \cdot x = c \cdot a_0 \cdot a \cdot a_0 \cdot d \cdot x$	$[14 \rightarrow 66, \text{flip}]$
488	$a \cdot a_0 \cdot x \cdot b_0 \cdot a \cdot c_0 = b_0 \cdot b \cdot x \cdot a \cdot a_0 \cdot d_0$	$[12 \rightarrow 81]$
550	$a \cdot c_0 \cdot x \cdot d_0 \cdot a \cdot a_0 = d_0 \cdot b \cdot x \cdot a \cdot c_0 \cdot b_0$	$[8 \rightarrow 96]$
645	$x \cdot b_0 \cdot y \cdot z \cdot a \cdot a_0 \cdot x = b_0 \cdot x \cdot y \cdot z \cdot b \cdot x \cdot b_0$	$[5 \rightarrow 137 :6]$
648	$b_0 \cdot x \cdot y \cdot z \cdot b \cdot x \cdot b_0 = x \cdot b_0 \cdot y \cdot z \cdot a \cdot a_0 \cdot x$	$[\text{flip } 645]$
944	$x \cdot y \cdot z \cdot u \cdot z \cdot y \cdot x = z \cdot x \cdot y \cdot u \cdot y \cdot x \cdot z$	$[27 \rightarrow 138]$
1006,1005	$d_0 \cdot d \cdot b \cdot d \cdot d_0 \cdot b_0 = d \cdot d_0 \cdot a \cdot c_0 \cdot c \cdot a_0$	$[10 \rightarrow 236, \text{flip}]$
1263	$a_0 \cdot d_0 \cdot x \cdot d_0 \cdot b_0 = c_0 \cdot b_0 \cdot x \cdot b_0 \cdot d_0$	$[3,2,309, \text{flip}]$
1265	$a_0 \cdot d_0 \cdot x \cdot y \cdot d_0 \cdot b_0 = c_0 \cdot b_0 \cdot x \cdot y \cdot b_0 \cdot d_0$	$[5 \rightarrow 1263 :6]$
1483	$c_0 \cdot b_0 \cdot b \cdot x \cdot a \cdot a_0 \cdot d_0 = a \cdot c_0 \cdot a_0 \cdot x \cdot b_0 \cdot c_0 \cdot a$	$[488 \rightarrow 138]$
1484	$a \cdot c_0 \cdot a_0 \cdot x \cdot b_0 \cdot c_0 \cdot a = c_0 \cdot b_0 \cdot b \cdot x \cdot a \cdot a_0 \cdot d_0$	$[\text{flip } 1483]$
1541	$a_0 \cdot d_0 \cdot b \cdot x \cdot a \cdot c_0 \cdot b_0 = a \cdot a_0 \cdot c_0 \cdot x \cdot d_0 \cdot a_0 \cdot a$	$[550 \rightarrow 138]$
1542	$a \cdot a_0 \cdot c_0 \cdot x \cdot d_0 \cdot a_0 \cdot a = a_0 \cdot d_0 \cdot b \cdot x \cdot a \cdot c_0 \cdot b_0$	$[\text{flip } 1541]$
3341	$a_0 \cdot d_0 \cdot x \cdot a \cdot c_0 \cdot b_0 = c_0 \cdot b_0 \cdot x \cdot a \cdot a_0 \cdot d_0$	$[29 \rightarrow 1265 :15]$
3343	$a_0 \cdot d_0 \cdot x \cdot y \cdot z \cdot d_0 \cdot b_0 = c_0 \cdot b_0 \cdot x \cdot y \cdot z \cdot b_0 \cdot d_0$	$[5 \rightarrow 1265 :6]$
5263,5262	$a_0 \cdot d_0 \cdot x \cdot y \cdot a \cdot c_0 \cdot b_0 = c_0 \cdot b_0 \cdot x \cdot y \cdot a \cdot a_0 \cdot d_0$	$[5 \rightarrow 3341 :6]$
5268	$a \cdot a_0 \cdot c_0 \cdot x \cdot d_0 \cdot a_0 \cdot a = c_0 \cdot b_0 \cdot b \cdot x \cdot a \cdot a_0 \cdot d_0$	$[1542 :5263]$
7031	$x \cdot y \cdot z \cdot u \cdot z \cdot y \cdot x = z \cdot y \cdot x \cdot u \cdot x \cdot y \cdot z$	$[138 \rightarrow 944]$

9084	$a_0 \cdot d \cdot d_0 \cdot a \cdot c_0 \cdot c \cdot a_0 = c_0 \cdot c \cdot a_0 \cdot a \cdot a_0 \cdot d \cdot d_0$	
		$[1005 \rightarrow 3343 : 17,430]$
10077	$c_0 \cdot a_0 \cdot x \cdot b_0 \cdot c_0 \cdot a = a_0 \cdot c_0 \cdot x \cdot d_0 \cdot a_0 \cdot a$	$[3,1484,5268]$
10079	$a_0 \cdot c_0 \cdot d \cdot d_0 \cdot a_0 \cdot a = c_0 \cdot a_0 \cdot c \cdot a_0 \cdot c_0 \cdot a$	$[16 \rightarrow 10077, \text{flip}]$
10101	$c_0 \cdot d \cdot d_0 \cdot a_0 \cdot a = c_0 \cdot c \cdot c_0 \cdot a_0 \cdot a$	$[3,27,10079, \text{flip}]$
10103	$d \cdot d_0 \cdot a_0 \cdot a = c \cdot c_0 \cdot a_0 \cdot a$	$[3,2,10101, \text{flip}]$
10151	$a_0 \cdot a \cdot d \cdot d_0 \cdot a \cdot a_0 = a \cdot a_0 \cdot c \cdot c_0 \cdot a_0 \cdot a$	$[10103 \rightarrow 22, \text{flip}]$
10421	$a \cdot d \cdot d_0 \cdot a \cdot a_0 = a \cdot c \cdot c_0 \cdot a \cdot a_0$	$[3,22,10151, \text{flip}]$
10423	$d \cdot d_0 \cdot a \cdot a_0 = c \cdot c_0 \cdot a \cdot a_0$	$[3,2,10421, \text{flip}]$
10477	$d \cdot d_0 \cdot a \cdot a_0 \cdot x = c \cdot c_0 \cdot a \cdot a_0 \cdot x$	$[10423 \rightarrow 5 : 6,6,6,6,6, \text{flip}]$
10555	$b_0 \cdot x \cdot d \cdot d_0 \cdot b \cdot x \cdot b_0 = x \cdot b_0 \cdot c \cdot c_0 \cdot a \cdot a_0 \cdot x$	
		$[10477 \rightarrow 645, \text{flip}]$
13651	$x \cdot d \cdot d_0 \cdot b \cdot x \cdot b_0 = x \cdot c \cdot c_0 \cdot b \cdot x \cdot b_0$	$[3,648,10555, \text{flip}]$
13665	$d \cdot d_0 \cdot b \cdot x \cdot b_0 = c \cdot c_0 \cdot b \cdot x \cdot b_0$	$[3,2,13651, \text{flip}]$
13671	$d \cdot d_0 \cdot b \cdot x \cdot y \cdot b_0 = c \cdot c_0 \cdot b \cdot x \cdot y \cdot b_0$	$[5 \rightarrow 13665 : 6]$
13744,13743	$d \cdot d_0 \cdot a \cdot c_0 \cdot c \cdot a_0 = c \cdot c_0 \cdot a \cdot c_0 \cdot c \cdot a_0$	
		$[27 \rightarrow 13671 : 1006,11,30]$
13747	$c_0 \cdot c \cdot a_0 \cdot a \cdot a_0 \cdot d \cdot d_0 = a_0 \cdot c \cdot c_0 \cdot a \cdot c_0 \cdot c \cdot a_0$	
		$[9084 : 13744, \text{flip}]$
18197	$c \cdot a_0 \cdot a \cdot a_0 \cdot d \cdot d_0 = c \cdot a_0 \cdot a \cdot a_0 \cdot c \cdot c_0$	$[3,7031,13747, \text{flip}]$
18201	$a_0 \cdot a \cdot a_0 \cdot d \cdot d_0 = a_0 \cdot a \cdot a_0 \cdot c \cdot c_0$	$[3,2,18197, \text{flip}]$
18205	$a \cdot a_0 \cdot d \cdot d_0 = a \cdot a_0 \cdot c \cdot c_0$	$[3,2,18201, \text{flip}]$
18225	$a_0 \cdot d \cdot d_0 = a_0 \cdot c \cdot c_0$	$[3,2,18205, \text{flip}]$
18263	$d \cdot d_0 = c \cdot c_0$	$[3,2,18225, \text{flip}]$
18265	$\square$	$[18263,1]$

We close this chapter with one more theorem on the quotient condition. For motivation, see Thm. MED-1 on p. 5.

Theorem MED-7. Cancellative medial algebras satisfy the quotient condition.

Let $a, b, c, d, a_0, b_0, c_0, d_0$ be constants.

$$\left.\begin{array}{l} \text{cancellation} \\ (xy)(zu) = (xz)(yu) \\ aa_0 = bb_0 \\ ca_0 = db_0 \\ ac_0 = bd_0 \end{array}\right\} \Rightarrow \{cc_0 = dd_0\}.$$

Proof (found by Otter 3.0.4 on gyro at 9.76 seconds).

1	$d \cdot d_0 = c \cdot c_0 \rightarrow \square$
3	$x \cdot y = z, \; x \cdot u = z \rightarrow y = u$
4	$x \cdot y = z, \; u \cdot y = z \rightarrow x = u$
5	$(x \cdot y) \cdot (z \cdot u) = (x \cdot z) \cdot (y \cdot u)$
6	$b \cdot b_0 = a \cdot a_0$

8	$d \cdot b_0 = c \cdot a_0$	
10	$b \cdot d_0 = a \cdot c_0$	
12	$((x \cdot y) \cdot (z \cdot u)) \cdot (v \cdot w) = ((x \cdot z) \cdot v) \cdot ((y \cdot u) \cdot w)$	$[5 \to 5]$
14	$((x \cdot y) \cdot z) \cdot ((u \cdot v) \cdot w) = ((x \cdot u) \cdot (y \cdot v)) \cdot (z \cdot w)$	[flip 12]
17	$(b \cdot x) \cdot (b_0 \cdot y) = (a \cdot a_0) \cdot (x \cdot y)$	$[6 \to 5, \text{flip}]$
21	$(d \cdot x) \cdot (b_0 \cdot y) = (c \cdot a_0) \cdot (x \cdot y)$	$[8 \to 5, \text{flip}]$
49	$(a \cdot c_0) \cdot (b_0 \cdot x) = (a \cdot a_0) \cdot (d_0 \cdot x)$	$[10 \to 17]$
88	$(x \cdot y) \cdot ((a \cdot a_0) \cdot (d_0 \cdot z)) = (x \cdot (a \cdot c_0)) \cdot (y \cdot (b_0 \cdot z))$	$[49 \to 5]$
107	$((x \cdot y) \cdot (z \cdot u)) \cdot ((v \cdot w) \cdot v_6) = ((x \cdot v) \cdot (z \cdot w)) \cdot ((y \cdot u) \cdot v_6)$	
		$[5 \to 14]$
370	$(x \cdot (a \cdot c_0)) \cdot (y \cdot (b_0 \cdot z)) = (x \cdot (a \cdot a_0)) \cdot (y \cdot (d_0 \cdot z))$	$[5 \to 88, \text{flip}]$
488	$(x \cdot c_0) \cdot (b_0 \cdot y) = (x \cdot a_0) \cdot (d_0 \cdot y)$	$[3,370,107]$
506	$d \cdot d_0 = c \cdot c_0$	$[4,21,488]$
508	$\square$	$[506,1]$

6. Lattice-like Algebras

6.1 Equational Theory of Lattices

The following six equations form a basis for lattice theory (LT).

$$x \wedge y = y \wedge x, \qquad x \vee y = y \vee x, \qquad \text{(commutativity)}$$
$$(x \wedge y) \wedge z = x \wedge (y \wedge z), \qquad (x \vee y) \vee z = x \vee (y \vee z), \qquad \text{(associativity)}$$
$$x \wedge (x \vee y) = x, \qquad x \vee (x \wedge y) = x. \qquad \text{(absorption)}$$

Note that the basis is self-dual, that is, it contains all of its dual equations. Three MACE runs easily show the basis to be independent. The two idempotence laws,

$$x \wedge x = x, \qquad x \vee x = x, \qquad \text{(idempotence)}$$

can be trivially derived from the two absorption laws, and we generally include the idempotence laws when using this basis.

For historical interest, we start with SAM's lemma [16], which was first proved in 1966 with the interactive program Semi-Automated Mathematics (SAM V). (Note that the representation we use is quite different from the standard automated theorem-proving benchmark form of SAM's lemma, which is relational and usually relies on hyperresolution.)

Theorem LT-2. SAM's lemma.

Let L be a modular lattice with 0 and 1. For all $x, y \in L$, if z_1 is a complement of $x \vee y$, and if z_2 is a complement of $x \wedge y$, then

$$(z_1 \vee (x \wedge z_2)) \wedge (z_1 \vee (y \wedge z_2)) = z_1.$$

To help Otter find a short and quick proof, we reformulate the conclusion, from $\alpha = \beta$ into $\alpha = x$, $\beta = y \rightarrow x = y$. Also, in keeping with the theme of this work, we use an equational form of modularity,

$$(x \wedge y) \vee (x \wedge z) = x \wedge (y \vee (x \wedge z)),$$

instead of the more usual implicational form.

Proof (found by Otter 3.0.4 on gyro at 23.51 seconds).

4	$x \wedge y = y \wedge x$	
6,5	$(x \wedge y) \wedge z = x \wedge (y \wedge z)$	
9	$x \vee y = y \vee x$	
11,10	$(x \vee y) \vee z = x \vee (y \vee z)$	
12	$x \wedge (x \vee y) = x$	
14	$x \vee (x \wedge y) = x$	
17,16	$0 \wedge x = 0$	
19,18	$0 \vee x = x$	
20	$1 \wedge x = x$	
25,24	$(x \wedge y) \vee (x \wedge z) = x \wedge (y \vee (x \wedge z))$	
28	$C_1 \wedge (A \vee B) = 0$	
30	$C_2 \vee (A \wedge B) = 1$	
32	$C_2 \wedge (A \wedge B) = 0$	
34	$C_1 \vee (A \wedge C_2) = D$	
37,36	$C_1 \vee (B \wedge C_2) = E$	
38	$D \wedge E \neq C_1$	
39	$E \wedge D \neq C_1$	$[4 \rightarrow 38]$
40	$C_1 \vee (C_2 \wedge B) = E$	$[4 \rightarrow 36]$
44	$(A \vee B) \wedge C_1 = 0$	$[4 \rightarrow 28]$
46	$C_1 \wedge ((A \vee B) \wedge x) = 0$	$[28 \rightarrow 5 : 17, \text{flip}]$
51	$x \wedge (y \wedge z) = y \wedge (z \wedge x)$	$[4 \rightarrow 5]$
54	$x \wedge (y \wedge z) = z \wedge (x \wedge y)$	$[\text{flip } 51]$
55	$C_2 \vee (B \wedge A) = 1$	$[4 \rightarrow 30]$
60,59	$A \wedge (B \wedge C_2) = 0$	$[4 \rightarrow 32 : 6]$
88,87	$(A \wedge C_2) \vee C_1 = D$	$[34 \rightarrow 9, \text{flip}]$
156	$C_1 \wedge E = C_1$	$[40 \rightarrow 12]$
166	$(x \vee y) \wedge x = x$	$[4 \rightarrow 12]$
168	$x \wedge ((x \vee y) \wedge z) = x \wedge z$	$[12 \rightarrow 5, \text{flip}]$
175,174	$E \wedge C_1 = C_1$	$[4 \rightarrow 156]$
196	$x \vee (y \wedge x) = x$	$[4 \rightarrow 14]$
201,200	$(x \wedge y) \vee x = x$	$[9 \rightarrow 14]$
217,216	$x \wedge 0 = 0$	$[4 \rightarrow 16]$
219,218	$x \vee 0 = x$	$[9 \rightarrow 18]$
231,230	$x \wedge 1 = x$	$[4 \rightarrow 20]$
296	$(A \vee B) \wedge (C_1 \vee ((A \vee B) \wedge x)) = (A \vee B) \wedge x$	$[44 \rightarrow 24 : 19, \text{flip}]$
306	$(E \wedge x) \vee C_1 = E \wedge (x \vee C_1)$	$[174 \rightarrow 24 : 175]$
312	$x \wedge (y \vee (x \wedge z)) = x \wedge (z \vee (x \wedge y))$	$[9 \rightarrow 24 : 25]$
349	$C_1 \wedge (x \wedge (A \vee B)) = 0$	$[4 \rightarrow 46]$
1295	$C_1 \wedge A = 0$	$[12 \rightarrow 349]$
1300,1299	$A \wedge C_1 = 0$	$[4 \rightarrow 1295]$
3663	$B \wedge (A \vee (B \wedge C_2)) = B$	$[55 \rightarrow 312 : 231, \text{flip}]$
3671	$A \wedge D = A \wedge C_2$	$[34 \rightarrow 312 : 1300, 219]$
3801	$D \wedge A = A \wedge C_2$	$[4 \rightarrow 3671]$

3854	$A \vee (B \wedge C_2) = A \vee B$	$[3663 \to 196 :11,201, \text{flip}]$
3916	$(B \wedge C_2) \vee A = A \vee B$	$[9 \to 3854]$
4122,4121	$(A \vee B) \wedge (B \wedge C_2) = B \wedge C_2$	$[3916 \to 166]$
4395	$(A \vee B) \wedge E = B \wedge C_2$	$[4121 \to 296 :37,4122]$
4420	$A \wedge E = 0$	$[4395 \to 168 :60, \text{flip}]$
4468	$E \wedge (x \wedge A) = 0$	$[4420 \to 54 :217, \text{flip}]$
4721	$E \wedge ((x \wedge A) \vee C_1) = C_1$	$[4468 \to 306 :19, \text{flip}]$
4972	$E \wedge D = C_1$	$[3801 \to 4721 :88]$
4974	$\square$	$[4972,39]$

6.1.1 Quasilattices

One obtains a basis for quasilattice theory (QLT) if the absorption laws of LT are replaced by a pair of link laws. The following set is a basis for QLT.

$$x \wedge y = y \wedge x, \qquad\qquad x \vee y = y \vee x,$$
$$(x \wedge y) \wedge z = x \wedge (y \wedge z), \qquad (x \vee y) \vee z = x \vee (y \vee z),$$
$$x \wedge x = x, \qquad\qquad x \vee x = x,$$
$$(x \wedge (y \vee z)) \vee (x \wedge y) = x \wedge (y \vee z), \quad (x \vee (y \wedge z)) \wedge (x \vee y) = x \vee (y \wedge z).$$

6.1.2 Weakly Associative Lattices

The following set is a basis for the variety of weakly associative lattices (WAL).

$x \wedge x = x,$	$x \vee x = x,$	(W1,W1')
$x \wedge y = y \wedge x,$	$x \vee y = y \vee x,$	(W2,W2')
$((x \vee z) \wedge (y \vee z)) \wedge z = z,$	$((x \wedge z) \vee (y \wedge z)) \vee z = z.$	(W3,W3')

We show here that by adding an absorption law to WAL, we obtain LT. The proof can be compared with the (mathematician's) proof presented in [40].

Theorem WAL-1. A relationship between WAL and LT.

$$\{\text{WAL}, \ x \wedge (y \vee (x \vee z)) = x\} \Rightarrow \{\text{LT}\}.$$

It is sufficient to derive the associative laws for $\wedge$ and $\vee$.

Proof (found by Otter 3.0.4 on gyro at 267.46 seconds).

1	$x = x$
2	$(A \wedge B) \wedge C = A \wedge (B \wedge C), \ (A \vee B) \vee C = A \vee (B \vee C) \ \to \ \square$
3	$x \wedge x = x$
5	$x \wedge y = y \wedge x$
6	$((x \vee y) \wedge (z \vee y)) \wedge y = y$
8	$x \vee x = x$
10	$x \vee y = y \vee x$
11	$((x \wedge y) \vee (z \wedge y)) \vee y = y$

13	$x \wedge (y \vee (x \vee z)) = x$	
15	$x \wedge (y \vee (z \vee x)) = x$	$[10 \to 13]$
17	$x \wedge (y \vee x) = x$	$[8 \to 13]$
19	$x \wedge ((x \vee y) \vee z) = x$	$[10 \to 13]$
21	$x \wedge (x \vee y) = x$	$[8 \to 13]$
31	$(x \vee y) \wedge y = y$	$[3 \to 6]$
33	$x \wedge ((y \vee x) \wedge (z \vee x)) = x$	$[5 \to 6]$
35	$(x \vee y) \wedge x = x$	$[5 \to 21]$
39	$(x \vee (y \wedge (x \vee z))) \vee (x \vee z) = x \vee z$	$[21 \to 11]$
49	$((x \wedge (y \vee z)) \vee z) \vee (y \vee z) = y \vee z$	$[17 \to 11]$
53	$(x \wedge y) \vee y = y$	$[8 \to 11]$
65	$x \vee (y \vee (x \vee z)) = y \vee (x \vee z)$	$[13 \to 53]$
67	$(x \wedge y) \vee x = x$	$[5 \to 53]$
70,69	$x \vee (y \wedge x) = x$	$[10 \to 53]$
71	$(x \wedge y) \wedge (z \vee y) = x \wedge y$	$[53 \to 13]$
73	$x \wedge (y \wedge x) = y \wedge x$	$[53 \to 35]$
80,79	$x \vee (x \wedge y) = x$	$[10 \to 67]$
85	$(x \wedge y) \wedge x = x \wedge y$	$[67 \to 21]$
93	$((x \vee (y \wedge z)) \wedge z) \wedge (y \wedge z) = y \wedge z$	$[69 \to 6]$
110,109	$x \vee (y \vee (z \vee x)) = y \vee (z \vee x)$	$[15 \to 53]$
111	$(x \wedge y) \wedge (x \vee z) = x \wedge y$	$[67 \to 19]$
153	$(x \wedge y) \wedge (x \wedge (z \vee (x \wedge y))) = x \wedge y$	$[79 \to 33]$
157	$(x \wedge y) \wedge ((z \vee (x \wedge y)) \wedge x) = x \wedge y$	$[79 \to 33]$
195	$(x \vee y) \vee (x \vee (z \vee y)) = x \vee (z \vee y)$	$[15 \to 39]$
197	$(x \vee y) \vee (x \vee (y \vee z)) = x \vee (y \vee z)$	$[13 \to 39]$
231	$(x \vee y) \vee (z \wedge y) = x \vee y$	$[71 \to 69]$
434,433	$((x \wedge y) \vee z) \vee (x \vee z) = x \vee z$	$[111 \to 49]$
539	$x \vee (y \wedge (x \wedge z)) = x$	$[79 \to 231 :80]$
541	$x \vee (y \wedge (z \wedge x)) = x$	$[69 \to 231 :70]$
629	$x \vee ((y \wedge x) \wedge z) = x$	$[85 \to 541]$
727	$x \vee (y \vee z) = x \vee (z \vee y)$	$[10 \to 65 :110]$
1036	$(x \wedge y) \wedge ((z \wedge x) \wedge y) = (z \wedge x) \wedge y$	$[629 \to 93]$
1421	$(x \vee y) \vee z = z \vee (y \vee x)$	$[10 \to 727]$
1423	$x \vee (y \vee z) = (z \vee y) \vee x$	$[\text{flip } 1421]$
2968,2967	$(x \wedge y) \wedge (x \wedge (z \vee y)) = x \wedge y$	$[231 \to 153]$
3194,3193	$(x \wedge (y \wedge z)) \wedge (z \wedge x) = x \wedge (y \wedge z)$	$[541 \to 157]$
3196,3195	$(x \wedge (y \wedge z)) \wedge (y \wedge x) = x \wedge (y \wedge z)$	$[539 \to 157]$
3200,3199	$(x \wedge y) \wedge ((z \vee y) \wedge x) = x \wedge y$	$[231 \to 157]$
4876,4875	$((x \vee y) \vee z) \vee (z \vee y) = z \vee (x \vee y)$	$[1423 \to 195]$
5085	$x \vee ((x \wedge y) \vee z) = x \vee z$	$[433 \to 197 :4876,434]$
5196,5195	$(x \vee y) \vee (x \vee z) = (x \vee y) \vee z$	$[35 \to 5085]$
5198,5197	$(x \vee y) \vee (y \vee z) = (x \vee y) \vee z$	$[31 \to 5085]$
5282,5281	$(x \vee y) \vee z = x \vee (y \vee z)$	$[197 :5196,5198]$
5737	$(A \wedge B) \wedge C = A \wedge (B \wedge C) \ \to \ \square$	$[2 :5282 :1]$

5741	$C \wedge (A \wedge B) = A \wedge (B \wedge C) \;\rightarrow\; \Box$	$[5 \rightarrow 5737]$
8492	$x \wedge ((y \vee x) \wedge z) = z \wedge x$	$[3199 \rightarrow 1036 : 3194,3200]$
8495,8494	$x \wedge (y \wedge (z \vee x)) = y \wedge x$	$[2967 \rightarrow 1036 : 3196,2968]$
8573	$(x \wedge y) \wedge (y \wedge z) = z \wedge (x \wedge y)$	$[69 \rightarrow 8492]$
8580	$(x \wedge (y \vee z)) \wedge z = x \wedge z$	$[73 \rightarrow 8492 : 8495,\text{flip}]$
8607	$x \wedge (y \wedge z) = (y \wedge z) \wedge (z \wedge x)$	$[\text{flip } 8573]$
8842,8841	$(x \wedge y) \wedge (y \wedge z) = x \wedge (y \wedge z)$	$[79 \rightarrow 8580]$
8853	$x \wedge (y \wedge z) = y \wedge (z \wedge x)$	$[8607 : 8842]$
8854	$\Box$	$[8853,5741]$

6.1.3 Near Lattices and Transitive Near Lattices

The following set is a basis for the variety of near lattices (NL).

$$x \wedge x = x, \qquad x \vee x = x,$$
$$x \wedge y = y \wedge x, \qquad x \vee y = y \vee x,$$
$$x \wedge (x \vee y) = x, \qquad x \vee (x \wedge y) = x.$$

To obtain a basis for the variety of transitive near lattices (TNL), we can adjoin the following pair of axioms to NL.

$$x \wedge (y \vee (x \vee z)) = x, \qquad x \vee (y \wedge (x \wedge z)) = x.$$

6.2 Distributivity and Modularity

We start with a previously known simple basis for distributive lattices, which is given without proof in [5, p. 35].

Theorem LT-3. Sholander's basis for distributive lattices.

$$\left\{ \begin{array}{l} x \wedge (x \vee y) = x \\ x \wedge (y \vee z) = (z \wedge x) \vee (y \wedge x) \end{array} \right\} \Rightarrow \left\{ \begin{array}{l} y \wedge x = x \wedge y \\ (x \wedge y) \wedge z = x \wedge (y \wedge z) \\ x \wedge (x \vee y) = x \\ y \vee x = x \vee y \\ (x \vee y) \vee z = x \vee (y \vee z) \\ x \vee (x \wedge y) = x \end{array} \right\}.$$

Proof (found by Otter 3.0.4 on gyro at 71.49 seconds).

1	$x = x$
2	$B \wedge A = A \wedge B,\; (A \wedge B) \wedge C = A \wedge (B \wedge C),\; A \wedge (A \vee B) = A,$
	$B \vee A = A \vee B,\; (A \vee B) \vee C = A \vee (B \vee C),$
	$A \vee (A \wedge B) = A \;\rightarrow\; \Box$
4,3	$x \wedge (x \vee y) = x$
5	$x \wedge (y \vee z) = (z \wedge x) \vee (y \wedge x)$
6	$B \wedge A = A \wedge B,\; (A \wedge B) \wedge C = A \wedge (B \wedge C),\; B \vee A = A \vee B,$

	$(A \lor B) \lor C = A \lor (B \lor C), \ A \lor (A \land B) = A \ \rightarrow \ \square \ [2:4:1]$
7	$(x \land y) \lor (z \land y) = y \land (z \lor x)$ $\hfill$ [flip 5]
8	$(x \land y) \lor (y \land y) = y$ $\hfill$ [3 → 5, flip]
19,18	$(x \land y) \land y = x \land y$ $\hfill$ [8 → 3]
20	$((x \land y) \lor (z \land y)) \land (z \lor x) = y \land (z \lor x)$ $\hfill$ [5 → 18]
22	$(x \land y) \lor (z \land y) = y \land (z \lor (x \land y))$ $\hfill$ [18 → 7]
23	$x \lor (y \land (x \lor z)) = (x \lor z) \land (y \lor x)$ $\hfill$ [3 → 7]
26,25	$(x \land (y \lor z)) \lor y = (y \lor z) \land (y \lor x)$ $\hfill$ [3 → 7]
33,32	$(x \land y) \land (y \land (z \lor x)) = x \land y$ $\hfill$ [7 → 3]
36	$(x \land y) \land ((y \land (z \lor x)) \land (u \lor (x \land y))) = x \land y$ $\hfill$ [32 → 32 :33]
55	$x \land (y \lor (z \land x)) = x \land (y \lor z)$ $\hfill$ [7 → 22, flip]
61	$x \land ((y \land x) \lor x) = x \land x$ $\hfill$ [8 → 55, flip]
67	$(x \land x) \lor (y \land x) = x \land x$ $\hfill$ [5 → 61 :19]
76,75	$x \land x = x$ $\hfill$ [8 → 67, flip]
78,77	$x \land (y \lor x) = x$ $\hfill$ [7 → 67 :76]
81	$x \lor (y \land x) = x$ $\hfill$ [67 :76,76]
95	$(x \land y) \lor y = y$ $\hfill$ [8 :76]
97	$x \lor y = y \lor x$ $\hfill$ [5 → 75 :78,4]
108	$(x \land (y \lor z)) \lor z = (y \lor z) \land (z \lor x)$ $\hfill$ [77 → 7]
119,118	$x \lor x = x$ $\hfill$ [75 → 81]
120	$(x \lor (y \land z)) \lor (z \land (x \lor y)) = x \lor (y \land z)$ $\hfill$ [55 → 81]
132	$((x \land y) \land z) \land (z \land y) = (x \land y) \land z$ $\hfill$ [81 → 32]
138	$x \land y = y \land x$ $\hfill$ [7 → 118 :119]
141	$(x \land y) \land x = y \land x$ $\hfill$ [118 → 20 :119,119]
158,157	$(x \lor y) \land (y \lor x) = x \lor y$ $\hfill$ [77 → 23, flip]
164	$(x \lor y) \land x = x$ $\hfill$ [3 → 23 :119,119, flip]
171,170	$(x \land y) \lor (z \land y) = y \land (x \lor z)$ $\hfill$ [7 → 97, flip]
184	$x \land (y \lor z) = x \land (z \lor y)$ $\hfill$ [7 :171]
190	$(x \lor (y \land z)) \land z = z \land (x \lor y)$ $\hfill$ [55 → 138, flip]
194	$(x \land (y \lor z)) \land (z \land x) = z \land x$ $\hfill$ [32 → 138, flip]
196	$x \land (y \land x) = y \land x$ $\hfill$ [18 → 138, flip]
206	$(x \land y) \lor x = x$ $\hfill$ [138 → 95]
208	$(x \land y) \land (x \land (z \lor y)) = y \land x$ $\hfill$ [138 → 32]
211,210	$x \lor (x \land y) = x$ $\hfill$ [138 → 81]
214	$(A \land B) \land C = A \land (B \land C), \ (A \lor B) \lor C = A \lor (B \lor C) \ \rightarrow \ \square$
	$\hfill$ [6 :211 :138,97,1]
219	$x \land (y \lor (x \lor z)) = x$ $\hfill$ [164 → 55 :78, flip]
224	$(x \lor (y \land z)) \land (x \lor z) = (x \lor y) \land (x \lor z)$ $\hfill$ [55 → 25 :26, flip]
249,248	$x \land (x \land y) = x \land y$ $\hfill$ [206 → 164]
269	$(x \land y) \land ((y \land (z \lor x)) \land x) = x \land y$ $\hfill$ [210 → 36]
282,281	$(x \land y) \lor (y \land x) = x \land y$ $\hfill$ [141 → 210]
315	$(x \land y) \land (z \lor x) = x \land y$ $\hfill$ [206 → 219]
525	$(x \lor y) \land z = z \land (y \lor x)$ $\hfill$ [138 → 184]
526	$x \land (y \lor z) = (z \lor y) \land x$ $\hfill$ [flip 525]

671	$((x \vee y) \wedge z) \vee y = (x \vee y) \wedge (y \vee z)$	$[138 \rightarrow 108]$
713	$(x \wedge y) \vee z = (z \vee x) \wedge (z \vee (x \wedge y))$	$[315 \rightarrow 25]$
724	$(x \vee y) \wedge (x \vee (y \wedge z)) = (y \wedge z) \vee x$	[flip 713]
798,797	$(x \vee (y \wedge z)) \vee (y \wedge (x \vee z)) = x \vee (z \wedge y)$	$[138 \rightarrow 120]$
1260	$(x \wedge y) \wedge ((z \wedge y) \wedge x) = (z \wedge y) \wedge x$	$[138 \rightarrow 132]$
1284	$(x \wedge y) \wedge z = z \wedge (y \wedge x)$	$[281 \rightarrow 525 : 282]$
2656,2655	$(x \wedge y) \wedge ((y \wedge z) \wedge x) = (y \wedge z) \wedge x$	$[210 \rightarrow 194]$
2710,2709	$(x \wedge (y \vee z)) \wedge z = z \wedge x$	$[269 : 2656]$
3035	$x \wedge (y \wedge (z \vee x)) = x \wedge y$	$[18 \rightarrow 208 : 2710, 33, \text{flip}]$
3069	$x \wedge (y \vee (z \wedge (u \vee x))) = ((u \vee x) \wedge (y \vee z)) \wedge x$	
		$[190 \rightarrow 2709, \text{flip}]$
3076,3075	$((x \vee y) \wedge z) \wedge y = y \wedge z$	$[138 \rightarrow 2709]$
3098,3097	$x \wedge (y \vee (z \wedge (u \vee x))) = x \wedge (y \vee z)$	$[3069 : 3076]$
3101	$x \vee (y \wedge z) = x \vee (z \wedge y)$	$[2709 \rightarrow 120 : 3098, 798, 2710]$
3127,3126	$(x \wedge y) \wedge (z \wedge x) = (x \wedge y) \wedge z$	$[210 \rightarrow 3035]$
3135,3134	$(x \wedge y) \wedge (z \wedge y) = (x \wedge y) \wedge z$	$[81 \rightarrow 3035]$
3144	$(x \wedge y) \wedge z = (z \wedge y) \wedge x$	$[1260 : 3127, 3135]$
3369	$x \vee (y \vee z) = x \vee (z \vee y)$	$[157 \rightarrow 3101 : 158]$
3452,3451	$(x \wedge y) \wedge z = x \wedge (y \wedge z)$	$[1284 \rightarrow 3144, \text{flip}]$
3685	$(A \vee B) \vee C = A \vee (B \vee C) \;\; \rightarrow \;\; \square$	$[214 : 3452 : 1]$
3692	$(B \vee A) \vee C = A \vee (B \vee C) \;\; \rightarrow \;\; \square$	$[97 \rightarrow 3685]$
3730,3729	$(x \vee y) \wedge (x \vee (z \wedge y)) = x \vee (z \wedge y)$	$[196 \rightarrow 224 : 76, \text{flip}]$
3733	$(x \vee (y \vee z)) \wedge (x \vee y) = x \vee y$	$[164 \rightarrow 224 : 76, \text{flip}]$
3746,3745	$x \vee (y \wedge z) = (x \vee y) \wedge (x \vee z)$	$[138 \rightarrow 224 : 3730]$
3747	$(x \vee y) \wedge (x \vee z) = (y \wedge z) \vee x$	$[724 : 3746, 249]$
3801	$(x \vee y) \vee z = z \vee (y \vee x)$	$[97 \rightarrow 3369]$
5932,5931	$(x \vee y) \wedge (y \vee (x \vee z)) = y \vee x$	$[526 \rightarrow 3733]$
5957	$(x \vee y) \vee (y \vee z) = (y \vee z) \vee x$	$[3733 \rightarrow 671 : 5932]$
6203	$((x \wedge y) \vee z) \vee x = x \vee z$	$[3747 \rightarrow 671 : 5932]$
6252	$(x \vee y) \vee (x \vee z) = (x \vee z) \vee y$	$[164 \rightarrow 6203]$
6290	$(x \vee y) \vee z = (x \vee z) \vee (x \vee y)$	[flip 6252]
6533,6532	$(x \vee y) \vee (x \vee z) = (x \vee y) \vee z$	$[3801 \rightarrow 5957]$
6552	$(x \vee y) \vee z = (x \vee z) \vee y$	$[6290 : 6533]$
6570	$(x \vee y) \vee z = y \vee (x \vee z)$	$[97 \rightarrow 6552, \text{flip}]$
6572	$\square$	$[6570, 3692]$

The M_5-N_5 argument. The class T of all one-element lattices defined by the single absorption law $x = y$ is called the trivial variety. This is obviously the smallest lattice variety. If a lattice contains more than one element, it has the two-element chain as a sublattice, and hence the class D of all distributive lattices is the unique nontrivial variety containing T.

If a lattice L is nondistributive, it must contain either M_5 or N_5 (Fig. 6.1) as a sublattice. Both algebras are subdirectly irreducible.

M_5 is the smallest nondistributive modular lattice; similarly, N_5 is the smallest nonmodular lattice. Hence a lattice L is modular if and only if it

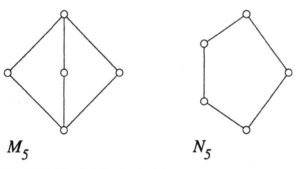

M_5 N_5

Fig. 6.1. Nondistributive Lattices

has no sublattice isomorphic to N_5 (see, e.g., [14, Theorem 1, p. 70]). Thus, to prove that an identity $f = g$ of type $\langle 2, 2 \rangle$ implies modularity, we need only show that $f = g$ fails in N_5. In that case, we say that $\{f = g\} \Rightarrow$ {modularity} by an N_5 argument. Similarly, to prove that an identity $f = g$ implies distributivity, we need only show that $f = g$ fails in the two lattices M_5 and N_5. In that case, we say that $\{f = g\} \Rightarrow$ {distributivity} by an M_5-N_5 argument.

    For an elegant proof of the M_5-N_5 argument, one simply takes a look at the corresponding free lattices: the free lattice on three generators contains N_5, and the free modular lattice on three generators contains M_5. In principle, these proofs can be modified into strict first-order proofs by considering the elements of the respective free lattices that are the culprits and force the desired equality from the given properties. However, this is more easily said than done. In some situations, such first-order proofs are even more difficult, if not impossible, to carry out.

    Here is a simple procedure to generate such first-order problems: Take a favorite identity $f = g$ of type $\langle 2, 2 \rangle$ implying, say, distributivity in lattices, and let the identity be regular. Then by a structure theorem of H. Lakser, R. Padmanabhan, and C. R. Platt [29],

$$\{f = g\} \Rightarrow \{\text{distributivity}\} \pmod{\text{QLT}}.$$

This proof now is higher order because we use the description of subdirectly irreducible quasilattices in terms of those lattices, apply the M_5-N_5 argument to the LT part, and then conclude the above implication for QLT. Obviously this idea generates a nice collection of problems for a first-order theorem prover such as Otter. We give here an example theorem with two proofs—Otter's and the M_5-N_5 proof.

Theorem QLT-1. A new form of distributivity for QLT.

$$\left\{ \begin{array}{l} \text{QLT} \\ x \wedge (y \vee (x \wedge z)) = x \wedge (y \vee z) \end{array} \right\} \Rightarrow \{x \wedge (y \vee z) = (x \wedge y) \vee (x \wedge z)\}.$$

Proof (found by Otter 3.0.4 on gyro at 4.96 seconds).

2	$x \wedge x = x$	
4	$x \wedge y = y \wedge x$	
6,5	$(x \wedge y) \wedge z = x \wedge (y \wedge z)$	
7	$x \vee x = x$	
9	$x \vee y = y \vee x$	
11,10	$(x \vee y) \vee z = x \vee (y \vee z)$	
12	$(x \wedge (y \vee z)) \vee (x \wedge y) = x \wedge (y \vee z)$	
15,14	$(x \vee (y \wedge z)) \wedge (x \vee y) = x \vee (y \wedge z)$	
17,16	$x \wedge (y \vee (x \wedge z)) = x \wedge (y \vee z)$	
18	$A \wedge (B \vee C) \neq (A \wedge B) \vee (A \wedge C)$	
19	$(A \wedge B) \vee (A \wedge C) \neq A \wedge (B \vee C)$	[flip 18]
20	$x \wedge (y \wedge z) = y \wedge (x \wedge z)$	[4 → 5 :6]
22,21	$x \wedge (x \wedge y) = x \wedge y$	[2 → 5, flip]
29,28	$x \vee (x \vee y) = x \vee y$	[7 → 10, flip]
30	$x \vee (y \vee z) = y \vee (z \vee x)$	[9 → 10]
37,36	$x \wedge (y \wedge x) = x \wedge y$	[4 → 21]
41,40	$x \vee (y \vee x) = x \vee y$	[9 → 28]
44	$x \wedge (y \wedge (x \wedge z)) = x \wedge (y \wedge z)$	[36 → 5 :6,6, flip]
49,48	$x \vee (y \vee (x \vee z)) = x \vee (y \vee z)$	[40 → 10 :11,11, flip]
51,50	$(x \wedge (y \vee z)) \vee (x \wedge z) = x \wedge (z \vee y)$	[9 → 12]
52	$((x \vee y) \wedge z) \vee (z \wedge x) = z \wedge (x \vee y)$	[4 → 12]
70	$x \wedge (y \wedge z) = x \wedge (z \wedge y)$	[4 → 20 :6]
100	$(x \vee ((x \vee y) \wedge z)) \wedge (x \vee y) = x \vee ((x \vee y) \wedge z)$	[28 → 14]
104	$(x \vee (x \wedge y)) \wedge x = x \vee (x \wedge y)$	[7 → 14]
123	$x \vee (y \vee z) = z \vee (y \vee x)$	[9 → 30]
154,153	$x \wedge (y \vee (z \wedge x)) = x \wedge (y \vee z)$	[36 → 16 :17, flip]
159,158	$x \wedge (y \vee (z \wedge (x \wedge u))) = x \wedge (y \vee (z \wedge u))$	[20 → 16]
161,160	$x \wedge (y \vee (z \vee (x \wedge u))) = x \wedge (y \vee (z \vee u))$	[16 → 16 :17, flip]
167	$x \wedge ((x \wedge y) \vee y) = x \wedge y$	[7 → 16 :22, flip]
170,169	$(x \vee (y \wedge z)) \wedge y = y \wedge (x \vee z)$	[4 → 16]
172	$x \vee (x \wedge y) = x \wedge (x \vee y)$	[104 :170, flip]
175,174	$x \vee ((x \vee y) \wedge z) = (x \vee y) \wedge (x \vee z)$	[100 :170, flip]
207	$(x \vee (y \wedge z)) \wedge (x \vee ((y \wedge z) \vee y)) = x \vee (y \wedge z)$	
		[14 → 167 :11,49,15]
213	$x \wedge (y \vee z) = x \wedge (z \vee y)$	[123 → 167 :161,29,41]
230,229	$x \wedge (y \wedge ((x \wedge z) \vee z)) = y \wedge (x \wedge z)$	[167 → 20, flip]
245	$x \vee (y \wedge x) = x \wedge (x \vee y)$	[4 → 172]
250,249	$(x \wedge y) \vee x = x \wedge (x \vee y)$	[9 → 172]
254,253	$(x \vee (y \wedge z)) \wedge (x \vee (y \wedge (y \vee z))) = x \vee (y \wedge z)$	[207 :250]
269	$(B \wedge A) \vee (A \wedge C) \neq A \wedge (B \vee C)$	[4 → 19]
276	$(x \vee y) \wedge z = z \wedge (y \vee x)$	[4 → 213]
280	$x \wedge (y \vee z) = (z \vee y) \wedge x$	[flip 276]
350	$(x \wedge y) \vee (y \wedge x) = y \wedge x$	[36 → 245 :6,230,37]

356	$x \vee ((y \wedge z) \vee (y \wedge (x \vee z))) = x \vee (y \wedge z)$	
		$[16 \to 245 :11,11,250,254]$
361,360	$(x \wedge y) \vee y = y \wedge (y \vee x)$	$[9 \to 245]$
615	$x \vee (y \wedge z) = (y \wedge z) \vee ((z \wedge y) \vee x)$	$[350 \to 123]$
621,620	$(x \wedge y) \vee ((y \wedge x) \vee z) = (y \wedge x) \vee z$	$[350 \to 10, \text{flip}]$
623	$x \vee (y \wedge z) = (z \wedge y) \vee x$	$[615 :621]$
866,865	$x \wedge ((y \wedge (x \wedge z)) \vee u) = x \wedge ((y \wedge z) \vee u)$	$[44 \to 50 :159,51, \text{flip}]$
869	$x \wedge ((y \wedge x) \vee z) = x \wedge (y \vee z)$	$[36 \to 50 :154,51, \text{flip}]$
880,879	$(x \wedge y) \vee (x \wedge (z \vee y)) = x \wedge (y \vee z)$	$[9 \to 50]$
884	$x \vee (y \wedge (z \vee x)) = x \vee (y \wedge z)$	$[356 :880]$
957	$(C \wedge A) \vee (B \wedge A) \neq A \wedge (B \vee C)$	$[623 \to 269]$
977	$(x \wedge y) \vee ((x \vee z) \wedge y) = y \wedge (x \vee z)$	$[623 \to 52]$
1053,1052	$x \wedge (y \wedge ((z \wedge x) \vee u)) = x \wedge (y \wedge (z \vee u))$	$[70 \to 869 :6,866,6]$
1114,1113	$x \vee (y \wedge z) = (x \vee z) \wedge (x \vee y)$	$[280 \to 884 :175, \text{flip}]$
1133,1132	$x \wedge ((x \vee y) \wedge (y \vee z)) = x \wedge (y \vee z)$	$[977 :1114,361,6,1053,29]$
1134	$A \wedge (C \vee B) \neq A \wedge (B \vee C)$	$[957 :1114,361,6,1053,1133]$
1135	$\square$	$[1134,213]$

A higher-order proof (Thm. QLT-1). It is clear that the given identity $x \wedge (y \vee (x \wedge z)) = x \wedge (y \vee z)$ implies distributivity in lattices because it fails in both M_5 and N_5.

Hence, by the structure theorem mentioned above, the only subdirectly irreducible quasilattices satisfying the given identity are shown in Fig. 6.2, where the element ∞ is an absorbing element for the binary operations, that is, $x \vee \infty = x \wedge \infty = \infty$.

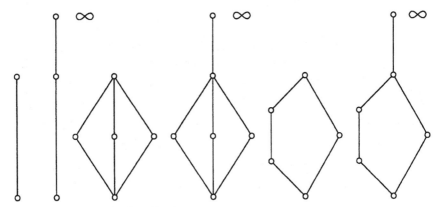

Fig. 6.2. Irreducible Quasilattices

Thus, the only subdirectly irreducible quasilattices satisfying the given identity are distributive, whatever may be the form of distributivity. Q.E.D.

6.2.1 Lattices

Theorem LT-4. The distributive law implies its dual in lattice theory.

$$\left\{ \begin{array}{l} \text{LT} \\ x \wedge (y \vee z) = (x \wedge y) \vee (x \wedge z) \end{array} \right\} \Rightarrow \{x \vee (y \wedge z) = (x \vee y) \wedge (x \vee z)\}.$$

Proof (found by Otter 3.0.4 on gyro at 4.08 seconds).

4	$x \wedge y = y \wedge x$	
6,5	$(x \wedge y) \wedge z = x \wedge (y \wedge z)$	
9	$x \vee y = y \vee x$	
10	$(x \vee y) \vee z = x \vee (y \vee z)$	
12	$x \wedge (x \vee y) = x$	
14	$x \vee (x \wedge y) = x$	
16	$x \wedge (y \vee z) = (x \wedge y) \vee (x \wedge z)$	
17	$(x \wedge y) \vee (x \wedge z) = x \wedge (y \vee z)$	[flip 16]
19	$A \vee (B \wedge C) \neq (A \vee B) \wedge (A \vee C)$	
20	$x \wedge (y \vee x) = x$	[9 → 12]
22	$(x \vee y) \wedge x = x$	[4 → 12]
35	$x \vee (y \wedge x) = x$	[4 → 14]
39	$x \wedge (y \wedge x) = x \wedge y$	[14 → 20 :6]
53	$(x \wedge y) \vee y = y$	[9 → 35]
55	$x \vee ((y \wedge x) \vee z) = x \vee z$	[35 → 10, flip]
70	$(x \vee (y \vee z)) \wedge (x \vee y) = x \vee y$	[10 → 22]
91,90	$x \vee ((x \vee y) \wedge z) = (x \vee y) \wedge (x \vee z)$	[22 → 17]
145	$A \vee (C \wedge B) \neq (A \vee B) \wedge (A \vee C)$	[4 → 19]
346,345	$x \vee (y \wedge (x \vee z)) = x \vee (y \wedge z)$	[17 → 55]
502,501	$(x \vee y) \wedge (x \vee (z \wedge y)) = x \vee (z \wedge y)$	[53 → 70]
816	$x \vee (y \wedge z) = (x \vee z) \wedge (x \vee y)$	[39 → 90 :91,346,502, flip]
818	□	[816,145]

It was previously known that the following self-dual equation can be used to express distributivity in lattice theory:

$$(x \wedge y) \vee (y \wedge z) \vee (z \wedge x) = (x \vee y) \wedge (y \vee z) \wedge (z \vee x).$$

The following theorem gives us an alternative equation.

Theorem LT-5. A new self-dual form of distributivity for lattice theory.

$$\left\{ \begin{array}{l} \text{LT} \\ (((x \wedge y) \vee z) \wedge y) \vee (z \wedge x) = (((x \vee y) \wedge z) \vee y) \wedge (z \vee x) \end{array} \right\} \Rightarrow$$

$$\{x \vee (y \wedge z) = (x \vee y) \wedge (x \vee z)\}.$$

Proof (found by Otter 3.0.4 on gyro at 11.74 seconds).

3,2	$x \wedge x = x$	
4	$x \wedge y = y \wedge x$	
6,5	$(x \wedge y) \wedge z = x \wedge (y \wedge z)$	
8,7	$x \vee x = x$	
9	$x \vee y = y \vee x$	
11,10	$(x \vee y) \vee z = x \vee (y \vee z)$	
12	$x \wedge (x \vee y) = x$	
15,14	$x \vee (x \wedge y) = x$	
16	$(((x \wedge y) \vee z) \wedge y) \vee (z \wedge x) = (((x \vee y) \wedge z) \vee y) \wedge (z \vee x)$	
20	$A \vee (B \wedge C) \neq (A \vee B) \wedge (A \vee C)$	
21	$x \wedge (y \wedge z) = y \wedge (x \wedge z)$	$[4 \to 5 :6]$
22	$x \wedge (x \wedge y) = x \wedge y$	$[2 \to 5, \text{flip}]$
29,28	$x \wedge (y \vee x) = x$	$[9 \to 12]$
33,32	$(x \vee y) \wedge x = x$	$[4 \to 12]$
36	$x \vee (y \vee z) = y \vee (x \vee z)$	$[9 \to 10 :11]$
38,37	$x \vee (x \vee y) = x \vee y$	$[7 \to 10, \text{flip}]$
47	$x \vee (y \wedge x) = x$	$[4 \to 14]$
51	$(x \wedge y) \vee x = x$	$[9 \to 14]$
53	$x \vee ((x \wedge y) \vee z) = x \vee z$	$[14 \to 10, \text{flip}]$
56,55	$x \wedge (y \wedge x) = x \wedge y$	$[14 \to 28 :6]$
57	$x \wedge (y \vee (z \vee x)) = x$	$[10 \to 28]$
61	$(x \vee y) \wedge y = y$	$[4 \to 28]$
67	$(((x \wedge y) \vee z) \wedge x) \vee (z \wedge y) = (((y \vee x) \wedge z) \vee x) \wedge (z \vee y)$	
		$[4 \to 16]$
76	$(((((x \vee y) \wedge z) \vee x) \wedge z) \vee x = (x \vee z) \wedge (x \vee y)$	
		$[12 \to 16 :11,33,38]$
89	$(((x \vee (y \vee x)) \wedge z) \vee x) \wedge (z \vee (x \vee y)) = x \vee (z \wedge (x \vee y))$	
		$[32 \to 16 :33,11, \text{flip}]$
92,91	$x \vee (y \vee x) = x \vee y$	$[32 \to 14 :11]$
94,93	$(x \vee y) \wedge (x \wedge z) = x \wedge z$	$[32 \to 5, \text{flip}]$
95	$(((x \vee y) \wedge z) \vee x) \wedge (z \vee (x \vee y)) = x \vee (z \wedge (x \vee y))$ $[89 :92]$	
110,109	$x \vee ((y \wedge x) \vee z) = x \vee z$	$[47 \to 10, \text{flip}]$
118	$(((x \vee y) \wedge z) \vee y) \wedge (z \wedge x) = z \wedge x$	$[16 \to 61 :6,94]$
122	$(((x \vee y) \wedge z) \vee y) \wedge (z \vee (x \vee y)) = y \vee (z \wedge (x \vee y))$	
		$[61 \to 16 :33,11,8, \text{flip}]$
127	$(B \wedge C) \vee A \neq (A \vee B) \wedge (A \vee C)$	$[9 \to 20]$
141,140	$(x \vee y) \wedge (z \wedge y) = z \wedge y$	$[61 \to 21, \text{flip}]$
142	$(x \vee y) \wedge (z \wedge x) = z \wedge x$	$[32 \to 21, \text{flip}]$
207,206	$(x \vee y) \wedge (z \vee (y \vee x)) = x \vee y$	$[91 \to 57]$
559,558	$((x \wedge y) \vee z) \wedge (u \vee (y \vee z)) = (x \wedge y) \vee z$	$[109 \to 57]$
696,695	$(x \vee (y \wedge z)) \wedge (z \wedge y) = z \wedge y$	$[55 \to 140 :56]$
747,746	$((x \wedge y) \vee z) \wedge (y \wedge x) = y \wedge x$	$[55 \to 142 :56]$
936	$(x \wedge y) \vee (z \vee ((((y \vee x) \wedge z) \vee x) \wedge (z \vee y))) = (x \wedge y) \vee z$	

$$[67 \to 53 :11,11,15]$$

1222,1221 $(x \wedge (y \vee z)) \vee y = (y \vee (x \wedge (y \vee z))) \wedge (y \vee z)$ $\quad [55 \to 76 :747]$

1224,1223 $((x \vee y) \wedge z) \vee x = (x \vee ((x \vee y) \wedge z)) \wedge (x \vee y)$ $\quad [22 \to 76 :33]$

1226,1225 $(x \vee ((x \vee y) \wedge z)) \wedge (x \vee y) = (x \vee z) \wedge (x \vee y)$

$$[4 \to 76 :1222,6,696,1224]$$

1234,1233 $x \vee ((x \vee y) \wedge z) = (x \vee z) \wedge (x \vee y)$ $\quad [9 \to 76 :1224,1226,6,141]$

1246,1245 $x \vee (y \wedge (x \vee z)) = (x \vee y) \wedge (x \vee z)$ $\quad [95 :1224,1234,6,3,6,29, \text{flip}]$

1248,1247 $((x \vee y) \wedge z) \vee x = (x \vee z) \wedge (x \vee y)$ $\quad [1223 :1234,6,3]$

1254,1253 $(x \wedge y) \vee ((z \vee x) \wedge (z \vee y)) = (x \wedge y) \vee z$ $\quad [936 :1246,110]$

2199 $(x \wedge y) \vee (((y \vee z) \wedge x) \vee z) = ((y \vee z) \wedge x) \vee z$ $\quad [118 \to 51]$

2238,2237 $x \vee (y \wedge (z \vee x)) = (x \vee y) \wedge (x \vee z)$ $\quad [9 \to 122 :1248,6,207, \text{flip}]$

2265,2264 $((x \vee y) \wedge z) \vee y = (y \vee z) \wedge (y \vee x)$ $\quad [36 \to 122 :559,2238]$

2280 $(x \wedge y) \vee z = (z \vee x) \wedge (z \vee y)$ $\quad [2199 :2265,1254,2265]$

2281 $\square$ $\quad [2280,127]$

Theorem LT-6. McKenzie's basis for the variety generated by N_5. (Suggested by David Kelly.)

Proof (found by Otter 3.0.4 on gyro at 101.45 seconds).

1 $\quad x = x$

2 $\quad x \vee x = x$

5,4 $\quad x \wedge x = x$

6 $\quad x \vee y = y \vee x$

7 $\quad x \wedge y = y \wedge x$

9,8 $\quad (x \vee y) \vee z = x \vee (y \vee z)$

11,10 $\quad (x \wedge y) \wedge z = x \wedge (y \wedge z)$

13,12 $\quad x \wedge (x \vee y) = x$

15,14 $\quad x \vee (x \wedge y) = x$

16 $\quad x \wedge (y \vee (z \wedge (x \vee u))) =$
$\quad\quad (x \wedge (y \vee (x \wedge z))) \vee (x \wedge ((x \wedge y) \vee (z \wedge u)))$

17 $\quad (x \wedge (y \vee (x \wedge z))) \vee (x \wedge ((x \wedge y) \vee (z \wedge u))) =$
$\quad\quad x \wedge (y \vee (z \wedge (x \vee u)))$ $\quad [\text{flip } 16]$

19 $\quad x \vee (y \wedge (z \vee (x \wedge u))) =$
$\quad\quad (x \vee (y \wedge (x \vee z))) \wedge (x \vee ((x \vee y) \wedge (z \vee u)))$

20 $\quad (x \vee (y \wedge z)) \wedge (z \vee (x \wedge y)) = (z \wedge (x \vee (y \wedge z))) \vee (x \wedge (y \vee z))$

21 $\quad (x \wedge (y \vee (z \wedge x))) \vee (y \wedge (z \vee x)) = (y \vee (z \wedge x)) \wedge (x \vee (y \wedge z))$
$\quad\quad [\text{flip } 20]$

23 $\quad A \wedge ((B \vee C) \wedge (B \vee D)) \neq (A \wedge ((B \vee C) \wedge (B \vee D))) \wedge ((A \wedge$
$\quad\quad (B \vee (C \wedge D))) \vee ((A \wedge C) \vee (A \wedge D)))$

24 $\quad A \wedge ((B \vee C) \wedge ((B \vee D) \wedge ((A \wedge (B \vee (C \wedge D))) \vee ((A \wedge C) \vee$
$\quad\quad (A \wedge D))))) \neq A \wedge ((B \vee C) \wedge (B \vee D))$ $\quad [\text{copy},23 :11,11, \text{flip}]$

25 $\quad (x \vee (y \wedge (x \vee z))) \wedge (x \vee ((x \vee y) \wedge (z \vee u))) =$
$\quad\quad x \vee (y \wedge (z \vee (x \wedge u)))$ $\quad [\text{flip } 19]$

26 $\quad x \wedge (y \vee x) = x$ $\quad [6 \to 12]$

28	$(x \vee y) \wedge x = x$	$[7 \to 12]$
30	$x \vee (y \vee z) = y \vee (x \vee z)$	$[6 \to 8 :9]$
32,31	$x \vee (x \vee y) = x \vee y$	$[2 \to 8, \text{flip}]$
33	$x \vee (y \vee z) = y \vee (z \vee x)$	$[6 \to 8]$
36	$x \vee (y \vee z) = z \vee (x \vee y)$	$[\text{flip } 33]$
40,39	$x \vee (y \wedge x) = x$	$[7 \to 14]$
43	$(x \wedge y) \vee x = x$	$[6 \to 14]$
45	$x \vee ((x \wedge y) \vee z) = x \vee z$	$[14 \to 8, \text{flip}]$
48,47	$x \wedge (y \wedge x) = x \wedge y$	$[14 \to 26 :11]$
50,49	$x \wedge (y \vee (z \vee x)) = x$	$[8 \to 26]$
51	$(x \vee y) \wedge y = y$	$[7 \to 26]$
55	$x \vee (y \vee x) = x \vee y$	$[28 \to 14 :9]$
59	$(x \wedge y) \vee y = y$	$[6 \to 39]$
62,61	$x \vee ((y \wedge x) \vee z) = x \vee z$	$[39 \to 8, \text{flip}]$
63	$(x \vee y) \wedge (x \wedge z) = x \wedge z$	$[28 \to 10, \text{flip}]$
69	$x \wedge (y \wedge z) = y \wedge (x \wedge z)$	$[7 \to 10 :11]$
71,70	$x \wedge (x \wedge y) = x \wedge y$	$[4 \to 10, \text{flip}]$
76	$x \wedge (y \wedge z) = y \wedge (z \wedge x)$	$[7 \to 10]$
77	$x \wedge (y \wedge z) = z \wedge (x \wedge y)$	$[\text{flip } 76]$
78	$x \vee (y \wedge (z \wedge x)) = x$	$[10 \to 39]$
83,82	$(x \wedge (y \wedge z)) \vee (x \wedge y) = x \wedge y$	$[10 \to 43]$
86	$(x \vee (y \vee z)) \wedge z = z$	$[8 \to 51]$
90	$(x \wedge (y \wedge z)) \vee z = z$	$[10 \to 59]$
92	$(x \wedge y) \vee (y \vee z) = y \vee z$	$[59 \to 8, \text{flip}]$
104	$(x \wedge (y \vee (z \wedge x))) \vee (x \wedge ((x \wedge y) \vee (z \wedge u))) =$	
	$\quad x \wedge (y \vee (z \wedge (x \vee u)))$	$[7 \to 17]$
112	$(x \wedge (y \vee (z \vee (x \wedge u)))) \vee (x \wedge ((x \wedge (y \vee z)) \vee (u \wedge v))) =$	
	$\quad x \wedge (y \vee (z \vee (u \wedge (x \vee v))))$	$[8 \to 17 :9]$
126	$(x \wedge (y \vee (x \wedge z))) \vee (x \wedge ((y \wedge x) \vee (z \wedge u))) =$	
	$\quad x \wedge (y \vee (z \wedge (x \vee u)))$	$[7 \to 17]$
128	$(x \wedge (y \vee (x \wedge (z \vee u)))) \vee (x \wedge ((x \wedge y) \vee u)) =$	
	$\quad x \wedge (y \vee ((z \vee u) \wedge (x \vee u)))$	$[51 \to 17]$
132	$(x \wedge (y \vee (x \wedge z))) \vee (x \wedge ((x \wedge y) \vee z)) = x \wedge (y \vee z)$	
		$[26 \to 17 :50]$
167,166	$x \wedge (y \wedge (z \wedge x)) = x \wedge (y \wedge z)$	$[10 \to 47]$
168	$(x \wedge y) \vee (y \wedge x) = x \wedge y$	$[47 \to 39]$
175,174	$x \wedge (y \wedge (x \wedge z)) = x \wedge (y \wedge z)$	$[47 \to 10 :11,11, \text{flip}]$
189,188	$x \wedge (y \wedge (z \vee x)) = x \wedge y$	$[14 \to 49 :11]$
192	$x \wedge (y \vee (x \vee z)) = x$	$[6 \to 49]$
210	$(x \vee y) \wedge (y \vee x) = x \vee y$	$[55 \to 26]$
242,241	$x \vee (y \wedge (x \wedge z)) = x$	$[43 \to 19 :11,15,40,11,13]$
273	$x \vee (y \vee (z \wedge x)) = x \vee y$	$[12 \to 78 :9]$
295	$(x \vee (y \vee z)) \wedge y = y$	$[6 \to 86]$
305	$(x \wedge y) \vee (z \vee y) = z \vee y$	$[26 \to 90]$

361        $(x \wedge y) \vee (y \wedge (x \wedge z)) = y \wedge x$
           $[43 \rightarrow 21 :48,11,11,11,189,11,83,11,11,11,167,242,11,5]$
409        $(x \vee (y \wedge z)) \wedge ((z \vee (x \wedge y)) \wedge (x \wedge (y \vee z))) = x \wedge (y \vee z)$
           $[21 \rightarrow 51 :11]$
481,480 $x \vee (y \vee (x \wedge z)) = y \vee x$ $[14 \rightarrow 30, \text{flip}]$
487        $x \vee ((y \wedge (x \vee z)) \vee z) = x \vee z$ $[59 \rightarrow 30, \text{flip}]$
527        $x \vee (y \vee z) = z \vee (y \vee x)$ $[6 \rightarrow 33]$
576,575 $x \wedge (y \wedge (x \vee z)) = x \wedge y$ $[45 \rightarrow 192 :11]$
578,577 $((x \wedge y) \vee z) \wedge (x \vee z) = (x \wedge y) \vee z$ $[45 \rightarrow 26]$
626        $x \vee (y \wedge ((x \wedge z) \vee (x \wedge u))) = x$ $[14 \rightarrow 25 :40,13, \text{flip}]$
644        $((x \wedge y) \vee (x \wedge ((x \wedge y) \vee z))) \wedge ((x \wedge y) \vee (x \wedge (z \vee u))) =$
           $(x \wedge y) \vee (x \wedge (z \vee (x \wedge (y \wedge u))))$ $[43 \rightarrow 25 :11]$
686        $(x \wedge y) \vee (x \wedge (z \vee (x \wedge (y \wedge u)))) = ((x \wedge y) \vee (x \wedge ((x \wedge y) \vee$
           $z))) \wedge ((x \wedge y) \vee (x \wedge (z \vee u)))$ $[\text{flip } 644]$
824,823 $(x \vee y) \wedge (z \wedge (x \wedge u)) = z \wedge (x \wedge u)$ $[63 \rightarrow 69, \text{flip}]$
835        $x \wedge (((x \wedge y) \vee z) \wedge y) = x \wedge y$ $[28 \rightarrow 69, \text{flip}]$
840        $(x \vee (y \wedge z)) \wedge (y \wedge (z \vee x)) = y \wedge (z \vee x)$ $[409 :824]$
1109 $x \wedge (y \wedge (z \vee (y \wedge x))) = x \wedge y$ $[168 \rightarrow 49 :11]$
1121,1120 $((x \wedge y) \vee z) \wedge (u \wedge (y \vee z)) = ((x \wedge y) \vee z) \wedge u$ $[61 \rightarrow 188]$
1123,1122 $(x \vee y) \wedge (z \wedge (y \vee x)) = (x \vee y) \wedge z$ $[55 \rightarrow 188]$
1129,1128 $(x \vee y) \wedge (z \wedge (y \vee (u \vee x))) = (x \vee y) \wedge z$ $[36 \rightarrow 188]$
1156,1155 $(x \wedge y) \vee (y \wedge (z \vee x)) = y \wedge (z \vee x)$ $[188 \rightarrow 59]$
1271 $x \wedge (y \vee z) = (z \vee y) \wedge x$ $[210 \rightarrow 77 :1123]$
1272 $(x \vee y) \wedge z = z \wedge (y \vee x)$ $[\text{flip } 1271]$
1341,1340 $(x \vee (y \wedge z)) \wedge (u \wedge (z \vee x)) = (x \vee (y \wedge z)) \wedge u$ $[273 \rightarrow 188]$
1351 $(x \vee (y \wedge z)) \wedge y = y \wedge (z \vee x)$ $[840 :1341]$
1916 $A \wedge ((B \vee C) \wedge ((B \vee D) \wedge ((A \wedge D) \vee ((A \wedge C) \vee (A \wedge (B \vee$
           $(C \wedge D)))))))) \neq A \wedge ((B \vee C) \wedge (B \vee D))$ $[527 \rightarrow 24]$
1971,1970 $(x \wedge y) \vee (z \vee (y \wedge (x \vee u))) = z \vee (y \wedge (x \vee u))$ $[575 \rightarrow 305]$
2231,2230 $(x \wedge y) \vee (x \wedge (z \vee (y \vee u))) = x \wedge (z \vee (y \vee u))$ $[295 \rightarrow 82]$
2243,2242 $(x \wedge (y \wedge z)) \vee (x \wedge (y \vee u)) = x \wedge (y \vee u)$ $[63 \rightarrow 82]$
2247,2246 $(x \wedge y) \vee (x \wedge (z \vee y)) = x \wedge (z \vee y)$ $[51 \rightarrow 82]$
3172,3171 $(x \wedge y) \vee (x \wedge ((x \wedge (y \wedge z)) \vee (y \wedge u))) = x \wedge y$
           $[82 \rightarrow 104 :48,175,2243,576]$
3853 $(x \wedge ((x \wedge y) \vee z)) \vee (x \wedge ((x \wedge ((x \wedge y) \vee z)) \vee (y \wedge u))) =$
           $x \wedge (z \vee (y \wedge (x \vee u)))$ $[55 \rightarrow 112 :1971]$
4053,4052 $(x \wedge (y \vee (x \wedge z))) \vee (x \wedge z) = (y \vee (x \wedge z)) \wedge x$ $[51 \rightarrow 361]$
5139 $(x \wedge y) \vee (z \wedge (u \vee (x \wedge (z \vee y)))) = (x \wedge y) \vee (z \wedge (u \vee (z \wedge x)))$
           $[126 \rightarrow 273 :9,1971,9,1971]$
5159 $(x \wedge y) \vee (z \wedge (u \vee (z \wedge x))) = (x \wedge y) \vee (z \wedge (u \vee (x \wedge (z \vee y))))$
           $[\text{flip } 5139]$
5530,5529 $x \vee (y \wedge ((y \wedge x) \vee z)) = x \vee (y \wedge (x \vee z))$
           $[128 \rightarrow 487 :578,62, \text{flip}]$
6026 $x \wedge (((y \wedge z) \vee (y \wedge u)) \wedge y) = x \wedge ((y \wedge z) \vee (y \wedge u))$

$$[626 \to 26 :11]$$

6072,6071 $x \wedge (y \wedge ((x \wedge z) \vee (x \wedge u))) = y \wedge ((x \wedge z) \vee (x \wedge u))$ $[626 \to 51]$

6739,6738 $(x \vee (y \wedge (x \vee z))) \wedge y = y \wedge (x \vee z)$ $[92 \to 132 :4053,32]$

6830,6829 $x \wedge (y \vee (x \wedge (y \vee z))) = x \wedge (y \vee z)$

$$[132 \to 132 :71,9,2247,5530,71,4053,6739,9,2247,5530, \text{flip}]$$

6857 $((x \wedge y) \vee z) \wedge ((x \wedge (y \vee (x \wedge z))) \vee (x \wedge ((x \wedge y) \vee (z \vee u)))) =$
$((x \wedge y) \vee z) \wedge x$ $[132 \to 104 :1121,3172,9, \text{flip}]$

7082,7081 $(x \wedge y) \vee (x \wedge ((x \wedge y) \vee z)) = x \wedge ((x \wedge y) \vee z)$ $[835 \to 82]$

7141 $x \wedge ((x \wedge ((x \wedge y) \vee z)) \vee (y \wedge u)) =$
$x \wedge (z \vee (y \wedge (x \vee u)))$ $[3853 :7082]$

7189 $(x \wedge y) \vee (x \wedge (z \vee (x \wedge (y \wedge u)))) = ((x \wedge y) \vee z) \wedge ((x \wedge y) \vee$
$(x \wedge (z \vee u)))$ $[686 :7082,11,6072]$

8277 $(x \wedge y) \vee (y \wedge (z \vee (y \wedge x))) = y \wedge (z \vee (y \wedge x))$ $[1109 \to 59]$

9238,9237 $((x \wedge y) \vee z) \wedge x = x \wedge (y \vee z)$ $[6 \to 1351]$

9241,9240 $x \wedge ((x \wedge y) \vee z) = x \wedge (y \vee z)$ $[1272 \to 1351]$

9397,9396 $x \wedge ((y \wedge z) \vee (y \wedge u)) = x \wedge (y \wedge (z \vee (y \wedge u)))$ $[6026 :9238, \text{flip}]$

9603,9602 $x \wedge (y \vee (z \wedge (x \vee u))) = x \wedge (y \vee z)$ $[7141 :9241,9241,9,481, \text{flip}]$

9965,9964 $((x \wedge y) \vee z) \wedge (x \wedge (y \vee (z \vee u))) = x \wedge (y \vee z)$

$$[6857 :9241,9397,9,2231,6830,9238]$$

10254 $(x \wedge y) \vee (x \wedge (z \vee (x \wedge (y \wedge u)))) =$
$((x \wedge y) \vee z) \wedge (x \wedge (y \vee (x \wedge (z \vee u))))$ $[7189 :9397]$

10762,10761 $(x \wedge y) \vee (z \wedge (u \vee (z \wedge x))) = (x \wedge y) \vee (z \wedge (u \vee x))$

$$[5159 :9603]$$

10966,10965 $x \wedge (y \vee (x \wedge z)) = x \wedge (y \vee z)$ $[8277 :10762,1156, \text{flip}]$

11208,11207 $(x \wedge y) \vee (x \wedge (z \vee (y \wedge u))) = x \wedge (y \vee z)$

$$[10254 :10966,10966,9965]$$

11311,11310 $x \wedge ((y \wedge z) \vee (y \wedge u)) = x \wedge (y \wedge (z \vee u))$ $[9396 :10966]$

11477 $A \wedge ((B \vee C) \wedge (B \vee D)) \neq A \wedge ((B \vee C) \wedge (B \vee D))$

$$[1916 :11208,11311,1129,167]$$

11478 □ $[11477,1]$

6.2.2 Quasilattices

Theorem QLT-2. The distributive law implies its dual in QLT.

$$\left\{ \begin{array}{l} \text{QLT} \\ x \wedge (y \vee z) = (x \wedge y) \vee (x \wedge z) \end{array} \right\} \Rightarrow \{x \vee (y \wedge z) = (x \vee y) \wedge (x \vee z)\}.$$

Proof (found by Otter 3.0.4 on gyro at 7.28 seconds).

2	$x \wedge x = x$
4	$x \wedge y = y \wedge x$
5	$(x \wedge y) \wedge z = x \wedge (y \wedge z)$
8,7	$x \vee x = x$
9	$x \vee y = y \vee x$
11,10	$(x \vee y) \vee z = x \vee (y \vee z)$

14	$(x \vee (y \wedge z)) \wedge (x \vee y) = x \vee (y \wedge z)$	
16	$x \wedge (y \vee z) = (x \wedge y) \vee (x \wedge z)$	
18,17	$(x \wedge y) \vee (x \wedge z) = x \wedge (y \vee z)$	[flip 16]
19	$A \vee (B \wedge C) \neq (A \vee B) \wedge (A \vee C)$	
23	$x \wedge (x \wedge y) = x \wedge y$	$[2 \to 5, \text{flip}]$
30	$x \vee (x \vee y) = x \vee y$	$[7 \to 10, \text{flip}]$
43,42	$x \vee (y \vee x) = x \vee y$	$[9 \to 30]$
51,50	$x \vee (y \vee (x \vee z)) = x \vee (y \vee z)$	$[42 \to 10 :11,11, \text{flip}]$
58	$(x \vee ((x \vee y) \wedge z)) \wedge (x \vee y) = x \vee ((x \vee y) \wedge z)$	$[30 \to 14]$
108	$x \vee (x \wedge y) = x \wedge (x \vee y)$	$[2 \to 17]$
116	$x \wedge (y \vee (x \wedge z)) = x \wedge (y \vee z)$	$[23 \to 17 :18, \text{flip}]$
122,121	$(x \wedge y) \vee x = x \wedge (y \vee x)$	$[2 \to 17]$
123	$x \wedge (y \vee z) = x \wedge (z \vee y)$	$[9 \to 17 :18]$
163	$A \vee (C \wedge B) \neq (A \vee B) \wedge (A \vee C)$	$[4 \to 19]$
181,180	$(x \vee (y \wedge z)) \wedge (x \vee (y \wedge (z \vee y))) = x \vee (y \wedge z)$	
		$[14 \to 108 :8,11,51,122, \text{flip}]$
183	$x \vee (y \wedge x) = x \wedge (x \vee y)$	$[4 \to 108]$
226	$(x \vee y) \wedge z = z \wedge (y \vee x)$	$[4 \to 123]$
230	$x \wedge (y \vee z) = (z \vee y) \wedge x$	[flip 226]
916,915	$(x \vee (y \wedge z)) \wedge y = y \wedge (x \vee z)$	$[4 \to 116]$
918,917	$x \vee ((x \vee y) \wedge z) = (x \vee y) \wedge (x \vee z)$	$[58 :916, \text{flip}]$
919	$x \vee (y \wedge (z \vee x)) = x \vee (y \wedge z)$	$[116 \to 183 :11,18,43,11,122,181]$
1440	$x \vee (y \wedge z) = (x \vee z) \wedge (x \vee y)$	$[230 \to 919 :918, \text{flip}]$
1442	$\square$	$[1440,163]$

Theorem QLT-3. A self-dual form of distributivity for QLT.

$$\left\{ \begin{array}{l} \text{QLT} \\ (((x \wedge y) \vee z) \wedge y) \vee (z \wedge x) = (((x \vee y) \wedge z) \vee y) \wedge (z \vee x) \end{array} \right\} \Rightarrow$$

$$\{x \vee (y \wedge z) = (x \vee y) \wedge (x \vee z)\}.$$

(Otter's original proof, which was found with the standard strategy with `max_weight`=24, `change_limit_after`=50, and `new_max_weight`=17, is 183 steps and was found in about 15 minutes. We asked Larry Wos to try some of his new methods for shortening proofs by using the resonance strategy [82], and he found the following 113-step proof.)

Proof (found by Otter 3.0.4 on gyro at 165.96 seconds).

3,2	$x \wedge x = x$
4	$x \wedge y = y \wedge x$
6,5	$(x \wedge y) \wedge z = x \wedge (y \wedge z)$
8,7	$x \vee x = x$
9	$x \vee y = y \vee x$
11,10	$(x \vee y) \vee z = x \vee (y \vee z)$

12	$(x \wedge (y \vee z)) \vee (x \wedge y) = x \wedge (y \vee z)$	
15,14	$(x \vee (y \wedge z)) \wedge (x \vee y) = x \vee (y \wedge z)$	
16	$(((x \wedge y) \vee z) \wedge y) \vee (z \wedge x) = (((x \vee y) \wedge z) \vee y) \wedge (z \vee x)$	
18	$A \wedge (B \vee C) \neq (A \wedge B) \vee (A \wedge C)$	
19	$(A \wedge B) \vee (A \wedge C) \neq A \wedge (B \vee C)$	[flip 18]
20	$x \wedge (y \wedge z) = y \wedge (x \wedge z)$	$[4 \to 5 :6]$
22,21	$x \wedge (x \wedge y) = x \wedge y$	$[2 \to 5, \text{flip}]$
23	$x \wedge (y \wedge z) = y \wedge (z \wedge x)$	$[4 \to 5]$
26	$x \wedge (y \wedge z) = z \wedge (x \wedge y)$	[flip 23]
30,29	$x \wedge (y \wedge x) = y \wedge x$	$[2 \to 20, \text{flip}]$
33,32	$x \wedge (y \wedge (x \wedge z)) = x \wedge (y \wedge z)$	$[20 \to 21]$
57	$x \vee (y \vee z) = y \vee (x \vee z)$	$[9 \to 10 :11]$
59,58	$x \vee (x \vee y) = x \vee y$	$[7 \to 10, \text{flip}]$
60	$x \vee (y \vee z) = y \vee (z \vee x)$	$[9 \to 10]$
63	$x \vee (y \vee z) = z \vee (x \vee y)$	[flip 60]
67,66	$x \vee (y \vee x) = y \vee x$	$[7 \to 57, \text{flip}]$
70,69	$x \vee (y \vee (x \vee z)) = x \vee (y \vee z)$	$[57 \to 58]$
74	$x \vee (y \vee z) = z \vee (y \vee x)$	$[9 \to 60]$
80	$x \vee (y \vee (z \vee u)) = y \vee (u \vee (x \vee z))$	$[57 \to 63 :11]$
81	$x \vee (y \vee (z \vee u)) = z \vee (x \vee (u \vee y))$	[flip 80]
94	$(x \wedge (y \vee (z \vee u))) \vee (x \wedge u) = x \wedge (u \vee (y \vee z))$	$[60 \to 12]$
99	$(x \wedge (y \wedge (z \vee u))) \vee (x \wedge (y \wedge z)) = x \wedge (y \wedge (z \vee u))$	$[5 \to 12 :6,6]$
101	$((x \vee y) \wedge z) \vee (z \wedge x) = z \wedge (x \vee y)$	$[4 \to 12]$
103	$x \vee (y \vee ((x \vee y) \wedge x)) = x \vee y$	$[2 \to 12 :11,3]$
105	$(x \wedge ((y \wedge x) \vee z)) \vee (y \wedge x) = x \wedge ((y \wedge x) \vee z)$	$[29 \to 12]$
110,109	$(x \wedge (y \vee z)) \vee (y \wedge x) = x \wedge (y \vee z)$	$[4 \to 12]$
114,113	$(x \wedge y) \vee (x \wedge (y \vee z)) = x \wedge (y \vee z)$	$[9 \to 12]$
122	$((x \wedge y) \vee z) \wedge (z \vee x) = z \vee (x \wedge y)$	$[9 \to 14]$
126,125	$(x \vee (y \wedge z)) \wedge (y \vee x) = x \vee (y \wedge z)$	$[9 \to 14]$
127	$(x \vee (x \wedge y)) \wedge x = x \vee (x \wedge y)$	$[7 \to 14]$
130,129	$(x \vee y) \wedge (x \vee (y \wedge z)) = x \vee (y \wedge z)$	$[4 \to 14]$
134	$((x \vee y) \wedge x) \vee (y \wedge x) = ((x \wedge y) \vee x) \wedge (y \vee x)$	$[2 \to 16 :8]$
136	$(x \wedge ((y \wedge x) \vee z)) \vee (z \wedge y) = (((y \vee x) \wedge z) \vee x) \wedge (z \vee y)$	$[4 \to 16]$
147	$(x \wedge (y \wedge (z \vee u))) \vee (x \wedge (y \wedge u)) = x \wedge (y \wedge (z \vee u))$	$[66 \to 99 :67]$
150	$(x \wedge (y \vee z)) \vee (x \wedge ((y \vee z) \wedge y)) = x \wedge (y \vee z)$	$[2 \to 99 :3]$
163,162	$((x \vee y) \wedge z) \vee (z \wedge y) = z \wedge (x \vee y)$	$[66 \to 101 :67]$
165,164	$(x \wedge y) \vee (y \wedge x) = y \wedge x$	$[7 \to 101 :8]$
166	$(((x \wedge y) \vee z) \wedge y) \vee (x \wedge y) = y \wedge ((x \wedge y) \vee z)$	$[29 \to 101]$
170	$((x \vee y) \wedge z) \vee (x \wedge z) = z \wedge (x \vee y)$	$[4 \to 101]$
173,172	$((x \vee y) \wedge x) \vee x = x \wedge (x \vee y)$	$[2 \to 101]$
175,174	$(x \vee y) \wedge (y \vee x) = x \vee y$	$[94 \to 101 :67,59,67,67,3]$
176	$(x \wedge y) \vee ((y \vee z) \wedge x) = x \wedge (y \vee z)$	$[9 \to 101]$
179,178	$x \vee (y \vee ((y \vee x) \wedge x)) = y \vee x$	$[66 \to 103 :11,70,67]$
187,186	$((x \vee y) \wedge x) \vee (x \vee y) = x \vee y$	$[63 \to 103]$

193,192	$x \vee (y \vee (z \wedge (z \vee x))) = y \vee (z \vee x)$	[103 → 81 :173, flip]
196	$(x \wedge ((y \wedge x) \vee z)) \vee (x \wedge y) = x \wedge ((y \wedge x) \vee z)$	[4 → 105]
199,198	$(x \wedge y) \vee (y \wedge ((x \wedge y) \vee z)) = y \wedge ((x \wedge y) \vee z)$	[9 → 105]
205,204	$(x \wedge y) \vee (y \wedge (x \vee z)) = y \wedge (x \vee z)$	[4 → 113]
208	$(x \wedge y) \vee (x \wedge (z \vee y)) = x \wedge (z \vee y)$	[66 → 113 :67]
212	$(x \wedge y) \vee ((x \wedge (y \vee z)) \vee u) = (x \wedge (y \vee z)) \vee u$	[113 → 10, flip]
214	$((x \wedge y) \vee z) \wedge (z \vee y) = z \vee (x \wedge y)$	[29 → 122 :30]
216	$((x \wedge y) \vee z) \wedge (x \vee z) = z \vee (x \wedge y)$	[9 → 122]
217	$((x \wedge y) \vee x) \wedge x = x \vee (x \wedge y)$	[7 → 122]
224	$x \vee (y \wedge (x \vee y)) = x \vee y$	[122 → 16 :11,110,67,187,3]
227,226	$((x \wedge y) \vee z) \wedge (z \vee x) = (x \wedge y) \vee z$	[122 → 12 :11,179, flip]
232	$(x \vee y) \wedge (y \vee (x \wedge z)) = y \vee (x \wedge z)$	[9 → 129]
234	$(x \vee y) \wedge (x \vee (z \wedge y)) = x \vee (z \wedge y)$	[29 → 129 :30]
236	$((x \wedge y) \vee x) \wedge (x \wedge (y \vee z)) = x \wedge (y \vee z)$	[113 → 129 :114]
243,242	$(x \wedge y) \vee x = x \vee (y \wedge x)$	[127 → 134 :6,30,11,165,22,3, flip]
247,246	$x \vee (y \wedge x) = x \wedge (x \vee y)$	[4 → 134 :163,243,126, flip]
254	$x \wedge ((x \vee y) \wedge (y \vee z)) = x \wedge (y \vee z)$	[236 :243,247,6,33]
257,256	$x \vee (x \wedge y) = (x \vee y) \wedge x$	[217 :243,247,6,30, flip]
265,264	$(x \wedge y) \vee x = x \wedge (x \vee y)$	[242 :247]
268	$(x \wedge ((y \wedge x) \vee z)) \vee (y \wedge z) = (((y \vee x) \wedge z) \vee x) \wedge (z \vee y)$	
		[4 → 136]
275	$(x \wedge (y \wedge (z \vee u))) \vee (y \wedge (x \wedge u)) = y \wedge (x \wedge (z \vee u))$	[20 → 147]
287	$(x \wedge (y \vee z)) \vee (x \wedge ((z \vee y) \wedge y)) = x \wedge (y \vee z)$	[9 → 150]
291	$((x \vee y) \wedge z) \vee (y \wedge z) = z \wedge (x \vee y)$	[4 → 162]
294,293	$((x \vee y) \wedge y) \vee y = y \wedge (x \vee y)$	[2 → 162]
296,295	$(x \wedge y) \vee ((z \vee y) \wedge x) = x \wedge (z \vee y)$	[9 → 162]
304	$x \vee (y \wedge z) = (y \wedge z) \vee ((z \wedge y) \vee x)$	[164 → 74]
307,306	$(x \wedge y) \vee (z \vee (y \wedge x)) = z \vee (y \wedge x)$	[164 → 57, flip]
309,308	$(x \wedge y) \vee ((y \wedge x) \vee z) = (y \wedge x) \vee z$	[164 → 10, flip]
311	$x \vee (y \wedge z) = (z \wedge y) \vee x$	[304 :309]
314	$(((x \wedge y) \vee z) \wedge y) \vee (y \wedge x) = y \wedge ((x \wedge y) \vee z)$	[4 → 166]
316	$(x \wedge y) \vee ((x \vee z) \wedge y) = y \wedge (x \vee z)$	[9 → 170]
322	$(x \vee y) \wedge (z \wedge (y \vee x)) = z \wedge (y \vee x)$	[174 → 26, flip]
330,329	$((x \vee y) \wedge y) \vee (y \vee x) = x \vee y$	[174 → 176 :175]
333	$x \vee (y \vee (z \wedge (x \vee z))) = y \vee (x \vee z)$	[178 → 81 :294, flip]
359,358	$x \wedge ((y \wedge x) \vee x) = x \wedge (y \vee x)$	[136 → 196 :294,6,175, flip]
368	$((x \wedge y) \vee y) \wedge (y \wedge (x \vee z)) = y \wedge (x \vee z)$	[204 → 129 :205]
370	$x \wedge ((y \vee z) \wedge ((y \wedge x) \vee x)) = x \wedge (y \vee z)$	[204 → 14 :6,205]
391,390	$((x \wedge y) \vee z) \wedge (z \vee y) = (x \wedge y) \vee z$	[214 → 113 :330, flip]
393	$x \wedge ((y \vee z) \wedge (y \vee (x \wedge (y \vee z)))) = x \wedge (y \vee z)$	[204 → 216 :6,110]
397	$x \wedge (y \vee (z \wedge (z \vee x))) = x \wedge (y \vee z)$	[162 → 216 :11,247,6,130,296]
402	$((x \wedge y) \vee z) \wedge (x \vee z) = (x \wedge y) \vee z$	[216 → 208 :330, flip]
407,406	$(x \wedge y) \vee y = y \wedge (x \vee y)$	[198 → 224 :359, flip]
411,410	$x \wedge ((y \vee z) \wedge (y \vee x)) = x \wedge (y \vee z)$	[370 :407,33]

413,412 $x \wedge ((y \vee x) \wedge (y \vee z)) = x \wedge (y \vee z)$ [368 :407,6,33]

440,439 $x \wedge ((x \vee y) \wedge ((y \wedge x) \vee z)) = x \wedge ((y \wedge x) \vee z)$

[198 → 232 :247,6,33,199]

446,445 $(x \vee y) \wedge (y \vee (z \wedge (x \vee y))) = y \vee (z \wedge (x \vee y))$ [66 → 234]

448,447 $(x \vee y) \wedge (x \vee (z \wedge (x \vee y))) = x \vee (z \wedge (x \vee y))$ [58 → 234]

449    $x \wedge (y \vee (x \wedge (y \vee z))) = x \wedge (y \vee z)$ [393 :448]

457    $x \wedge ((y \wedge x) \vee (x \wedge z)) = (y \wedge x) \vee (x \wedge z)$ [246 → 232 :6,440]

480,479 $x \wedge ((x \vee y) \wedge (z \vee (u \vee y))) = x \wedge (z \vee (u \vee y))$ [192 → 254 :193]

532    $(((x \vee y) \wedge z) \vee y) \wedge (z \vee x) = (x \wedge z) \vee (y \wedge ((x \wedge y) \vee z))$

[9 → 268, flip]

539,538 $(x \wedge y) \vee (z \wedge (x \vee z)) = (x \wedge y) \vee z$

[226 → 287 :257,11,247,67,15,227]

589    $(C \wedge A) \vee (A \wedge B) \neq A \wedge (B \vee C)$ [311 → 19]

602    $x \wedge (y \vee (x \wedge (z \vee y))) = x \wedge (z \vee y)$ [322 → 316 :407,6,446,6,175]

668,667 $(x \vee y) \wedge ((z \wedge y) \vee x) = (z \wedge y) \vee x$ [4 → 390]

673,672 $(x \wedge y) \vee ((z \vee y) \wedge z) = (x \wedge y) \vee z$ [390 → 291 :11,257,59,668]

676    $(x \wedge y) \vee (z \wedge (y \vee z)) = (x \wedge y) \vee z$

[390 → 287 :257,11,247,67,15,391]

698,697 $(x \vee y) \wedge (z \wedge ((x \wedge u) \vee y)) = z \wedge ((x \wedge u) \vee y)$ [402 → 26, flip]

704,703 $(x \wedge y) \vee (y \vee z) = (y \vee z) \wedge (x \vee (y \vee z))$ [406 → 212 :539,407]

712,711 $x \wedge (y \wedge (z \vee (x \wedge y))) = x \wedge (y \wedge (z \vee y))$ [234 → 410 :6,6]

723    $(x \vee y) \wedge (z \wedge (x \vee z)) = (x \vee y) \wedge z$ [410 → 275 :3,165, flip]

765,764 $x \wedge (y \vee (x \vee z)) = x \wedge ((x \vee z) \wedge (y \vee x))$ [449 → 397 :6,712,480,6]

779    $(((x \vee y) \wedge z) \vee y) \wedge (x \vee z) = (x \wedge z) \vee (y \wedge ((x \wedge y) \vee z))$

[9 → 532]

785    $(x \wedge y) \vee (z \wedge (u \wedge (x \vee u))) = (x \wedge y) \vee (z \wedge u)$ [5 → 538 :712]

796,795 $x \wedge ((y \vee x) \wedge ((y \wedge z) \vee u)) = x \wedge ((y \wedge z) \vee u)$

[538 → 410 :6,698,411,6, flip]

807    $x \wedge (((y \vee z) \wedge x) \vee z) = x \wedge (y \vee z)$ [311 → 602]

867,866 $x \vee ((y \wedge x) \vee z) = (x \vee z) \wedge (y \vee (x \vee z))$ [676 → 333 :704]

928,927 $(x \wedge y) \vee (z \wedge (u \wedge (x \vee z))) = (x \wedge y) \vee (u \wedge z)$ [20 → 785]

940,939 $(x \vee y) \wedge (((x \vee z) \wedge y) \vee z) = (x \vee y) \wedge ((x \wedge y) \vee z)$

[779 → 807 :11,265,867,480,765,928,673, flip]

944,943 $x \wedge ((y \wedge x) \vee z) = x \wedge (y \vee z)$ [723 → 807 :6,940,796,6,413]

951,950 $(x \wedge y) \vee (y \wedge z) = y \wedge (x \vee (y \wedge z))$ [457 :944, flip]

952    $x \wedge (y \vee (x \wedge z)) = x \wedge (z \vee y)$ [314 :951,11,307,944]

957    $A \wedge (C \vee (A \wedge B)) \neq A \wedge (B \vee C)$ [589 :951]

958    $\square$ [957,952]

Theorem QLT-4. Bowden's inequality gives distributivity in QLT.

The inequality $x \vee (y \wedge z) \geq (x \vee y) \wedge z$ is written as the following equation.

$$(x \vee (y \wedge z)) \vee ((x \vee y) \wedge z) = x \vee (y \wedge z).$$

Proof (found by Otter 3.0.4 on gyro at 8.18 seconds).

3,2	$x \wedge x = x$	
4	$x \wedge y = y \wedge x$	
6,5	$(x \wedge y) \wedge z = x \wedge (y \wedge z)$	
8,7	$x \vee x = x$	
9	$x \vee y = y \vee x$	
11,10	$(x \vee y) \vee z = x \vee (y \vee z)$	
12	$(x \wedge (y \vee z)) \vee (x \wedge y) = x \wedge (y \vee z)$	
14	$(x \vee (y \wedge z)) \wedge (x \vee y) = x \vee (y \wedge z)$	
16	$(x \vee (y \wedge z)) \vee ((x \vee y) \wedge z) = x \vee (y \wedge z)$	
17	$x \vee ((y \wedge z) \vee ((x \vee y) \wedge z)) = x \vee (y \wedge z)$	[copy,16 :11]
21	$A \vee (B \wedge C) \neq (A \vee B) \wedge (A \vee C)$	
23	$x \wedge (x \wedge y) = x \wedge y$	[2 → 5, flip]
25	$x \wedge (y \wedge z) = y \wedge (z \wedge x)$	[4 → 5]
28	$x \wedge (y \wedge z) = z \wedge (x \wedge y)$	[flip 25]
29	$x \vee (y \vee z) = y \vee (x \vee z)$	[9 → 10 :11]
30	$x \vee (x \vee y) = x \vee y$	[7 → 10, flip]
32	$x \vee (y \vee z) = y \vee (z \vee x)$	[9 → 10]
39,38	$x \wedge (y \wedge x) = x \wedge y$	[4 → 23]
43,42	$x \vee (y \vee x) = x \vee y$	[9 → 30]
46	$x \wedge (y \wedge (x \wedge z)) = x \wedge (y \wedge z)$	[38 → 5 :6,6, flip]
48	$x \vee (y \vee (z \vee x)) = x \vee (y \vee z)$	[10 → 42]
52	$(x \wedge (y \vee z)) \vee (x \wedge z) = x \wedge (z \vee y)$	[9 → 12]
54	$((x \vee y) \wedge z) \vee (z \wedge x) = z \wedge (x \vee y)$	[4 → 12]
56	$x \vee (y \vee ((x \vee y) \wedge x)) = x \vee y$	[2 → 12 :11,3]
60	$(x \wedge ((x \wedge y) \vee z)) \vee (x \wedge y) = x \wedge ((x \wedge y) \vee z)$	[23 → 12]
63,62	$(x \wedge (y \vee z)) \vee (y \wedge x) = x \wedge (y \vee z)$	[4 → 12]
68	$(x \wedge (y \vee z)) \vee ((x \wedge y) \vee u) = (x \wedge (y \vee z)) \vee u$	[12 → 10, flip]
125	$x \vee (y \vee z) = z \vee (y \vee x)$	[9 → 32]
160,159	$x \vee (y \vee ((x \vee y) \wedge y)) = x \vee y$	[2 → 17 :3]
161	$x \vee ((y \wedge z) \vee ((y \vee x) \wedge z)) = x \vee (y \wedge z)$	[9 → 17]
230	$x \vee (((y \vee x) \wedge y) \vee y) = y \vee x$	[32 → 56]
249	$(x \vee y) \wedge (y \vee x) = x \vee y$	[2 → 52 :11,160, flip]
255	$(x \wedge (y \vee z)) \vee (z \wedge x) = x \wedge (z \vee y)$	[4 → 52]
290,289	$(x \vee y) \wedge (z \wedge (y \vee x)) = (x \vee y) \wedge z$	[249 → 46 :39, flip]
293	$x \wedge (y \vee z) = (z \vee y) \wedge x$	[249 → 28 :290]
322,321	$(x \vee y) \wedge (y \vee (x \wedge z)) = y \vee (x \wedge z)$	[14 → 293, flip]
350	$(x \wedge y) \vee (y \wedge x) = y \wedge x$	[7 → 54 :8]
365,364	$((x \vee y) \wedge x) \vee x = x \wedge (x \vee y)$	[2 → 54]
368	$x \vee (y \wedge (y \vee x)) = y \vee x$	[230 :365]
405,404	$(x \wedge y) \vee (z \vee (y \wedge x)) = (x \wedge y) \vee z$	[350 → 48 :43, flip]
408	$x \vee (y \wedge z) = (y \wedge z) \vee ((z \wedge y) \vee x)$	[350 → 125]
411	$x \vee (y \wedge z) = (z \wedge y) \vee x$	[350 → 29 :405]
413,412	$(x \wedge y) \vee ((y \wedge x) \vee z) = (y \wedge x) \vee z$	[350 → 10, flip]

414	$(x \wedge y) \vee z = z \vee (y \wedge x)$	[flip 408 :413]
430	$x \vee (y \wedge (x \vee y)) = y \vee x$	[9 → 368]
521,520	$(x \wedge y) \vee ((x \vee z) \wedge y) = y \wedge (x \vee z)$	[54 → 411, flip]
536	$x \vee (y \wedge (z \vee x)) = x \vee (z \wedge y)$	[161 :521]
590	$(x \wedge (y \vee x)) \vee y = x \vee y$	[9 → 430]
675	$(x \wedge y) \vee (((x \wedge y) \vee z) \wedge x) = x \wedge ((x \wedge y) \vee z)$	[414 → 60]
1481	$((x \vee y) \wedge (x \vee z)) \vee x = (x \vee y) \wedge (x \vee z)$	[364 → 68 :63, flip]
1510	$x \vee (y \wedge (x \vee z)) = x \vee (z \wedge y)$	[9 → 536]
1522,1521	$x \vee ((x \vee y) \wedge z) = x \vee (y \wedge z)$	[293 → 536]
1535,1534	$(x \wedge (y \vee z)) \vee z = z \vee (y \wedge x)$	[9 → 536]
1550	$x \vee (y \wedge z) = x \vee (z \wedge (x \vee y))$	[flip 1510]
1559,1558	$(x \wedge y) \vee (z \wedge x) = x \wedge ((x \wedge y) \vee z)$	[675 :1522]
1617,1616	$x \wedge (y \vee (z \wedge x)) = x \wedge (y \vee z)$	[255 :1559,1535]
1634,1633	$(x \wedge (y \vee z)) \vee y = y \vee (z \wedge x)$	[536 → 590 :6,322,1617,1535]
1645,1644	$x \vee (y \wedge (x \vee z)) = (x \vee z) \wedge (x \vee y)$	[1481 :1634]
1653	$x \vee (y \wedge z) = (x \vee y) \wedge (x \vee z)$	[1550 :1645]
1655	□	[1653,21]

Corollary LT-7. Bowden's inequality for lattice theory.

If a lattice satisfies Bowden's inequality, it is distributive [5, p. 36].

The next two theorems were previously known by model-theoretic arguments using N_5 and the Plonka-sums argument [29]. These are the first equational proofs known to us.

Problem QLT-5. Self-dual modularity axiom for quasilattices.

$$\left\{ \begin{array}{l} \text{QLT} \\ (x \wedge y) \vee (z \wedge (x \vee y)) = (x \vee y) \wedge (z \vee (x \wedge y)) \end{array} \right\} \Rightarrow$$

$$\{x \wedge (y \vee (x \wedge z)) = (x \wedge y) \vee (x \wedge z)\}.$$

(Otter's original proof, which was found with the standard strategy with `max_weight=23` and the flag `control_memory`, is 125 steps and was found in 914 seconds. Larry Wos, starting with that proof, found the following 30-step proof by using the resonance strategy [82]).

Proof (found by Otter 3.0.4 on gyro at 12.04 seconds).

2	$x \vee y = y \vee x$
3	$x = x$
5,4	$x \wedge x = x$
6	$x \wedge y = y \wedge x$
8,7	$(x \wedge y) \wedge z = x \wedge (y \wedge z)$
10,9	$x \vee x = x$
13,12	$(x \vee y) \vee z = x \vee (y \vee z)$
15,14	$(x \wedge (y \vee z)) \vee (x \wedge y) = x \wedge (y \vee z)$
16	$(x \vee (y \wedge z)) \wedge (x \vee y) = x \vee (y \wedge z)$
18	$(x \wedge y) \vee (z \wedge (x \vee y)) = (x \vee y) \wedge (z \vee (x \wedge y))$

19	$A \wedge (B \vee (A \wedge C)) \neq (A \wedge B) \vee (A \wedge C)$	
20	$(A \wedge B) \vee (A \wedge C) \neq A \wedge (B \vee (A \wedge C))$	[flip 19]
22	$x \vee (y \vee ((x \vee y) \wedge x)) = x \vee y$	$[4 \rightarrow 14 \ :13,5]$
31,30	$x \vee (((y \vee x) \wedge y) \vee y) = y \vee x$	$[2 \rightarrow 22 \ :13]$
38	$(x \wedge y) \vee (x \wedge (y \vee z)) = x \wedge (y \vee z)$	$[14 \rightarrow 30 \ :5,10,15]$
46	$(x \wedge (y \vee z)) \vee (x \wedge z) = x \wedge (y \vee z)$	$[30 \rightarrow 14 \ :31]$
50	$(x \vee y) \wedge (x \vee (y \wedge z)) = x \vee (y \wedge z)$	$[16 \rightarrow 6, \text{flip}]$
53,52	$(x \vee y) \wedge (y \vee (x \wedge z)) = y \vee (x \wedge z)$	$[2 \rightarrow 50]$
78	$(x \vee (y \wedge z)) \wedge (x \vee z) = x \vee (z \wedge y)$	$[6 \rightarrow 16]$
79	$(x \wedge y) \vee (y \wedge (x \vee z)) = y \wedge (x \vee z)$	$[6 \rightarrow 38]$
100	$x \vee (y \wedge z) = (x \vee (z \wedge y)) \wedge (x \vee y)$	[flip 78]
219	$x \wedge (x \wedge y) = x \wedge y$	$[4 \rightarrow 7, \text{flip}]$
231	$(x \vee (y \wedge z)) \wedge (x \vee z) = x \vee (y \wedge z)$	$[6 \rightarrow 100, \text{flip}]$
238,237	$(x \vee (y \wedge z)) \wedge (z \vee x) = x \vee (y \wedge z)$	$[2 \rightarrow 231]$
283,282	$x \wedge (y \wedge x) = x \wedge y$	$[6 \rightarrow 219]$
284	$(x \wedge (y \vee (x \wedge z))) \vee (x \wedge z) = x \wedge (y \vee (x \wedge z))$	$[219 \rightarrow 46]$
292	$(x \vee (y \wedge z)) \wedge ((x \vee z) \wedge u) = (x \vee (y \wedge z)) \wedge u$	$[231 \rightarrow 7, \text{flip}]$
300	$(x \vee (y \wedge z)) \wedge ((z \vee x) \wedge u) = (x \vee (y \wedge z)) \wedge u$	$[2 \rightarrow 292]$
382	$((x \wedge y) \vee (z \wedge u)) \wedge (y \vee u) = (x \wedge y) \vee (z \wedge u)$	
		$[237 \rightarrow 300 \ :238, \text{flip}]$
438	$(x \vee y) \wedge ((z \wedge x) \vee (u \wedge y)) = (z \wedge x) \vee (u \wedge y)$	$[6 \rightarrow 382]$
441,440	$(x \vee y) \wedge ((z \wedge y) \vee (u \wedge x)) = (z \wedge y) \vee (u \wedge x)$	$[2 \rightarrow 438]$
447	$(x \wedge y) \vee (z \wedge (u \wedge (x \vee y))) = (x \vee y) \wedge ((z \wedge u) \vee (x \wedge y))$	
		$[7 \rightarrow 18]$
448	$x \wedge (y \vee x) = x \vee (y \wedge x)$	$[79 \rightarrow 18 \ :53]$
452	$(x \vee y) \wedge ((z \wedge u) \vee (x \wedge y)) = (x \wedge y) \vee (z \wedge (u \wedge (x \vee y)))$	
		[flip 447]
456,455	$x \vee (y \wedge x) = x \wedge (x \vee y)$	$[2 \rightarrow 448, \text{flip}]$
463,462	$(x \wedge y) \vee (z \wedge (u \wedge (x \vee y))) = (x \vee y) \wedge ((x \wedge y) \vee (z \wedge u))$	
		$[2 \rightarrow 452, \text{flip}]$
474	$(x \wedge y) \vee (z \wedge x) = x \wedge ((x \vee y) \wedge (z \vee (x \wedge y)))$	
		$[455 \rightarrow 18 \ :283,463,441,456,283,8]$
520	$(x \wedge y) \vee (x \wedge z) = x \wedge ((x \vee y) \wedge (z \vee (x \wedge y)))$	$[6 \rightarrow 474]$
524,523	$(x \wedge y) \vee (x \wedge z) = x \wedge ((x \vee z) \wedge (y \vee (x \wedge z)))$	$[2 \rightarrow 520]$
527,526	$x \wedge ((x \vee y) \wedge (z \vee (x \wedge y))) = x \wedge (z \vee (x \wedge y))$	$[284 \ :524,13,10]$
532	$A \wedge (B \vee (A \wedge C)) \neq A \wedge (B \vee (A \wedge C))$	$[20 \ :524,527]$
533	$\square$	$[532,3]$

Theorem QLT-6. Another modularity axiom for quasilattices.

$$\left\{ \begin{array}{l} \text{QLT} \\ ((x \vee y) \wedge z) \vee y = ((z \vee y) \wedge x) \vee y \end{array} \right\} \Rightarrow \{x \wedge (y \vee (x \wedge z)) = (x \wedge y) \vee (x \wedge z)\}.$$

Proof (found by Otter 3.0.4 on gyro at 475.97 seconds).

2	$x \wedge x = x$

4	$x \wedge y = y \wedge x$	
6,5	$(x \wedge y) \wedge z = x \wedge (y \wedge z)$	
8,7	$x \vee x = x$	
9	$x \vee y = y \vee x$	
11,10	$(x \vee y) \vee z = x \vee (y \vee z)$	
12	$(x \wedge (y \vee z)) \vee (x \wedge y) = x \wedge (y \vee z)$	
14	$(x \vee (y \wedge z)) \wedge (x \vee y) = x \vee (y \wedge z)$	
16	$((x \vee y) \wedge z) \vee y = ((z \vee y) \wedge x) \vee y$	
17	$A \wedge (B \vee (A \wedge C)) \neq (A \wedge B) \vee (A \wedge C)$	
18	$(A \wedge B) \vee (A \wedge C) \neq A \wedge (B \vee (A \wedge C))$	[flip 17]
19	$x \wedge (y \wedge z) = y \wedge (x \wedge z)$	$[4 \to 5 :6]$
21,20	$x \wedge (x \wedge y) = x \wedge y$	$[2 \to 5, \text{flip}]$
22	$x \wedge (y \wedge z) = y \wedge (z \wedge x)$	$[4 \to 5]$
26	$x \vee (y \vee z) = y \vee (x \vee z)$	$[9 \to 10 :11]$
28,27	$x \vee (x \vee y) = x \vee y$	$[7 \to 10, \text{flip}]$
29	$x \vee (y \vee z) = y \vee (z \vee x)$	$[9 \to 10]$
32	$x \vee (y \vee z) = z \vee (x \vee y)$	[flip 29]
35	$x \wedge (y \wedge x) = x \wedge y$	$[4 \to 20]$
40,39	$x \vee (y \vee x) = x \vee y$	$[9 \to 27]$
44,43	$x \wedge (y \wedge (x \wedge z)) = x \wedge (y \wedge z)$	$[35 \to 5 :6,6, \text{flip}]$
47	$x \vee (y \vee (x \vee z)) = x \vee (y \vee z)$	$[39 \to 10 :11,11, \text{flip}]$
52,51	$(x \wedge (y \vee z)) \vee (x \wedge z) = x \wedge (z \vee y)$	$[9 \to 12]$
59	$(x \wedge ((y \wedge x) \vee z)) \vee (x \wedge y) = x \wedge ((y \wedge x) \vee z)$	$[35 \to 12]$
65	$(x \wedge (x \vee y)) \vee x = x \wedge (x \vee y)$	$[2 \to 12]$
73	$x \wedge (y \wedge z) = x \wedge (z \wedge y)$	$[4 \to 19 :6]$
79	$x \wedge (y \wedge z) = z \wedge (y \wedge x)$	$[4 \to 22]$
94	$x \vee (y \vee z) = x \vee (z \vee y)$	$[9 \to 26 :11]$
99	$(x \vee (y \wedge (z \wedge u))) \wedge (x \vee u) = x \vee (u \wedge (y \wedge z))$	$[22 \to 14]$
101	$(x \vee (y \wedge (z \wedge u))) \wedge (x \vee (y \wedge z)) = x \vee (y \wedge (z \wedge u))$	$[5 \to 14 :6]$
108	$((x \wedge y) \vee z) \wedge (z \vee x) = z \vee (x \wedge y)$	$[9 \to 14]$
109	$x \wedge (y \wedge ((x \wedge y) \vee x)) = x \wedge y$	$[7 \to 14 :6,8]$
121	$(x \vee y) \wedge (x \vee (y \wedge z)) = x \vee (y \wedge z)$	$[4 \to 14]$
186	$((x \vee y) \wedge (z \wedge u)) \vee y = (((u \wedge z) \vee y) \wedge x) \vee y$	$[73 \to 16]$
197	$((x \vee y) \wedge x) \vee y = x \vee y$	$[2 \to 16 :11,8,11,8, \text{flip}]$
203,202	$((x \vee y) \wedge z) \vee y = y \vee ((z \vee y) \wedge x)$	$[9 \to 16, \text{flip}]$
222,221	$x \vee ((y \vee x) \wedge y) = y \vee x$	$[197 :203]$
226	$x \vee (((y \wedge z) \vee x) \wedge u) = x \vee ((u \vee x) \wedge (z \wedge y))$	$[186 :203,203]$
258	$x \vee ((x \vee y) \wedge y) = y \vee x$	$[9 \to 221]$
282,281	$(x \vee y) \wedge (y \vee x) = x \vee y$	$[221 \to 14 :40,222]$
288	$x \wedge (y \vee z) = x \wedge (z \vee y)$	$[221 \to 12 :52,222]$
330,329	$x \vee ((x \vee y) \wedge y) = x \vee y$	$[258 \to 27 :40, \text{flip}]$
392	$x \wedge (y \vee z) = (z \vee y) \wedge ((y \vee z) \wedge x)$	$[281 \to 79]$
399,398	$(x \vee y) \wedge ((y \vee x) \wedge z) = (x \vee y) \wedge z$	$[281 \to 5, \text{flip}]$
400	$(x \vee y) \wedge z = z \wedge (y \vee x)$	$[\text{flip } 392 :399]$

401	$x \wedge (y \vee z) = (z \vee y) \wedge x$	$[392 : 399]$
641	$x \vee (x \wedge y) = x \wedge (x \vee y)$	$[401 \rightarrow 65 : 203,8]$
695,694	$x \vee (y \wedge x) = x \wedge (x \vee y)$	$[4 \rightarrow 641]$
699,698	$(x \wedge y) \vee x = x \wedge (x \vee y)$	$[9 \rightarrow 641]$
705	$x \wedge (y \wedge (x \vee y)) = x \wedge y$	$[109 : 699,44]$
735	$(x \wedge y) \vee (y \wedge x) = x \wedge (y \wedge ((x \wedge y) \vee y))$	$[35 \rightarrow 694 : 6]$
744,743	$(x \wedge y) \vee y = y \wedge (y \vee x)$	$[9 \rightarrow 694]$
745	$(x \wedge y) \vee (y \wedge x) = x \wedge (y \wedge (y \vee x))$	$[735 : 744,21]$
854,853	$x \wedge (y \wedge (y \vee x)) = x \wedge y$	$[9 \rightarrow 705]$
872,871	$(x \wedge y) \vee (y \wedge x) = x \wedge y$	$[745 : 854]$
1269,1268	$(x \wedge y) \vee (z \vee (y \wedge x)) = (x \wedge y) \vee z$	$[871 \rightarrow 47 : 40, \text{flip}]$
1271	$x \vee (y \wedge z) = x \vee (z \wedge y)$	$[871 \rightarrow 94 : 872]$
1272	$x \vee (y \wedge z) = (z \wedge y) \vee x$	$[871 \rightarrow 32 : 1269]$
1341	$x \wedge (y \vee (z \wedge u)) = x \wedge ((u \wedge z) \vee y)$	$[1271 \rightarrow 288]$
1358	$x \wedge ((y \wedge z) \vee u) = x \wedge (u \vee (z \wedge y))$	$[\text{flip } 1341]$
1408	$(C \wedge A) \vee (A \wedge B) \neq A \wedge (B \vee (A \wedge C))$	$[1272 \rightarrow 18]$
1431	$(x \wedge ((y \wedge x) \vee z)) \vee (y \wedge x) = x \wedge ((y \wedge x) \vee z)$	$[4 \rightarrow 59]$
3062	$((x \wedge y) \vee z) \wedge (z \vee x) = (x \wedge y) \vee z$	$[108 \rightarrow 20 : 282, \text{flip}]$
3234,3233	$(x \vee y) \wedge (x \vee ((x \vee y) \wedge z)) = x \vee ((x \vee y) \wedge z)$	$[27 \rightarrow 121]$
5031	$x \vee ((x \vee (y \wedge z)) \wedge y) = x \vee (y \wedge z)$	$[329 \rightarrow 101 : 3234,330]$
5341,5340	$(x \wedge (y \vee z)) \vee y = y \vee ((x \vee y) \wedge z)$	$[400 \rightarrow 202]$
5400,5399	$(x \wedge y) \vee (y \wedge ((y \vee x) \wedge z)) = y \wedge ((x \wedge y) \vee z)$	$[1431 : 5341,695,6]$
7284,7283	$x \vee ((x \vee y) \wedge (z \wedge y)) = x \vee (y \wedge z)$	$[3062 \rightarrow 226 : 40,11,40, \text{flip}]$
7333	$x \vee (y \wedge ((x \vee y) \wedge z)) = x \vee (y \wedge z)$	$[99 \rightarrow 5031 : 28,7284]$
7550,7549	$(x \wedge y) \vee (y \wedge z) = y \wedge ((x \wedge y) \vee z)$	$[743 \rightarrow 7333 : 6,21,5400, \text{flip}]$
7555	$A \wedge ((C \wedge A) \vee B) \neq A \wedge (B \vee (A \wedge C))$	$[1408 : 7550]$
7556	$\square$	$[7555,1358]$

6.3 Uniqueness of Operations

6.3.1 Lattices

Theorem LT-8. Uniqueness of the meet operation in LT.

If $\wedge$, $\vee$, and $\cdot$ are binary operations such that both $\langle L; \vee, \wedge \rangle$ and $\langle L; \vee, \cdot \rangle$ are lattices, then $x \wedge y = x \cdot y$.

Proof (found by Otter 3.0.4 on gyro at 79.56 seconds).

4	$x \wedge y = y \wedge x$
5	$(x \wedge y) \wedge z = x \wedge (y \wedge z)$
9	$x \vee y = y \vee x$
12	$x \wedge (x \vee y) = x$
14	$x \vee (x \wedge y) = x$
18	$x \cdot y = y \cdot x$

19	$(x \cdot y) \cdot z = x \cdot (y \cdot z)$	
21	$x \cdot (x \vee y) = x$	
23	$x \vee (x \cdot y) = x$	
25	$A \wedge B \neq A \cdot B$	
33	$x \wedge (y \vee x) = x$	$[9 \to 12]$
37	$(x \vee y) \wedge x = x$	$[4 \to 12]$
41	$(x \wedge y) \vee (x \wedge (y \wedge z)) = x \wedge y$	$[5 \to 14]$
43	$x \vee (y \wedge x) = x$	$[4 \to 14]$
45	$(x \wedge y) \vee x = x$	$[9 \to 14]$
47	$x \cdot (y \vee x) = x$	$[9 \to 21]$
49	$(x \vee y) \cdot x = x$	$[18 \to 21]$
66	$x \vee (y \cdot x) = x$	$[18 \to 23]$
70	$(x \cdot y) \vee x = x$	$[9 \to 23]$
74	$B \wedge A \neq A \cdot B$	$[4 \to 25]$
83	$(x \vee y) \wedge y = y$	$[4 \to 33]$
110	$(x \wedge y) \vee y = y$	$[9 \to 43]$
126	$x \cdot (y \cdot (z \vee (x \cdot y))) = x \cdot y$	$[19 \to 47]$
133,132	$x \cdot (x \wedge y) = x \wedge y$	$[45 \to 49]$
167	$x \wedge (x \cdot y) = x \cdot y$	$[70 \to 37]$
171	$x \wedge (y \cdot x) = y \cdot x$	$[66 \to 83]$
188,187	$x \cdot (y \wedge x) = y \wedge x$	$[110 \to 49]$
456	$(x \wedge y) \vee (x \wedge (y \cdot z)) = x \wedge y$	$[167 \to 41]$
8044	$(x \wedge y) \vee (y \cdot x) = x \wedge y$	$[171 \to 456]$
8185	$x \wedge y = y \cdot x$	$[8044 \to 126 : 133,188]$
8187	$\square$	$[8185,74]$

6.3.2 Quasilattices

Because the meet operation is unique for lattices, the analogous conjecture arose for quasilattices. MACE easily provided a counterexample.

Example QLT-7. Uniqueness of the meet operation in QLT.

Let $\wedge$, $\vee$, and $\cdot$ be binary operations such that both $\langle QL; \vee, \wedge \rangle$ and $\langle QL; \vee, \cdot \rangle$ are quasilattices. Are the two meet operations necessarily the same?

Counterexample. The clauses

$$x \vee x = x$$
$$x \wedge x = x$$
$$x \cdot x = x$$
$$x \vee y = y \vee x$$
$$x \wedge y = y \wedge x$$
$$x \cdot y = y \cdot x$$
$$(x \vee y) \vee z = x \vee (y \vee z)$$
$$(x \wedge y) \wedge z = x \wedge (y \wedge z)$$

$$(x \cdot y) \cdot z = x \cdot (y \cdot z)$$
$$(x \wedge (y \vee z)) \vee (x \wedge y) = x \wedge (y \vee z)$$
$$(x \vee (y \wedge z)) \wedge (x \vee y) = x \vee (y \wedge z)$$
$$(x \cdot (y \vee z)) \vee (x \cdot y) = x \cdot (y \vee z)$$
$$(x \vee (y \cdot z)) \cdot (x \vee y) = x \vee (y \cdot z)$$
$$A \cdot B \neq A \wedge B$$

have the following model (found by MACE 1.2.0 on gyro at 0.36 seconds).

```
^ | 0 1 2            v | 0 1 2
--+------            --+------
0 | 0 0 0            0 | 0 0 0
1 | 0 1 0            1 | 0 1 0
2 | 0 0 2            2 | 0 0 2


. | 0 1 2
--+------
0 | 0 1 0
1 | 1 1 1
2 | 0 1 2            A: 0,  B: 1
```

6.3.3 Weakly Associative Lattices

Theorem WAL-2. Uniqueness of the meet operation in WAL.

If $\langle S; \vee, \wedge \rangle$ and $\langle S; \vee, \cdot \rangle$ are both weakly associative lattices, then $x \wedge y = x \cdot y$.

Proof (found by Otter 3.0.4 on gyro at 13.99 seconds).

2	$x \wedge x = x$	
4	$x \wedge y = y \wedge x$	
5	$((x \vee y) \wedge (z \vee y)) \wedge y = y$	
7	$x \vee x = x$	
9	$x \vee y = y \vee x$	
10	$((x \wedge y) \vee (z \wedge y)) \vee y = y$	
12	$x \cdot x = x$	
14	$x \cdot y = y \cdot x$	
15	$((x \vee y) \cdot (z \vee y)) \cdot y = y$	
17	$((x \cdot y) \vee (z \cdot y)) \vee y = y$	
19	$A \wedge B \neq A \cdot B$	
28	$(x \vee y) \wedge y = y$	$[2 \to 5]$
33	$(x \vee y) \wedge x = x$	$[9 \to 28]$
35	$x \wedge (y \vee x) = x$	$[4 \to 28]$
45	$(x \wedge y) \vee y = y$	$[7 \to 10]$
51	$x \wedge (x \vee y) = x$	$[4 \to 33]$
59	$(x \wedge y) \vee x = x$	$[4 \to 45]$

61	$x \vee (y \wedge x) = x$	$[9 \rightarrow 45]$
83	$(x \vee y) \cdot y = y$	$[12 \rightarrow 15]$
89	$x \vee (x \wedge y) = x$	$[9 \rightarrow 59]$
96,95	$(x \vee y) \vee x = x \vee y$	$[51 \rightarrow 61]$
99	$((x \vee (y \wedge z)) \cdot z) \cdot (y \wedge z) = y \wedge z$	$[61 \rightarrow 15]$
125	$x \vee (y \cdot x) = x$	$[83 \rightarrow 17 :96]$
153	$x \vee (x \cdot y) = x$	$[14 \rightarrow 125]$
161	$((x \vee (y \cdot z)) \wedge z) \wedge (y \cdot z) = y \cdot z$	$[125 \rightarrow 5]$
199	$(x \cdot y) \wedge x = x \cdot y$	$[153 \rightarrow 35]$
748,747	$(x \cdot y) \cdot (x \wedge y) = x \wedge y$	$[89 \rightarrow 99]$
1042,1041	$(x \wedge y) \wedge (x \cdot y) = x \wedge y$	$[747 \rightarrow 199 :748]$
1471	$x \wedge y = x \cdot y$	$[153 \rightarrow 161 :1042]$
1473	$\square$	$[1471,19]$

Note that Thm. LT-8 is a corollary of the Thm. WAL-2. Although Thm. LT-8 is weaker, Otter with the same strategy takes more time to find a proof. This situation occurs frequently, and part of the reason here is Otter's difficulty in handling associative-commutative operations.

6.3.4 Transitive Near Lattices

Example TNL-2. Uniqueness of the meet operation in TNL.

Let $\langle S; \vee, \wedge \rangle$ and $\langle S; \vee, \cdot \rangle$ be two transitive near lattices. Are the two meet operations necessarily the same?

Counterexample. The clauses

$$x \wedge x = x$$
$$x \wedge y = y \wedge x$$
$$x \wedge (x \vee y) = x$$
$$x \wedge (y \vee (x \vee z)) = x$$
$$x \vee x = x$$
$$x \vee y = y \vee x$$
$$x \vee (x \wedge y) = x$$
$$x \vee (y \wedge (x \wedge z)) = x$$
$$x \cdot x = x$$
$$x \cdot y = y \cdot x$$
$$x \cdot (x \vee y) = x$$
$$x \cdot (y \vee (x \vee z)) = x$$
$$x \vee x = x$$
$$x \vee y = y \vee x$$
$$x \vee (x \cdot y) = x$$
$$x \vee (y \cdot (x \cdot z)) = x$$
$$0 \wedge 1 \neq 0 \cdot 1$$

have the following model (found by MACE 1.2.0 on gyro at 0.45 seconds).

```
^ | 0 1 2 3 4          v | 0 1 2 3 4
--+----------          --+----------
0 | 0 2 2 0 4          0 | 0 3 0 3 0
1 | 2 1 2 1 4          1 | 3 1 1 3 1
2 | 2 2 2 2 2          2 | 0 1 2 3 4
3 | 0 1 2 3 4          3 | 3 3 3 3 3
4 | 4 4 2 4 4          4 | 0 1 4 3 4


. | 0 1 2 3 4
--+----------
0 | 0 4 2 0 4
1 | 4 1 2 1 4
2 | 2 2 2 2 2
3 | 0 1 2 3 4
4 | 4 4 2 4 4
```

6.4 Single Axioms

In [43], R. McKenzie presented the following self-dual basis, consisting of four absorption equations, for lattice theory.

$$y \vee (x \wedge (y \wedge z)) = y, \qquad y \wedge (x \vee (y \vee z)) = y, \qquad \text{(L1,L2)}$$
$$((x \wedge y) \vee (y \wedge z)) \vee y = y, \qquad ((x \vee y) \wedge (y \vee z)) \wedge y = y. \qquad \text{(L3,L4)}$$

The next theorem is a verification of the McKenzie basis.

Theorem LT-9. McKenzie's absorption basis for LT.

The four equations (L1, L2, L3, L4) are a basis for lattice theory. It is clear that they are a part of lattice theory, and by duality, it is sufficient to derive the set

$$\left\{ \begin{array}{l} x \wedge y = y \wedge x \\ (x \wedge y) \wedge z = x \wedge (y \wedge z) \\ x \wedge (x \vee y) = x \end{array} \right\}.$$

Proof (found by Otter 3.0.4 on gyro at 392.74 seconds).

1	$x = x$
2	$B \wedge A = A \wedge B, \ (A \wedge B) \wedge C = A \wedge (B \wedge C),$
	$A \wedge (A \vee B) = A \ \rightarrow \ \square$
4,3	$x \vee (y \wedge (x \wedge z)) = x$
5	$x \wedge (y \vee (x \vee z)) = x$
7	$((x \wedge y) \vee (y \wedge z)) \vee y = y$
9	$((x \vee y) \wedge (y \vee z)) \wedge y = y$
11	$x \wedge (y \vee x) = x$

$[3 \rightarrow 5]$

13	$x \vee (y \wedge x) = x$	$[5 \to 3]$
15	$(x \wedge (y \wedge z)) \wedge y = x \wedge (y \wedge z)$	$[3 \to 11]$
18,17	$(x \vee y) \vee y = x \vee y$	$[11 \to 13]$
19	$(x \vee (y \vee z)) \vee y = x \vee (y \vee z)$	$[5 \to 13]$
22,21	$(x \wedge y) \wedge y = x \wedge y$	$[13 \to 11]$
28,27	$(x \wedge y) \vee y = y$	$[11 \to 7 :18]$
29	$x \vee x = x$	$[5 \to 7 :28]$
31	$x \wedge x = x$	$[7 \to 11]$
34,33	$x \wedge (x \vee y) = x$	$[7 \to 5]$
37	$B \wedge A = A \wedge B, \ (A \wedge B) \wedge C = A \wedge (B \wedge C) \ \to \ \Box$	$[2 :34 :1]$
41,40	$x \vee (x \wedge y) = x$	$[31 \to 3]$
42	$x \vee (y \vee x) = y \vee x$	$[11 \to 27]$
48	$((x \wedge y) \vee (y \wedge z)) \wedge y = (x \wedge y) \vee (y \wedge z)$	$[7 \to 33]$
56	$(x \wedge ((y \wedge x) \vee z)) \wedge (y \wedge x) = y \wedge x$	$[13 \to 9]$
60	$(x \vee y) \wedge y = y$	$[29 \to 9 :22]$
62	$((x \vee (y \wedge z)) \wedge z) \wedge (y \wedge z) = y \wedge z$	$[27 \to 9]$
68	$(x \wedge y) \wedge x = x \wedge y$	$[40 \to 11]$
70	$(x \wedge ((x \wedge y) \vee z)) \wedge (x \wedge y) = x \wedge y$	$[40 \to 9]$
75,74	$x \wedge (y \wedge x) = y \wedge x$	$[13 \to 60]$
83,82	$(x \wedge y) \vee x = x$	$[40 \to 42 :41]$
84	$(x \wedge (y \wedge z)) \vee y = y$	$[3 \to 42 :4]$
120	$(x \vee y) \vee (y \wedge z) = x \vee y$	$[82 \to 19 :83]$
145,144	$(x \wedge (y \wedge z)) \wedge z = x \wedge (y \wedge z)$	$[74 \to 15 :75]$
152	$(x \wedge (y \wedge z)) \vee z = z$	$[74 \to 84]$
158	$((x \wedge y) \wedge z) \vee x = x$	$[68 \to 84]$
280	$((x \wedge y) \wedge z) \vee y = y$	$[68 \to 152]$
298	$(x \wedge y) \vee (z \vee x) = z \vee x$	$[60 \to 158]$
950	$(x \wedge y) \wedge ((z \wedge y) \wedge x) = (z \wedge y) \wedge x$	$[280 \to 56]$
955,954	$(x \wedge y) \wedge (y \wedge x) = y \wedge x$	$[82 \to 56]$
1234	$((x \vee y) \wedge z) \wedge (y \wedge z) = y \wedge z$	$[120 \to 62]$
1620	$(x \wedge (y \vee z)) \wedge (z \wedge x) = z \wedge x$	$[298 \to 56]$
1766	$(x \wedge y) \wedge (x \wedge (y \wedge z)) = x \wedge (y \wedge z)$	$[84 \to 70]$
2857,2856	$(x \wedge (y \wedge z)) \vee (z \wedge y) = z \wedge y$	$[954 \to 84]$
2860	$x \wedge y = y \wedge x$	$[954 \to 48 :2857,955,955,2857]$
2901	$x \wedge (y \wedge z) = x \wedge (z \wedge y)$	$[2860 \to 15 :145]$
2914	$(A \wedge B) \wedge C = A \wedge (B \wedge C) \ \to \ \Box$	$[2860 \to 37 :1]$
3316	$(x \wedge y) \wedge z = z \wedge (y \wedge x)$	$[2860 \to 2901]$
3342	$x \wedge (y \wedge z) = (z \wedge y) \wedge x$	$[\text{flip } 3316]$
10469,10468	$(x \wedge y) \wedge ((z \vee y) \wedge x) = y \wedge x$	$[3342 \to 1234]$
10470	$(x \wedge y) \wedge (y \wedge (z \vee x)) = x \wedge y$	$[3316 \to 1234]$
13162,13161	$(x \wedge y) \wedge (x \wedge (z \vee y)) = y \wedge x$	$[3342 \to 1620]$
19054,19053	$((x \wedge y) \wedge z) \wedge (z \wedge x) = (x \wedge y) \wedge z$	$[40 \to 10470]$
19056,19055	$((x \wedge y) \wedge z) \wedge (z \wedge y) = (x \wedge y) \wedge z$	$[13 \to 10470]$
23311	$(x \wedge (y \vee z)) \wedge z = z \wedge x$	$[13161 \to 950 :19054,13162]$

23318,23317	$((x \vee y) \wedge z) \wedge y = y \wedge z$	$[10468 \to 950 : 19056,10469]$
23552	$x \wedge ((y \vee x) \wedge z) = x \wedge z$	$[68 \to 23311 : 23318, \text{flip}]$
23932,23931	$(x \wedge y) \wedge (x \wedge z) = (x \wedge y) \wedge z$	$[40 \to 23552]$
23934,23933	$(x \wedge y) \wedge (y \wedge z) = (x \wedge y) \wedge z$	$[13 \to 23552]$
24033	$(x \wedge y) \wedge z = x \wedge (y \wedge z)$	$[1766 : 23932,23934]$
24035	$\square$	$[24033,2914]$

The following theorem gives us a new and simpler absorption basis, which is useful in the construction of short single axioms.

Theorem LT-10. An absorption 3-basis for LT.

$$\left\{ \begin{array}{ll} y \wedge (x \vee (y \vee z)) = y & \text{(L2)} \\ ((x \wedge y) \vee (y \wedge z)) \vee y = y & \text{(L3)} \\ ((y \vee x) \wedge (y \vee z)) \wedge y = y & \text{(L4')} \end{array} \right\}.$$

It is sufficient to prove (L1) and (L4) of the McKenzie basis.

Proof (found by Otter 3.0.4 on gyro at 7.85 seconds).

1	$x = x$	
2	$B \vee (A \wedge (B \wedge C)) = B, \ ((A \vee B) \wedge (B \vee C)) \wedge B = B \ \to \ \square$	
3	$x \wedge (y \vee (x \vee z)) = x$	
5	$((x \wedge y) \vee (y \wedge z)) \vee y = y$	
7	$((x \vee y) \wedge (x \vee z)) \wedge x = x$	
11	$((x \wedge y) \vee y) \vee y = y$	$[3 \to 5]$
14,13	$x \wedge (x \vee y) = x$	$[5 \to 3]$
37	$(x \vee y) \wedge x = x$	$[3 \to 7]$
41	$(x \vee x) \vee x = x$	$[7 \to 11]$
43	$(x \vee (x \wedge y)) \vee x = x$	$[7 \to 5]$
59	$x \wedge ((y \wedge x) \vee x) = (y \wedge x) \vee x$	$[11 \to 37]$
70,69	$x \vee x = x$	$[41 \to 37 : 14, \text{flip}]$
76,75	$x \wedge (y \vee x) = x$	$[41 \to 3 : 70,70]$
77	$(x \wedge y) \vee y = y$	$[59 : 76, \text{flip}]$
91	$x \vee (y \vee x) = y \vee x$	$[75 \to 77]$
99	$x \wedge (y \wedge x) = y \wedge x$	$[77 \to 37]$
109	$(x \wedge y) \wedge (z \vee y) = x \wedge y$	$[77 \to 3]$
116,115	$x \vee (x \wedge y) = x$	$[43 \to 37 : 14, \text{flip}]$
146,145	$((x \vee y) \wedge (y \vee z)) \wedge y = y$	$[91 \to 7]$
153	$B \vee (A \wedge (B \wedge C)) = B \ \to \ \square$	$[2 : 146 : 1]$
202	$x \vee (y \wedge x) = x$	$[99 \to 115]$
474	$(x \vee y) \vee (z \wedge y) = x \vee y$	$[109 \to 202]$
1504	$x \vee (y \wedge (x \wedge z)) = x$	$[115 \to 474 : 116]$
1506	$\square$	$[1504,153]$

6.4.1 Presence of Jónsson Polynomials

A term in three variables, say $p(x, y, z)$, is a *ternary majority polynomial* [20] (also Jónsson polynomial) for a theory if it satisfies the *majority properties*

$$p(x, x, y) = p(x, z, x) = p(u, x, x) = x.$$

For example, each of the following is a ternary majority polynomial for lattice theory:

$$(x \wedge y) \vee (y \wedge z) \vee (x \wedge z),$$
$$(x \wedge z) \vee (y \wedge (x \vee z)).$$

Henceforth, $p(t_1, t_2, t_3)$ should be read as an abbreviation for a majority term (any majority term admitted by the theory).

The interest in the existence of a majority term is that any pair of absorption equations can be transformed into an equivalent (modulo majority properties) single equation. Let $g(x) = x$ and $h(x) = x$ represent absorption equations; then $p(g(x), h(x), y) = x$ is equivalent to the pair. (The pair can also be combined as $p(y, g(x), h(x)) = x$ or $p(g(x), y, h(x)) = x$; this fact is relevant if the size of the resulting equation is a concern.) The only constraint on variables is that y not occur in $g(x)$ or $h(x)$; the other variables in $g(x)$ and $h(x)$ need not be distinct.

For example, with the lattice theory basis $\{(L2),(L3),(L4')\}$, representing the equations as $L2(y) = y$, $L3(y) = y$, and $L4'(y) = y$, we can apply the transformation twice to get

$$p(p(L2(y), L3(y), u), L4'(y), v) = y.$$

Therefore, this equation, along with the majority properties, is a basis for lattice theory. With the following reduction schema, we can combine this basis into a single equational axiom.

A Majority Reduction Schema

With the majority term transformation of the preceding subsection and the following theorem, we can construct a single axiom any theory that has a majority polynomial and a finite basis consisting exclusively of absorption equations. This was first proved in [53].

Theorem MAJ-2. A majority polynomial reduction schema.

$$\left\{ \begin{array}{l} p(x, y, y) = y \\ p(y, x, y) = y \\ p(y, y, x) = y \\ f(y) = y. \end{array} \right\} \Leftrightarrow \{p(p(x, y, y), u, p(p(x, y, y), f(y), z)) = y\}.$$

($\Rightarrow$) is trivial by simplification. The proof of ($\Leftarrow$) follows.

Proof (found by Otter 3.0.4 on gyro at 0.09 seconds).

1        $x = x$
2        $p(A, A, B) = A$, $p(A, B, A) = A$, $p(B, A, A) = A$,
             $f(A) = A \rightarrow \square$
3        $p(p(x, y, y), z, p(p(x, y, y), f(y), u)) = y$

8,7	$p(p(x,y,y),z,y) = y$	$[3 \rightarrow 3]$
10,9	$p(x,y,x) = x$	$[7 \rightarrow 7]$
11	$p(x,y,p(p(z,x,x),f(x),u)) = p(p(z,x,x),f(x),u)$	$[3 \rightarrow 7]$
13	$p(A,A,B) = A, \ p(B,A,A) = A, \ f(A) = A \ \rightarrow \ \Box$	$[2 :10 :1]$
15,14	$p(x,y,p(x,f(x),z)) = x$	$[7 \rightarrow 3 :8]$
17,16	$p(x,y,y) = y$	$[9 \rightarrow 3 :10]$
18	$p(A,A,B) = A, \ f(A) = A \ \rightarrow \ \Box$	$[13 :17 :1]$
19	$p(x,f(x),y) = x$	$[11 :17,15,17, \text{flip}]$
22,21	$f(x) = x$	$[16 \rightarrow 19]$
24,23	$p(x,x,y) = x$	$[19 :22]$
25	$\Box$	$[18 :24,22 :1,1]$

Continuing the lattice theory example of the preceding subsection, to build a single axiom, we can simply substitute the term

$$p(p(L2(y), L3(y), u), L4'(y), v)$$

for $f(y)$ in the reduction schema, taking care to keep the variables separate where necessary. We can program Otter to do this with the following input file.

```
op(400, xfx, [^,v]).

set(demod_inf).
assign(max_given, 1).

list(demodulators).
L2(Y) = Y ^ (X v (Y v Z)).
L3(Y) = ((X ^ Y) v (Y ^ Z)) v Y.
L4m(Y) = ((Y v X) ^ (Y v Z)) ^ Y.
f(Y) = p(p(L2(Y),L3(Y),U),L4m(Y), V).
p(x,y,z) = (x ^ y) v (z ^ (x v y)).   % real variables here
end_of_list.

list(sos).
p(p(X,Y,Y),W1,p(p(X,Y,Y),f(Y),W2)) = Y.
end_of_list.
```

The equation in list(sos) is simply rewritten with the demodulators. This gives us a single axiom of length 243, with 7 variables. There is one trick going on here: the upper-case "variables" are really constants to Otter, thus giving us control over identification of variables. Without the trick, we obtain an axiom with 12 variables (of the same length).

The motivation for the next several theorems is to find a shorter single axiom for lattice theory. (Parts of this work are also reported in [40]). We first give a new schema, found by automatically examining a large set of candidates, that handles two absorption equations.

Theorem MAJ-3. A majority schema for two absorption equations.

$$\left.\begin{cases} p(x,y,y)=y \\ p(y,x,y)=y \\ p(y,y,x)=y \\ f(y)=y \\ g(y)=y \end{cases}\right\} \Leftrightarrow \{p(p(x,y,y),p(x,p(y,z,f(y)),g(y)),u)=y\}.$$

($\Rightarrow$) is trivial by simplification. The proof of ($\Leftarrow$) follows.

Proof (found by Otter 3.0.4 on gyro at 0.78 seconds).

1	$x=x$	
2	$p(A,A,B)=A,\ p(A,B,A)=A,\ p(B,A,A)=A,$	
	$\quad f(A)=A,\ g(A)=A\ \rightarrow\ \square$	
4,3	$p(p(x,y,y),p(x,p(y,z,f(y)),g(y)),u)=y$	
5	$p(x,p(p(y,x,x),p(p(y,p(x,z,f(x)),g(x)),u,f(p(y,p(x,z,f(x)),$	
	$\quad g(x)))),g(p(y,p(x,z,f(x)),g(x)))),v)=$	
	$\quad p(y,p(x,z,f(x)),g(x))$	$[3\rightarrow3]$
7	$p(p(x,p(y,z,z),p(y,z,z)),p(x,z,g(p(y,z,z))),u)=p(y,z,z)$	
		$[3\rightarrow3]$
12,11	$p(p(x,y,y),p(x,p(z,p(y,u,f(y)),g(y)),g(y)),v)=y$	
		$[3\rightarrow7\ :4,4,4]$
25	$p(p(p(x,y,y),y,y),y,z)=y$	$[3\rightarrow11]$
29	$p(p(x,y,y),p(x,p(z,p(u,p(y,v,f(y)),g(y)),g(y)),g(y)),w)=y$	
		$[11\rightarrow7\ :12,12,12]$
42,41	$p(x,x,y)=x$	$[25\rightarrow25]$
43	$p(A,B,A)=A,\ p(B,A,A)=A,\ f(A)=A,\ g(A)=A\ \rightarrow\ \square$	
		$[2\ :42\ :1]$
64	$p(p(x,y,y),p(p(x,y,y),y,g(p(x,y,y))),z)=p(x,y,y)$	$[41\rightarrow7]$
68	$p(x,p(x,p(y,p(x,z,f(x)),g(x)),g(x)),u)=x$	$[41\rightarrow11]$
82	$p(p(x,p(y,z,f(y)),g(y)),p(y,p(p(p(x,y,y),p(x,p(y,z,f(y)),$	
	$\quad g(y)),u,f(p(x,p(y,z,f(y)),g(y)))),g(p(x,p(y,z,f(y)),$	
	$\quad g(y)))),v,f(p(p(x,y,y),p(x,p(y,z,f(y)),g(y)),u,$	
	$\quad f(p(x,p(y,z,f(y)),g(y)))),g(p(x,p(y,z,f(y)),g(y)))))),$	
	$\quad g(p(p(x,y,y),p(p(x,p(y,z,f(y)),g(y)),u,f(p(x,p(y,z,$	
	$\quad f(y)),g(y)))),g(p(x,p(y,z,f(y)),g(y)))))),w)=$	
	$\quad p(p(x,y,y),p(p(x,p(y,z,f(y)),g(y)),u,f(p(x,p(y,z,f(y)),$	
	$\quad g(y)))),g(p(x,p(y,z,f(y)),g(y))))$	$[3\rightarrow5]$
91,90	$p(x,p(y,z,f(y)),g(y))=y$	$[41\rightarrow5\ :4,42,\ \mathrm{flip}]$
95,94	$p(x,p(y,x,g(x)),z)=p(y,x,g(x))$	$[41\rightarrow5\ :42,42,91,42]$
103,102	$p(x,p(x,y,f(x)),z)=p(x,y,f(x))$	$[41\rightarrow5\ :42,42,91,42]$
107,106	$p(x,y,f(x))=x$	
		$[82\ :91,91,91,91,91,91,91,91,91,91,91,91,91,103,103,91,91,91,91]$
115,114	$p(x,y,g(y))=y$	$[68\ :107,95,95]$
117,116	$p(p(x,y,y),y,z)=y$	$[29\ :107,115,115,115]$
121,120	$p(x,y,y)=y$	$[64\ :117,117,\ \mathrm{flip}]$

122	$p(A, B, A) = A,\ f(A) = A,\ g(A) = A\ \rightarrow\ \Box$	[43 :121 :1]
124,123	$f(x) = x$	$[120 \rightarrow 106]$
125	$p(A, B, A) = A,\ g(A) = A\ \rightarrow\ \Box$	[122 :124 :1]
127,126	$p(x, y, x) = x$	[106 :124]
128	$g(A) = A\ \rightarrow\ \Box$	[125 :127 :1]
129	$g(x) = x$	$[114 \rightarrow 126, \text{flip}]$
131	$\Box$	[129,128]

With the absorption basis {L2,L3,L4'}, the schema of Thm. MAJ-3, and the majority term $(x \wedge z) \vee (y \wedge (x \vee z))$, we can substitute $p(\text{L3}(y), \text{L4}'(y), u)$ for $f(y)$ and L2(y) for $g(y)$, obtaining a single axiom for LT of length 119, again with 7 variables.

6.4.2 A Short Single Axiom for Lattices

To build a simpler single axiom for LT, we can use the fact that the three equations

$$\left\{ \begin{array}{l} (x \wedge y) \vee (x \wedge (x \vee y)) = x \\ (x \wedge x) \vee (y \wedge (x \vee x)) = x \\ (x \wedge y) \vee (y \wedge (x \vee y)) = y \end{array} \right\} \tag{B}$$

hold, given the reduction schema of Thm. MAJ-3 and the majority polynomial

$$p(x, y, z) = (x \wedge z) \vee (y \wedge (x \vee z)).$$

We first show that if we add (L3) and one other equation to (B), we obtain a basis for LT.

Lemma LT-11. Another absorption basis for LT.

$$\left\{ \begin{array}{l} (x \wedge y) \vee (x \wedge (x \vee y)) = x \\ (x \wedge x) \vee (y \wedge (x \vee x)) = x \\ (x \wedge y) \vee (y \wedge (x \vee y)) = y \\ ((x \wedge y) \vee (y \wedge z)) \vee y = y \quad \text{(L3)} \\ ((x \vee (y \vee z)) \wedge (u \vee y)) \wedge y = y \end{array} \right\} \Rightarrow$$

$$\left\{ \begin{array}{ll} y \wedge (x \vee (y \vee z)) = y & \text{(L2)} \\ ((y \vee x) \wedge (y \vee z)) \wedge y = y & \text{(L4')} \end{array} \right\}.$$

Proof (found by Otter 3.0.4 on gyro at 1.64 seconds).

1	$x = x$
2	$B \wedge (A \vee (B \vee C)) = B,\ ((B \vee A) \wedge (B \vee C)) \wedge B = B\ \rightarrow\ \Box$
3	$(x \wedge y) \vee (x \wedge (x \vee y)) = x$
5	$(x \wedge x) \vee (y \wedge (x \vee x)) = x$
7	$(x \wedge y) \vee (y \wedge (x \vee y)) = y$
9	$((x \wedge y) \vee (y \wedge z)) \vee y = y$

11	$((x \vee (y \vee z)) \wedge (u \vee y)) \wedge y = y$	
13	$((x \wedge y) \wedge (x \wedge (x \vee y))) \vee ((x \wedge y) \wedge x) = x \wedge y$	$[3 \rightarrow 3]$
26,25	$x \vee x = x$	$[7 \rightarrow 9]$
31	$(x \wedge x) \vee (y \wedge x) = x$	$[5 :26]$
42,41	$x \wedge x = x$	$[25 \rightarrow 7 :26]$
43	$x \vee (y \wedge x) = x$	$[31 :42]$
57	$((x \vee y) \wedge (z \vee x)) \wedge x = x$	$[25 \rightarrow 11]$
59	$((x \vee (y \vee z)) \wedge y) \wedge y = y$	$[25 \rightarrow 11]$
67	$(x \vee (x \wedge y)) \vee x = x$	$[11 \rightarrow 9]$
77	$(x \wedge y) \vee y = y$	$[43 \rightarrow 9]$
80,79	$(x \wedge y) \wedge y = x \wedge y$	$[77 \rightarrow 3 :26]$
82,81	$(x \vee (y \vee z)) \wedge y = y$	$[59 :80]$
105	$(x \vee (x \wedge y)) \wedge x = x \vee (x \wedge y)$	$[67 \rightarrow 3 :26]$
122,121	$(x \vee y) \wedge x = x$	$[77 \rightarrow 81]$
123	$x \vee (x \wedge y) = x$	$[105 :122, \text{flip}]$
127	$(x \wedge ((y \vee (x \vee z)) \wedge ((y \vee (x \vee z)) \vee x))) \vee (x \wedge (y \vee (x \vee z))) = x$	
		$[81 \rightarrow 13 :82,82]$
169	$(x \vee y) \vee x = x \vee y$	$[121 \rightarrow 123]$
174,173	$(x \vee (y \vee z)) \vee y = x \vee (y \vee z)$	$[81 \rightarrow 123]$
188,187	$x \wedge (y \vee (x \vee z)) = x$	$[127 :174,42,26]$
199	$((B \vee A) \wedge (B \vee C)) \wedge B = B \ \rightarrow \ \square$	$[2 :188 :1]$
414	$((x \vee y) \wedge (x \vee z)) \wedge x = x$	$[169 \rightarrow 57]$
416	$\square$	$[414,199]$

Theorem LT-12. A short single axiom for LT.

We use the majority polynomial $p(x, y, z) = (x \wedge z) \vee (y \wedge (x \vee z))$. With the axiom schema axiom of Thm. MAJ-3, we substitute the two additional equations from Lem. LT-11. (The three equations (B) need not be used in the construction, because they are already satisfied.) When the schema is expanded with the majority polynomial and written in terms of $\wedge$ and $\vee$, it has length 79, with 7 variables:

$$(((x \wedge y) \vee (y \wedge (x \vee y))) \wedge z) \vee (((x \wedge (((x_1 \wedge y) \vee (y \wedge x_2)) \vee$$
$$y)) \vee (((y \wedge (((x_1 \vee (y \vee x_2)) \wedge (x_3 \vee y)) \wedge y)) \vee (u \wedge (y \vee$$
$$(((x_1 \vee (y \vee x_2)) \wedge (x_3 \vee y)) \wedge y)))) \wedge (x \vee (((x_1 \wedge y) \vee (y \wedge$$
$$x_2)) \vee y)))) \wedge (((x \wedge y) \vee (y \wedge (x \vee y))) \vee z)) = y.$$

This is the shortest lattice theory single axiom known to us.

6.4.3 Weakly Associative Lattices

The results in this section are also reported in [40].

The term

$$p(x, y, z) = (x \wedge z) \vee (y \wedge (x \vee z))$$

is a majority polynomial for WAL as well as for LT, so if we can find an absorption basis for WAL, we can construct a single axiom in the same way as for LT in Thm. LT-12 above. In particular, we wish to find absorption equations that can be added to the three equations (B) (p. 143) to give WAL.

Lemma WAL-3. An absorption basis for WAL.

$$
\left\{
\begin{array}{l}
(x \wedge y) \vee (x \wedge (x \vee y)) = x \\
(x \wedge x) \vee (y \wedge (x \vee x)) = x \\
(x \wedge y) \vee (y \wedge (x \vee y)) = y \\
((x \vee y) \wedge (z \vee x)) \wedge x = x \\
((x \wedge y) \vee (z \wedge x)) \vee x = x
\end{array}
\right\}
\Rightarrow
\left\{
\begin{array}{l}
x \wedge x = x \\
x \wedge y = y \wedge x \\
x \vee x = x \\
x \vee y = y \vee x
\end{array}
\right\}.
$$

Because the two additional equations are commuted variants of (W3) and (W3') (p. 111), it is sufficient to derive commutativity and idempotence of the two operations.

Proof (found by Otter 3.0.4 on gyro at 2.69 seconds).

1	$x = x$	
2	$A \wedge A = A,\ B \wedge A = A \wedge B,\ A \vee A = A,\ B \vee A = A \vee B\ \rightarrow\ \square$	
3	$(x \wedge y) \vee (x \wedge (x \vee y)) = x$	
5	$(x \wedge x) \vee (y \wedge (x \vee x)) = x$	
7	$(x \wedge y) \vee (y \wedge (x \vee y)) = y$	
9	$((x \vee y) \wedge (z \vee x)) \wedge x = x$	
11	$((x \wedge y) \vee (z \wedge x)) \vee x = x$	
13	$(x \wedge (y \vee (z \wedge x))) \wedge (z \wedge x) = z \wedge x$	$[7 \rightarrow 9]$
15	$(x \wedge (y \vee (x \wedge z))) \wedge (x \wedge z) = x \wedge z$	$[3 \rightarrow 9]$
20,19	$x \vee (x \wedge (((x \vee y) \wedge (z \vee x)) \vee x)) = x$	$[9 \rightarrow 7]$
23	$((x \wedge y) \vee x) \vee x = x$	$[9 \rightarrow 11]$
26,25	$((x \vee y) \wedge x) \wedge x = x$	$[11 \rightarrow 9]$
29	$(((x \wedge y) \vee x) \wedge x) \vee (x \wedge x) = x$	$[23 \rightarrow 7]$
33	$(x \wedge (y \wedge x)) \wedge (y \wedge x) = y \wedge x$	$[7 \rightarrow 25]$
44,43	$(x \wedge x) \wedge x = x$	$[25 \rightarrow 33\ :26,26]$
46,45	$(x \wedge (y \vee x)) \wedge x = x$	$[25 \rightarrow 13\ :26,26]$
47	$(x \wedge x) \wedge (x \wedge x) = x \wedge x$	$[29 \rightarrow 13]$
62,61	$x \wedge (x \wedge (y \vee x)) = x \wedge (y \vee x)$	$[7 \rightarrow 45\ :46]$
63	$x \vee (x \wedge ((x \wedge (y \vee x)) \vee x)) = x$	$[45 \rightarrow 7]$
69	$((x \wedge x) \vee (x \wedge x)) \vee (x \wedge x) = x \wedge x$	$[47 \rightarrow 23]$
73	$((x \wedge x) \wedge (y \vee x)) \wedge x = x$	$[43 \rightarrow 15\ :44,44]$
88,87	$x \wedge x = x$	$[23 \rightarrow 73\ :44]$
94,93	$(x \vee x) \vee x = x$	$[69\ :88,88,88,88]$
105	$x \vee (y \wedge (x \vee x)) = x$	$[5\ :88]$
107	$B \wedge A = A \wedge B,\ A \vee A = A,\ B \vee A = A \vee B\ \rightarrow\ \square$	$[2\ :88\ :1]$
113,112	$x \vee x = x$	$[87 \rightarrow 19\ :94,88]$
114	$B \wedge A = A \wedge B,\ B \vee A = A \vee B\ \rightarrow\ \square$	$[107\ :113\ :1]$

115	$x \vee (y \wedge x) = x$	$[105 : 113]$
118,117	$(x \wedge (((x \vee y) \wedge (z \vee x)) \vee x)) \vee x = x$	$[19 \to 3 : 62,88]$
122,121	$x \wedge (y \wedge x) = y \wedge x$	$[115 \to 13 : 88]$
127	$(x \wedge y) \vee y = y$	$[115 \to 3 : 122,88]$
130,129	$(x \wedge y) \wedge y = x \wedge y$	$[127 \to 3 : 113]$
133	$(x \vee y) \wedge x = x$	$[25 : 130]$
135	$x \wedge (((x \vee y) \wedge (z \vee x)) \vee x) = x$	$[19 \to 63 : 46,20,46,118, \text{flip}]$
140,139	$x \wedge ((x \wedge y) \vee x) = (x \wedge y) \vee x$	$[23 \to 133]$
141	$x \vee ((x \wedge (y \vee x)) \vee x) = x$	$[63 : 140]$
143	$(x \vee (x \vee y)) \vee (x \vee y) = x \vee y$	$[133 \to 23]$
151	$((x \wedge y) \vee (z \wedge y)) \vee y = y$	$[121 \to 11]$
171	$(x \wedge (y \vee x)) \vee x = x$	$[141 \to 135 : 140]$
178,177	$x \wedge (y \vee x) = x$	$[171 \to 3 : 46,46,113, \text{flip}]$
179	$(x \wedge y) \vee x = x$	$[139 : 178, \text{flip}]$
188,187	$(x \vee y) \vee y = x \vee y$	$[177 \to 115]$
193	$(x \vee (y \wedge (z \vee x))) \vee (z \vee x) = z \vee x$	$[177 \to 151]$
197	$x \vee (x \vee y) = x \vee y$	$[143 : 188]$
199	$(x \wedge y) \wedge x = x \wedge y$	$[179 \to 3 : 113]$
211	$x \wedge (x \vee y) = x$	$[197 \to 3 : 113]$
215	$(x \vee y) \vee x = x \vee y$	$[211 \to 115]$
235	$x \vee (x \wedge y) = x$	$[199 \to 115]$
242,241	$(x \wedge y) \wedge (y \wedge x) = y \wedge x$	$[235 \to 13]$
285	$x \wedge y = y \wedge x$	$[241 \to 199 : 242,242]$
287,286	$(x \vee y) \vee (y \vee x) = y \vee x$	$[211 \to 193]$
348	$x \vee y = y \vee x$	$[286 \to 215 : 287,287]$
351	$\square$	$[114,285,348]$

Theorem WAL-4. A short single axiom for WAL.

We use the same majority polynomial and axiom schema as in Thm. LT-12, and we use the two additional absorption laws of Lem. WAL-3. When the axiom is written in terms of $\wedge$ and $\vee$, it has length 75, with 6 variables:

$$(((x \wedge y) \vee (y \wedge (x \vee y))) \wedge z) \vee (((x \wedge (((y \wedge x_1) \vee (x_2 \wedge y)) \vee y)) \vee (((y \wedge (((y \vee x_1) \wedge (x_2 \vee y)) \wedge y)) \vee (u \wedge (y \vee (((y \vee x_1) \wedge (x_2 \vee y)) \wedge y)))) \wedge (x \vee (((y \wedge x_1) \vee (x_2 \wedge y)) \vee y)))) \wedge (((x \wedge y) \vee (y \wedge (x \vee y))) \vee z)) = y.$$

This is the first WAL single axiom known to us.

6.5 Boolean Algebras

Boolean algebras (BA) are ordinarily considered to be of type $\langle 2,2,1,0,0 \rangle$; however, here we use type $\langle 2,1 \rangle$, with the following simple basis.

$$x + y = y + x \qquad \text{(commutativity)}$$
$$(x + y) + z = x + (y + z) \qquad \text{(associativity)}$$
$$n(x + n(y)) + n(n(x) + n(y)) = y \qquad \text{(Huntington axiom)}$$

6.5.1 Frink's Theorem

The following theorem is on Frink's implicational basis for Boolean algebra. Padmanabhan found the first first-order proof that the system is a basis for Boolean algebra [54] (previous proofs were model theoretic), and we include Otter's proof here.

Theorem BA-1. A first-order proof of Frink's theorem.

$$
\left\{
\begin{array}{l}
x + x = x \\
((x + y) + z) + u = (y + z) + x \;\rightarrow\; ((x + y) + z) + n(u) = 0 \\
((x + y) + z) + n(u) = 0 \;\rightarrow\; ((x + y) + z) + u = (y + z) + x
\end{array}
\right\} \Rightarrow
$$

$$
\left\{
\begin{array}{l}
n(x + n(y)) + n(n(x) + n(y)) = y \\
(x + y) + z = x + (y + z) \\
y + x = x + y
\end{array}
\right\}.
$$

The conclusion is Huntington's basis for Boolean algebra.

Proof (found by Otter 3.0.4 on gyro at 7.84 seconds).

1	$x = x$	
2	$((x + y) + z) + u = (y + z) + x \;\rightarrow\; ((x + y) + z) + n(u) = 0$	
3	$((x + y) + z) + n(u) = 0 \;\rightarrow\; ((x + y) + z) + u = (y + z) + x$	
4	$n(A + n(B)) + n(n(A) + n(B)) = B,\; B + A = A + B,$	
	$\quad (A + B) + C = A + (B + C) \;\rightarrow\; \square$	
6,5	$x + x = x$	
8,7	$x + n(x) = 0$	[2,5 :6,6,6,6]
9	$(x + y) + z = (y + z) + x$	[5 → 3 :8 :1]
12	$(x + y) + n(z) = 0 \;\rightarrow\; (x + y) + z = (x + y) + x$	[5 → 3 :6]
14	$(x + y) + z = (z + x) + y$	[flip 9]
22	$(n(x) + y) + x = 0 + y$	[7 → 9, flip]
25,24	$(x + y) + x = x + y$	[5 → 9, flip]
26	$(x + n(y + x)) + y = 0$	[7 → 9, flip]
34	$(x + y) + n(z) = 0 \;\rightarrow\; (x + y) + z = x + y$	[12 :25]
47,46	$0 + x = 0$	[7 → 24 :8]
49,48	$(x + y) + y = y + x$	[9 → 24]
51	$(n(x) + y) + x = 0$	[22 :47]
71	$x + y = y + x$	[5 → 14 :25,49]
74	$n(A + n(B)) + n(n(A) + n(B)) = B,$	
	$\quad (C + A) + B = A + (B + C) \;\rightarrow\; \square$	[14 → 4 :71]
92,91	$x + (x + y) = x + y$	[24 → 71, flip]
94,93	$(x + y) + z = x + (y + z)$	[14 → 71]

95	$x + (y + z) = y + (z + x)$	$[9 \to 71 : 94]$
96	$n(x) + x = 0$	$[7 \to 71, \text{flip}]$
98	$n(A + n(B)) + n(n(A) + n(B)) = B \to \square$	$[74 : 94 : 95]$
99	$n(x) + (y + x) = 0$	$[51 : 94]$
101	$x + (y + n(z)) = 0 \to x + (y + z) = x + y$	$[34 : 94, 94]$
103	$x + (n(y + x) + y) = 0$	$[26 : 94]$
106,105	$x + (y + x) = x + y$	$[24 : 94]$
107	$x + (y + z) = z + (x + y)$	$[14 : 94, 94]$
108	$n(x) + (x + y) = 0$	$[71 \to 99]$
110	$x + (y + n(x)) = 0$	$[71 \to 108 : 94]$
131	$x + (y + n(y + x)) = 0$	$[105 \to 103 : 94, 106]$
137	$x + (n(x + y) + y) = 0$	$[71 \to 103]$
167	$n(A + n(B)) + n(n(B) + n(A)) = B \to \square$	$[71 \to 98]$
234	$x + n(y) = 0 \to x + y = x$	$[91 \to 101 : 92, 6]$
287	$n(n(x)) + x = n(n(x))$	$[234, 96]$
289	$n(x) + y = 0 \to y + x = y$	$[71 \to 234]$
368,367	$x + n(n(x)) = n(n(x))$	$[71 \to 287]$
386	$x + n(n(n(x))) = 0$	$[287 \to 110]$
422,421	$n(n(x)) = x$	$[234, 386 : 368]$
424	$x + y = 0 \to x + n(y) = x$	$[421 \to 234]$
440	$x + y = 0 \to y + n(x) = y$	$[421 \to 289]$
544	$x + n(y + n(x)) = x$	$[424, 110]$
546	$n(x) + n(x + y) = n(x)$	$[424, 108]$
587	$n(x + n(y)) + y = y$	$[71 \to 544]$
663,662	$n(x + y) + n(y) = n(y)$	$[421 \to 587]$
700,699	$n(x + y) + (z + n(x)) = z + n(x)$	$[546 \to 107, \text{flip}]$
797	$n(x + y) + y = y + n(x)$	$[440, 137 : 94, 700, \text{flip}]$
801	$x + n(x + y) = x + n(y)$	$[440, 131 : 94, 663, \text{flip}]$
1090,1089	$n(x + y) + n(y + n(x)) = n(y)$	$[797 \to 801 : 663]$
1099	$B = B \to \square$	$[167 : 1090, 422]$
1100	$\square$	$[1099, 1]$

6.5.2 Robbins Algebra

This section is on the celebrated Robbins problem, whether a Robbins algebra is necessarily a Boolean algebra. Consider the equations

$$n(x + n(y)) + n(n(x) + n(y)) = y, \qquad \text{(Huntington axiom)}$$
$$n(n(x + y) + n(x + n(y))) = x. \qquad \text{(Robbins axiom)}$$

The Huntington axiom, along with commutativity and associativity of $+$, is a basis for Boolean algebra, but it is unknown whether the Huntington axiom can be replaced with the Robbins axiom.

Problem RBA-1. The Robbins question.

$$\left.\begin{cases} x + y = y + x \\ (x + y) + z = x + (y + z) \\ \text{Robbins axiom} \end{cases}\right\} \overset{?}{\Rightarrow} \{\text{Huntington axiom}\}.$$

Winker and Wos attacked the problem by finding weaker and weaker conditions that force a Robbins algebra to be Boolean [79]. We present two examples.

Theorem RBA-2. A Robbins algebra with an idempotent element is Boolean.

$$\left.\begin{cases} x + y = y + x \\ (x + y) + z = x + (y + z) \\ n(n(x + y) + n(x + n(y))) = x \\ \exists c \, (c + c = c) \end{cases}\right\} \Rightarrow \{n(x + n(y)) + n(n(x) + n(y)) = y\}.$$

Proof (found by Otter 3.0.4 on gyro at 21.99 seconds).

1	$x = x$	
2	$x + y = y + x$	
4,3	$(x + y) + z = x + (y + z)$	
6,5	$n(n(x + y) + n(x + n(y))) = x$	
7	$c + c = c$	
9	$n(A + n(B)) + n(n(A) + n(B)) \neq B$	
10	$c + (c + x) = c + x$	$[7 \to 3, \text{flip}]$
13	$x + (y + z) = y + (z + x)$	$[2 \to 3]$
14	$x + (y + z) = z + (x + y)$	$[\text{flip } 13]$
16,15	$c + (x + c) = c + x$	$[2 \to 10]$
25	$n(n(c) + n(c + n(c))) = c$	$[7 \to 5]$
27	$n(n(x + (y + z)) + n(x + (y + n(z)))) = x + y$	$[3 \to 5 : 4]$
29	$n(n(x + y) + n(y + n(x))) = y$	$[2 \to 5]$
33	$n(n(x + (n(y + z) + n(y + n(z)))) + n(x + y)) = x$	$[5 \to 5]$
35	$n(n(x + y) + n(n(y) + x)) = x$	$[2 \to 5]$
39	$n(n(x + n(y)) + n(x + y)) = x$	$[2 \to 5]$
49	$x + (y + z) = z + (y + x)$	$[2 \to 13]$
63	$n(n(c + n(c)) + n(c)) = c$	$[2 \to 25]$
65	$n(n(n(c) + (c + n(c))) + c) = n(c)$	$[25 \to 5]$
73	$n(n(A) + n(B)) + n(A + n(B)) \neq B$	$[2 \to 9]$
98	$n(n(x + (n(c + n(c)) + n(c))) + n(x + c)) = x$	$[63 \to 5]$
117	$n(n(x + (y + z)) + n(y + (x + n(z)))) = y + x$	$[49 \to 29 : 4]$
147	$n(n(x + y) + n(n(x) + y)) = y$	$[2 \to 29]$
159	$n(n(x + n(y)) + n(y + x)) = x$	$[2 \to 29]$
227	$n(n(n(x) + y) + n(y + x)) = y$	$[2 \to 35]$
289	$n(c + n(n(c) + (c + n(c)))) = n(c)$	$[25 \to 39]$
303	$n(n(x + n(y + z)) + n(z + (x + y))) = x$	$[14 \to 39]$
381	$n(n(n(x) + y) + n(x + y)) = y$	$[2 \to 147]$

459	$n(n(c + (x + n(c))) + n(c + x)) = c + x$	$[10 \rightarrow 159 : 4]$
646,645	$n(n(c) + n(n(c) + (c + n(c)))) = c$	$[65 \rightarrow 381 : 4,4,16]$
652,651	$n(n(c) + (c + n(c))) = n(c + n(c))$	$[65 \rightarrow 227 : 646,\ \text{flip}]$
681	$n(c + n(c + n(c))) = n(c)$	$[289 : 652]$
1451,1450	$n(n(c + (n(c + n(c)) + n(c))) + n(c)) = c$	$[7 \rightarrow 98]$
1583,1582	$n(n(x + c) + n(n(c + n(c)) + (x + n(c)))) = x$	$[25 \rightarrow 303]$
1644	$c + n(c + n(c)) = c$	$[681 \rightarrow 459 : 1451,\ \text{flip}]$
1646	$n(c + n(c)) + c = c$	$[2 \rightarrow 1644]$
1719,1718	$n(c + n(c)) + x = x$	$[1646 \rightarrow 117 : 1583,\ \text{flip}]$
1725,1724	$x + n(c + n(c)) = x$	$[1646 \rightarrow 27 : 1719,6,\ \text{flip}]$
1770	$n(x + n(x)) = n(c + n(c))$	$[1718 \rightarrow 33 : 6,1719]$
1957,1956	$n(n(n(n(x)) + x)) = n(n(x))$	$[1770 \rightarrow 227 : 1719]$
1959,1958	$n(n(x)) = x$	$[1770 \rightarrow 227 : 1725,1957]$
2091,2090	$n(n(x) + y) + n(x + y) = n(y)$	$[381 \rightarrow 1958,\ \text{flip}]$
2177	$B \neq B$	$[73 : 2091,1959]$
2178	$\square$	$[2177,1]$

Theorem RBA-3. A Robbins algebra with $c + d = c$ is Boolean.

$$\left. \begin{cases} x + y = y + x \\ (x + y) + z = x + (y + z) \\ n(n(x + y) + n(x + n(y))) = x \\ \exists c \exists d\ (c + d = c) \end{cases} \right\} \Rightarrow \{n(x + n(y)) + n(n(x) + n(y)) = y\}.$$

By Thm. RBA-2, it is sufficient to show $\exists e, e + e = e$.

Proof (found by Otter 3.0.4 on gyro at 1506.42 seconds).

1	$x + x \neq x$	
4,3	$x + y = y + x$	
6,5	$(x + y) + z = x + (y + z)$	
8,7	$x + (y + z) = y + (x + z)$	
9	$n(n(x + y) + n(x + n(y))) = x$	
12,11	$c + d = c$	
14	$n(n(x + (y + z)) + n(y + n(x + z))) = y$	$[7 \rightarrow 9]$
18	$n(n(x + y) + n(y + n(x))) = y$	$[3 \rightarrow 9]$
20	$n(n(x + y) + n(x + (n(y + z) + n(y + n(z))))) = x$	$[9 \rightarrow 9 : 4]$
22	$n(n(x + y) + n(n(y) + x)) = x$	$[3 \rightarrow 9]$
26	$c + (x + d) = x + c$	$[11 \rightarrow 7,\ \text{flip}]$
31,30	$c + (d + x) = c + x$	$[11 \rightarrow 5,\ \text{flip}]$
32	$n(n(x + c) + n(c + n(x + d))) = c$	$[26 \rightarrow 9]$
40	$n(n(c + x) + n(d + (n(c) + x))) = d + x$	$[30 \rightarrow 18 : 4,8]$
44	$n(n(c) + n(d + n(c))) = d$	$[11 \rightarrow 18]$
58	$n(n(x + y) + n(n(x) + y)) = y$	$[3 \rightarrow 18]$
62	$n(n(x + n(y)) + n(y + x)) = x$	$[3 \rightarrow 18]$
67,66	$n(d + n(d + (n(c) + n(c)))) = n(c)$	$[44 \rightarrow 18 : 4,8,4]$
77,76	$n(n(c + x) + n(d + n(c + x))) = d$	$[30 \rightarrow 14]$

80	$n(n(x+(y+z))+n(y+n(z+x)))=y$	$[3 \to 14]$
102	$n(n(x+(n(y+z)+n(z+n(y))))+n(z+x))=x$	$[18 \to 22]$
108	$n(d+n(c+n(d+n(c))))=n(d+n(c))$	$[44 \to 22 :4,4]$
112	$n(x+n(x+(y+n(n(y)+x))))=n(n(y)+x)$	$[22 \to 22 :8,4,4]$
130	$n(n(x+(y+z))+n(n(x+y)+z))=z$	$[5 \to 58]$
146	$n(x+n(y+(x+n(n(y)+x))))=n(n(y)+x)$	$[58 \to 58 :6,4]$
176	$n(n(x+n(y+z))+n(y+(z+x)))=x$	$[5 \to 62]$
242	$n(n(x+c)+n(c+n(d+x)))=c$	$[3 \to 32]$
294	$n(n(x+n(y+n(z)))+n(x+(y+n(z+(y+n(y+$	
	$n(z)))))))=x$	$[62 \to 20 :8,4,4]$
434	$n(n(c+n(d+x))+n(x+c))=c$	$[3 \to 242]$
529,528	$n(n(c)+n(d+(d+n(c))))=d+d$	$[11 \to 40 :4]$
636	$n(n(n(d+(n(c)+x))+(y+n(c+x)))+n(y+(d+x)))=y$	
		$[40 \to 80]$
814	$n(n(x+(y+z))+n(n(y+x)+z))=z$	$[7 \to 130]$
866	$n(n(x+n(y+z))+n(z+(y+x)))=x$	$[3 \to 176]$
1140	$n(n(d+(x+n(c+y)))+n(y+(c+x)))=d+x$	$[30 \to 866 :6]$
1933,1932	$n(n(c)+n(d+(n(x+c)+n(c+n(x)))))=d$	$[11 \to 102 :4]$
2194	$n(n(c+n(d+n(c)))+n(c+n(c+n(d+n(c)))))=c$	
		$[108 \to 434 :4]$
2356	$n(n(d+n(c+x))+n(c+(x+n(d+n(c+x)))))=d$	
		$[76 \to 112 :4,8,31,4,6,77]$
5429,5428	$n(d+n(d+(n(c+x)+n(c+(x+n(d+n(c+x)))))))=$	
	$n(c+x)$	$[76 \to 294 :8,6,8,31,8]$
6608	$n(c+(x+n(d+n(c+x))))=n(c+x)$	
		$[2356 \to 814 :8,4,5429, \text{flip}]$
6611,6610	$n(c+n(d+n(c)))=n(c)$	$[11 \to 6608 :31,12]$
6653,6652	$n(n(c)+n(c+n(c)))=c$	$[2194 :6611,6611]$
6721,6720	$n(c+n(c+(n(c)+n(c+(c+n(c+n(c)))))))=n(c)$	
		$[6652 \to 294 :8]$
6798	$n(n(c+n(c))+n(c+(c+n(c+n(c)))))=c$	
		$[6652 \to 146 :4,6653]$
7832	$n(c+(c+n(c+n(c))))=n(c)$	$[6798 \to 814 :8,4,6721, \text{flip}]$
7834	$d+n(c+n(c))=d$	$[7832 \to 1140 :4,4,1933, \text{flip}]$
8090	$d+d=d$	$[7834 \to 636 :4,67,529]$
8092	$\square$	$[8090,1]$

The preceding theorem is very difficult for Otter, and the proof shown is the result of a very specialized search strategy. Our first automatic proof (without hints or special strategies) was found with a prototype theorem prover with associative-commutative (AC) unification and matching; that proof was the first known first-order proof. Wos and McCune then found an automatic proof (unpublished) with Otter using the *tail strategy*, which penalizes equalities with complex right-hand sides, and a restriction strategy that discards equalities with more than three variables. Independently,

R. Veroff found an Otter proof with the *hints strategy* [75], using the AC proof as hints. The proof shown above was found with a strategy that discards equalities that match some simple patterns that do not appear in some of the other proofs; we have included this proof because it is shorter than the other proofs we have found. We are applying the strategies we have developed for this theorem to searches for a proof of the main conjecture, Prob. RBA-1.

6.5.3 Ternary Boolean Algebra

The results in this section are also reported in [59]. Padmanabhan's reduction schema [53] gives us the following single axiom for ternary Boolean algebra (TBA).

$$f(f(x, g(x), y), g(f(f(z, g(f(f(u, v, w), v6, f(u, v, v7))),$$

$$f(v, f(v7, v6, w), u)), g(v8), z)), z) = y.$$

The following theorem gives us a simpler single axiom for TBA; it was found by running Otter searches with many candidate axioms derived from the preceding axiom.

Theorem TBA-1. A short single axiom for TBA.

$$\left\{ \begin{array}{l} f(f(v, w, x), y, f(v, w, z)) = f(v, w, f(x, y, z)) \\ f(y, x, x) = x \\ f(x, y, g(y)) = x \\ f(x, x, y) = x \\ f(g(y), y, x) = x \end{array} \right\} \Leftrightarrow$$

$$\{f(f(x, g(x), y), g(f(f(z, u, v), w, f(z, u, v6))), f(u, f(v6, w, v), z)) = y\}.$$

Proof ($\Rightarrow$) found by Otter 3.0.4 on gyro at 3.78 seconds.

1	$x = x$	
3,2	$f(f(x, y, z), u, f(x, y, v)) = f(x, y, f(z, u, v))$	
4	$f(x, y, y) = y$	
7,6	$f(x, y, g(y)) = x$	
9,8	$f(x, x, y) = x$	
10	$f(g(x), x, y) = y$	
12	$f(f(A, g(A), B), g(f(f(C, D, E), F, f(C, D, G))),$	
	$f(D, f(G, F, E), C)) \neq B$	
13	$f(f(A, g(A), B), g(f(C, D, f(E, F, G))), f(D, f(G, F, E), C)) \neq B$	
		[copy,12 :3]
14	$f(x, y, f(z, f(x, y, u), u)) = f(x, y, u)$	[2 → 4]
17,16	$f(f(x, y, z), u, y) = f(x, y, f(z, u, y))$	[4 → 2]
18	$f(x, y, f(z, x, u)) = f(z, x, f(x, y, u))$	[4 → 2]
19	$f(x, y, f(y, z, u)) = f(y, z, f(x, y, u))$	[flip 18]

20	$f(x, y, f(z, f(x, y, z), u)) = f(x, y, z)$	$[2 \to 8]$
22	$f(x, y, x) = x$	$[8 \to 2 : 9, 9]$
25,24	$f(f(x, y, z), u, x) = f(x, y, f(z, u, x))$	$[22 \to 2]$
33	$f(x, y, f(z, x, g(y))) = x$	$[6 \to 14 : 7]$
46,45	$f(x, y, f(z, u, f(v, x, g(y)))) = f(x, y, f(z, u, x))$	$[33 \to 2 : 25,$ flip$]$
57	$f(x, y, z) = f(x, z, f(g(z), y, z))$	$[6 \to 16]$
60	$f(x, y, f(g(y), z, y)) = f(x, z, y)$	$[$flip $57]$
61	$f(x, y, f(z, g(y), x)) = x$	$[16 \to 33 : 46]$
71	$f(g(g(x)), x, y) = g(g(x))$	$[6 \to 61]$
80,79	$g(g(x)) = x$	$[4 \to 71,$ flip$]$
84,83	$f(x, g(x), y) = y$	$[79 \to 10]$
85	$f(x, g(y), f(z, y, x)) = x$	$[79 \to 61]$
90,89	$f(x, g(y), y) = x$	$[79 \to 6]$
91	$f(B, g(f(C, D, f(E, F, G))), f(D, f(G, F, E), C)) \neq B$	$[13 : 84]$
163,162	$f(x, y, f(z, x, y)) = f(z, x, y)$	$[4 \to 19,$ flip$]$
286	$f(x, g(y), f(y, x, z)) = x$	$[89 \to 20 : 90]$
374	$f(x, y, z) = f(y, z, x)$	$[286 \to 85 : 80, 17, 163]$
403,402	$f(g(x), y, x) = y$	$[83 \to 374,$ flip$]$
420	$f(x, y, z) = f(x, z, y)$	$[60 : 403]$
444	$f(B, g(f(D, f(E, F, G), C)), f(D, f(G, F, E), C)) \neq B$	$[374 \to 91]$
612	$f(x, y, z) = f(z, y, x)$	$[374 \to 420]$
1085	$B \neq B$	$[612 \to 444 : 90]$
1086	$\square$	$[1085, 1]$

Proof ($\Leftarrow$) found by Otter 3.0.4 on gyro at 1.73 seconds.

1	$x = x$	
2	$f(f(D, E, A), B, f(D, E, C)) = f(D, E, f(A, B, C)),$	
	$f(B, A, A) = A, \quad f(A, B, g(B)) = A,$	
	$f(A, A, B) = A, \quad f(g(B), B, A) = A \; \to \; \square$	
4,3	$f(f(x, g(x)), y, g(f(f(z, u, v), w, f(z, u, v_6)))),$	
	$f(u, f(v_6, w, v), z)) = y$	
5	$f(f(x, y, z), g(f(f(f(u, v, w), v_6, f(u, v, v_7))), f(v, f(v_7, v_6, w), u)) =$	
	$f(y, f(z, g(f(x, y, v_8)), v_8), x)$	$[3 \to 3]$
6	$f(f(x, g(x)), y, g(f(f(z, u, f(v, f(w, v_6, v_7)), v_8)), g(f(f(v_8, v, v_7), v_6,$	
	$f(v_8, v, w))), f(z, u, f(v_9, g(v_9), v_{10})))), f(u, v_{10}, z)) = y \; [3 \to 3]$	
8	$f(x, f(y, g(f(z, x, u)), u), z) = f(f(z, x, y), g(f(f(v, w, v_6), v_7,$	
	$f(v, w, v_8))), f(w, f(v_8, v_7, v_6), v))$	$[$flip $5]$
9	$f(x, f(y, g(f(z, x, u)), u), z) = f(x, f(y, g(f(z, x, v)), v), z) \; [5 \to 5]$	
10	$f(g(x), f(y, g(f(x, g(x), z)), z), x) = y$	$[3 \to 5,$ flip$]$
13,12	$f(f(x, g(x)), y, g(f(z, g(z), f(u, g(u), v)))), f(g(z), v, z)) = y$	
		$[6 \to 6 : 4]$
15,14	$f(g(x), y, x) = y$	$[12 \to 12 : 13,$ flip$]$
18	$f(f(x, g(x)), y, g(f(z, g(z), g(g(u))))), u) = y$	$[12 \to 12 : 15]$
20	$f(x, g(f(y, g(y), z)), z) = x$	$[10 : 15]$

27,26	$f(x, g(g(g(y))), y) = x$	$[14 \to 20]$
31,30	$g(f(x, g(x), y)) = g(y)$	$[14 \to 20]$
33,32	$f(x, g(y), y) = x$	$[20 :31]$
35,34	$f(x, g(x), y) = y$	$[18 :31,27]$
37,36	$f(x, g(f(f(y, z, u), v, f(y, z, w))), f(z, f(w, v, u), y)) = x$	$[3 :35]$
38	$f(x, f(y, g(f(z, x, u)), u), z) = f(z, x, y)$	$[8 :37]$
39	$f(x, y, z) = f(y, f(z, g(f(x, y, u)), u), x)$	$[5 :37]$
43,42	$g(g(x)) = x$	$[14 \to 34]$
47,46	$f(x, y, g(y)) = x$	$[42 \to 32]$
49,48	$f(g(x), x, y) = y$	$[42 \to 34]$
51,50	$f(x, y, g(x)) = y$	$[42 \to 14]$
52	$f(f(D, E, A), B, f(D, E, C)) = f(D, E, f(A, B, C))$,	
	$f(B, A, A) = A,\ f(A, A, B) = A \to \square$	$[2 :47,49 :1,1]$
56,55	$f(x, f(y, g(f(z, x, u)), u), z) = f(x, f(y, g(z), g(x)), z)$	$[46 \to 9, \text{flip}]$
59	$f(x, y, z) = f(y, f(z, g(x), g(y)), x)$	$[39 :56]$
60	$f(x, f(y, g(z), g(x)), z) = f(z, x, y)$	$[38 :56]$
67	$f(x, g(f(y, z, u)), f(u, z, y)) = x$	$[48 \to 36 :49,51]$
107	$f(x, y, z) = f(z, y, x)$	$[48 \to 67 :43]$
111	$f(f(x, y, z), g(f(z, y, x)), u) = u$	$[67 \to 107, \text{flip}]$
132,131	$f(f(x, y, z), u, f(x, y, v)) = f(y, f(z, u, v), x)$	$[36 \to 111]$
139	$f(E, f(A, B, C), D) = f(D, E, f(A, B, C)),\ f(B, A, A) = A$,	
	$f(A, A, B) = A \to \square$	$[52 :132]$
144	$f(x, f(y, g(x), g(z)), z) = f(x, f(y, g(z), g(x)), z)$	$[50 \to 55]$
149	$f(x, f(y, g(z), g(x)), z) = f(x, f(y, g(x), g(z)), z)$	$[\text{flip } 144]$
160	$f(x, y, z) = f(y, f(x, g(z), g(y)), z)$	$[107 \to 59]$
171	$f(x, f(y, g(z), g(x)), z) = f(y, x, z)$	$[\text{flip } 160]$
290,289	$f(x, y, y) = y$	$[50 \to 60 :33, \text{flip}]$
294,293	$f(x, f(y, g(x), g(z)), z) = f(x, z, y)$	$[107 \to 60]$
301	$f(E, f(A, B, C), D) = f(D, E, f(A, B, C))$,	
	$f(A, A, B) = A \to \square$	$[139 :290 :1]$
310,309	$f(x, f(y, g(z), g(x)), z) = f(x, z, y)$	$[149 :294]$
340	$f(x, y, z) = f(z, x, y)$	$[171 :310]$
363,362	$f(x, x, y) = x$	$[107 \to 289]$
364	$\square$	$[301 :363 :340,1]$

7. Independent Self-Dual Bases

We write about two different types of dual equation in this chapter. The first type applies to group-like algebras with one binary operation, and the dual of an equation is obtained by simultaneously flipping the arguments of all occurrences of the binary operation; informally, the dual of an equation is its mirror image with respect to the binary operation. The second (and more familiar) type applies to Boolean algebras with one or more pairs of operations; the dual of an equation is obtained by simultaneously replacing all occurrences of each operation that occurs in a pair with the other member of the pair.

For both types of duality, we write $\widetilde{E}$ for the dual of an equation E. A set S of equations is self-dual if $E \in S \Rightarrow \widetilde{E} \in S$ (modulo renaming of variables). An equation E is self-dual if $E \equiv \widetilde{E}$. For example, the associative law $(xy)z = x(yz)$ is self-dual, for if we write it backward and then rename the variables, the result is exactly what we started with.

The focus of this chapter is the existence of independent self-dual n-bases, for several values of n, for groups and subvarieties, and for Boolean algebra. Our interest in this area arises from work by Tarski.

Let V be a variety of algebras, and let $\nabla(V)$ denote the set of cardinalities of independent equational bases for V. If V is finitely based, then $\nabla(V)$ is a set of natural numbers. The following results were announced by Tarski [72]. (See also [73], [45], and [56].)

Tarski's Unbounded Theorem. *Let f be a term in which the variable x occurs at least twice. If the finitely based variety V satisfies $f = x$, then $\nabla(V)$ is an unbounded interval.*

Tarski's Interpolation Theorem. *If there exist two independent bases for an equational theory K with m and n identities, respectively, then there exists an independent basis for K with j identities for every j in the interval $[m, n]$.*

However, there is an example of a variety of algebras admitting a duality definable by an independent self-dual basis with 2 and 4 identities, but having no such basis with 3 identities (see [24]). Thus, the analog of Tarski's theorems need not be true if we insist that the equational basis enjoys some additional syntactic property (e.g., being self-dual).

We are interested in similar results in which the basis satisfies syntactic constraints such as self-duality. For groups, we consider cardinalities 2, 3, and 4; and for subvarieties of groups, we present a schema for cardinalities 2, 3, and 4. In [60], we extend these results to obtain the following.

Group Theory Self-Dual Basis Theorem. *Every finitely based variety of group theory has an independent self-dual basis with n identities for all $n \geq 2$.*

These give us new (and the only constructive) proofs of Tarski's theorem for the group case.

For the equational theory of Boolean algebras, we verify a previously known self-dual 6-basis and present new self-dual bases of cardinalities 2 and 3. These results, coupled with a blow-up technique of D. Kelly and Padmanabhan [24], yields the following theorem.

Boolean Algebra Self-Dual Basis Theorem. *The equational theory of Boolean algebras has an independent self-dual basis with n equations for all $n \geq 2$.*

7.1 Self-Dual Bases for Group Theory

Theorem DUAL-GT-1 presents an independent self-dual 2-basis for the variety of groups, and Thm. DUAL-GT-2 does the same for Abelian groups. These two 2-bases were found by techniques similar to those presented in [35]: a large number of candidate bases were generated, and each was given to Otter to search for a proof.

Theorem DUAL-GT-1. An independent self-dual 2-basis for GT.

$$\left\{ \begin{array}{l} ((x \cdot y) \cdot z) \cdot (y \cdot z)' = x \\ (z \cdot y)' \cdot (z \cdot (y \cdot x)) = x \end{array} \right\} \Leftrightarrow \left\{ \begin{array}{l} x \cdot x' = y \cdot y' \\ x \cdot (y \cdot y') = x \\ (x \cdot y) \cdot z = x \cdot (y \cdot z) \end{array} \right\}.$$

($\Leftarrow$) is trivial, because the right-hand side is a well-known basis for group theory. The proof of ($\Rightarrow$) is below. The 2-basis is independent because each equation by itself has a projection (nongroup) model. (Also, it is known that no single axiom for groups can be as small as either of these [28].)

Proof (found by Otter 3.0.4 on gyro at 0.53 seconds).

1	$x = x$	
2	$B \cdot B' = A \cdot A'$, $A \cdot (B \cdot B') = A$, $(A \cdot B) \cdot C = A \cdot (B \cdot C) \rightarrow \Box$	
3	$((x \cdot y) \cdot z) \cdot (y \cdot z)' = x$	
6,5	$(x \cdot y)' \cdot (x \cdot (y \cdot z)) = z$	
7	$(x \cdot y) \cdot ((z \cdot u)' \cdot y)' = (x \cdot z) \cdot u$	$[3 \rightarrow 3]$
8	$x \cdot (y \cdot (z \cdot y)')' = x \cdot z$	$[3 \rightarrow 3]$

12	$(x \cdot y) \cdot z = (x \cdot u) \cdot ((y \cdot z)' \cdot u)'$	[flip 7]
35	$((x \cdot y) \cdot (z \cdot y)') \cdot z = x$	[3 → 8, flip]
37	$(x \cdot (y \cdot x)')' = y$	[8 → 5 :6, flip]
43	$((x \cdot y)' \cdot x)' = y$	[37 → 37]
45	$x \cdot (y \cdot ((x \cdot y)' \cdot z)) = z$	[37 → 5]
48,47	$(x' \cdot (y \cdot z)')' = y \cdot (z \cdot x)$	[5 → 43]
52,51	$x \cdot ((y \cdot x)' \cdot (y \cdot z)) = z$	[43 → 5]
63	$(x \cdot y) \cdot (z \cdot y)' = (x \cdot (u \cdot z)') \cdot u$	[43 → 7]
75,74	$(x \cdot y) \cdot (z \cdot ((y \cdot z)' \cdot u)) = x \cdot u$	[7 → 37 :48]
98	$((x \cdot ((y \cdot z) \cdot (u \cdot z)')) \cdot u) \cdot y' = x$	[35 → 3]
119,118	$(x \cdot y) \cdot (z' \cdot y) = x \cdot z$	[45 → 7 :75]
137,136	$(x \cdot y) \cdot z = x \cdot (y \cdot z)$	[12 :119]
143,142	$x \cdot (y \cdot y') = x$ [98 :137,137,137,137,137,137,137,137,52]	
182	$x \cdot (y \cdot (z \cdot y)') = x \cdot ((u \cdot z)' \cdot u)$	[63 :137,137]
193	$B \cdot B' = A \cdot A' \; \to \; \square$	[2 :143,137 :1,1]
196	$x \cdot (y \cdot (x \cdot y)') = z \cdot z'$	[142 → 45]
200,199	$(x \cdot y)' \cdot x = y'$	[142 → 5]
203	$x \cdot x' = y \cdot (z \cdot (y \cdot z)')$	[flip 196]
205,204	$x \cdot (y \cdot (z \cdot y)') = x \cdot z'$	[182 :200]
210	$x \cdot x' = y \cdot y'$	[flip 203 :205]
211	$\square$	[210,193]

Theorem DUAL-GT-2. An independent self-dual 2-basis for Abelian GT.

$$\left\{ \begin{array}{l} (z \cdot (x \cdot y)) \cdot (y \cdot z)' = x \\ (z \cdot y)' \cdot ((y \cdot x) \cdot z) = x \end{array} \right\} \Leftrightarrow \left\{ \begin{array}{l} x \cdot x' = y \cdot y' \\ x \cdot (y \cdot y') = x \\ (x \cdot y) \cdot z = x \cdot (y \cdot z) \\ x \cdot y = y \cdot x \end{array} \right\}.$$

($\Leftarrow$) is trivial, because the right-hand side is a well-known basis for Abelian group theory. The proof of ($\Rightarrow$) is below. MACE shows the 2-basis to be independent with a 3-element nongroup model of the first equation.

Proof (found by Otter 3.0.4 on gyro at 1.47 seconds).

1	$x = x$	
2	$B \cdot B' = A \cdot A', \; A \cdot (B \cdot B') = A, \; (A \cdot B) \cdot C = A \cdot (B \cdot C),$	
	$B \cdot A = A \cdot B \; \to \; \square$	
3	$(x \cdot (y \cdot z)) \cdot (z \cdot x)' = y$	
5	$(x \cdot y)' \cdot ((y \cdot z) \cdot x) = z$	
7	$(x \cdot y) \cdot ((z \cdot u)' \cdot x)' = u \cdot (y \cdot z)$	[3 → 3]
17	$((x \cdot y)' \cdot y)' \cdot z = z \cdot x$	[3 → 5]
20	$x \cdot y = ((y \cdot z)' \cdot z)' \cdot x$	[flip 17]
23	$x \cdot (y \cdot (y \cdot z)')' = z \cdot x$	[5 → 3]
26	$x \cdot y = y \cdot (z \cdot (z \cdot x)')'$	[flip 23]
31	$((x \cdot y) \cdot (z \cdot x)') \cdot z = y$	[5 → 17, flip]
62,61	$((x \cdot (y \cdot z))' \cdot y)' = z \cdot x$	[31 → 31, flip]

95	$(((x \cdot y)' \cdot y)' \cdot z)' \cdot ((x \cdot u) \cdot z) = u$	[20 → 5]
126	$x \cdot ((y \cdot (z \cdot x)') \cdot z) = y$	[31 → 7 :62, flip]
171	$x \cdot (((((y \cdot x)' \cdot z)' \cdot z)' \cdot u) \cdot y) = u$	[20 → 126]
173	$x \cdot (((y \cdot x)' \cdot z) \cdot y) = ((z \cdot u)' \cdot u)'$	[17 → 126]
183,182	$(x \cdot (x \cdot y)')' = y$	[5 → 126]
184	$((x \cdot y)' \cdot y)' = z \cdot (((u \cdot z)' \cdot x) \cdot u)$	[flip 173]
185	$x \cdot y = y \cdot x$	[26 :183]
198	$(((x \cdot (y \cdot z)') \cdot y) \cdot (u \cdot z)) \cdot x' = u$	[126 → 3]
212,211	$((x \cdot y)' \cdot y)' \cdot z = x \cdot z$	[20 → 185]
223,222	$x \cdot (((y \cdot x)' \cdot z) \cdot y) = z$	[171 :212]
245	$(x \cdot y)' \cdot ((x \cdot z) \cdot y) = z$	[95 :212]
255	$((x \cdot y)' \cdot y)' = x$	[184 :223]
257	$(x \cdot (y \cdot x)')' = y$	[185 → 182]
294,293	$(x \cdot y) \cdot z = x \cdot (y \cdot z)$	[255 → 7]
330,329	$(x \cdot y)' \cdot (x \cdot (z \cdot y)) = z$	[245 :294]
353	$x \cdot (y \cdot x') = y$	[198 :294,294,294,330,294]
423	$B \cdot B' = A \cdot A',\ A \cdot (B \cdot B') = A \rightarrow \square$	[2 :294 :1,185]
458	$x \cdot (x' \cdot y) = y$	[185 → 353]
465,464	$(x \cdot y')' = y \cdot x'$	[353 → 182]
482,481	$x \cdot (y \cdot y') = x$	[257 :465,294]
492	$B \cdot B' = A \cdot A' \rightarrow \square$	[423 :482 :1]
526	$x \cdot x' = y \cdot y'$	[481 → 458]
527	$\square$	[526,492]

The next two theorems present a 3-basis and a 4-basis for (ordinary) groups; both are independent and self-dual. These bases were found by trial and error: we conjectured several variations, giving each to Otter to search for a proof.

Theorem DUAL-GT-3. An independent self-dual 3-basis for GT.

$$\left\{ \begin{array}{ll} x \cdot (x' \cdot y) = y & (1) \\ (x \cdot y') \cdot y = x & (2) \\ (x \cdot y) \cdot ((z \cdot z') \cdot u) = (x \cdot (v' \cdot v)) \cdot (y \cdot u) & (3) \end{array} \right\}.$$

The Otter proof below derives a well-known basis for GT. The basis is easily shown to be independent: the first two equations are not a basis because 3 variables are required, and MACE finds 3-element nongroup models of $\{(1),(3)\}$ and (dually) of $\{(2),(3)\}$.

Proof (found by Otter 3.0.4 on gyro at 0.52 seconds).

1	$x = x$	
2	$B \cdot B' = A \cdot A',\ A \cdot (B \cdot B') = A,\ (A \cdot B) \cdot C = A \cdot (B \cdot C) \rightarrow \square$	
4,3	$x \cdot (x' \cdot y) = y$	
6,5	$(x \cdot y') \cdot y = x$	
7	$(x \cdot y) \cdot ((z \cdot z') \cdot u) = (x \cdot (v' \cdot v)) \cdot (y \cdot u)$	
8	$(x \cdot (y' \cdot y)) \cdot (z \cdot u) = (x \cdot z) \cdot ((v \cdot v') \cdot u)$	[flip 7]

9	$x'' \cdot y = x \cdot y$	$[3 \to 3, \text{flip}]$
14,13	$x'' = x$	$[9 \to 5 : 6, \text{flip}]$
15	$x' \cdot (x \cdot y) = y$	$[9 \to 3]$
17	$(x \cdot y) \cdot y' = x$	$[13 \to 5]$
27	$(x \cdot y) \cdot z = (x \cdot (u' \cdot u)) \cdot (y \cdot z)$	$[5 \to 7]$
38	$(x \cdot (y' \cdot y)) \cdot (z \cdot u) = (x \cdot z) \cdot u$	$[\text{flip } 27]$
42	$x \cdot (y \cdot x)' = y'$	$[15 \to 17]$
50	$(x \cdot y)' = y' \cdot x'$	$[17 \to 42, \text{flip}]$
65	$(x \cdot (y \cdot z)) \cdot (z' \cdot y') = x$	$[50 \to 17]$
80	$(x \cdot ((y \cdot z) \cdot u)) \cdot (u' \cdot (z' \cdot y')) = x$	$[50 \to 65]$
91,90	$(x \cdot y) \cdot ((z \cdot z') \cdot u) = x \cdot (y \cdot u)$	$[15 \to 8 : 14, \text{flip}]$
99,98	$x \cdot (y' \cdot y) = x$	$[65 \to 8 : 14,91, \text{flip}]$
118,117	$(x \cdot y) \cdot z = x \cdot (y \cdot z)$	$[38 : 99, \text{flip}]$
120,119	$x \cdot (y \cdot y') = x$	$[80 : 118,118,118,118,4,4]$
121	$B \cdot B' = A \cdot A' \;\to\; \square$	$[2 : 120,118 : 1,1]$
125	$x \cdot x' = y \cdot y'$	$[119 \to 3]$
126	$\square$	$[125,121]$

Theorem DUAL-GT-4. An independent self-dual 4-basis for GT.

$$\left\{ \begin{array}{ll} x \cdot (x' \cdot y) = y & (1) \\ (y \cdot x') \cdot x = y & (2) \\ x \cdot x' = y' \cdot y & (3) \\ ((x \cdot (y \cdot z)) \cdot y) \cdot u = x \cdot (y \cdot ((z \cdot y) \cdot u)) & (4) \end{array} \right\}.$$

The Otter proof below derives a well-known basis for GT. To show independence, we note that the first three equations are not a basis because 3 variables are required, and MACE finds 3-element nongroup models of $\{(1),(2),(4)\}$, of $\{(1),(3),(4)\}$, and of $\{(2),(3),(4)\}$.

Proof (found by Otter 3.0.4 on gyro at 0.17 seconds).

1	$x = x$	
2	$B \cdot B' = A \cdot A',\; A \cdot (B \cdot B') = A,\; (A \cdot B) \cdot C = A \cdot (B \cdot C) \;\to\; \square$	
3	$x \cdot (x' \cdot y) = y$	
5	$(x \cdot y') \cdot y = x$	
7	$((x \cdot (y \cdot z)) \cdot y) \cdot u = x \cdot (y \cdot ((z \cdot y) \cdot u))$	
9	$x \cdot x' = y' \cdot y$	
17,16	$(x' \cdot x) \cdot y = y$	$[9 \to 5]$
18	$x \cdot (y' \cdot y) = x''$	$[9 \to 3]$
19	$x'' = x \cdot (y' \cdot y)$	$[\text{flip } 18]$
22,21	$x'' = x$	$[5 \to 16]$
27,26	$x \cdot (y' \cdot y) = x$	$[19 : 22, \text{flip}]$
28	$(x \cdot y) \cdot y' = x$	$[21 \to 5]$
35,34	$(x \cdot y) \cdot z = x \cdot (y \cdot z)$	$[16 \to 7 : 27,27,17]$
44,43	$x \cdot (y \cdot y') = x$	$[28 : 35]$
45	$B \cdot B' = A \cdot A' \;\to\; \square$	$[2 : 44,35 : 1,1]$

| 51 | $x \cdot x' = y \cdot y'$ | $[43 \to 3]$ |
| 52 | $\square$ | $[51,45]$ |

7.2 Self-Dual Schemas for Subvarieties of Group Theory

In this section, we give self-dual schemas for equational subvarieties of groups. In particular, we give bases containing meta-terms that are to be replaced with terms specifying the subvariety of interest. The resulting basis is self-dual. Any equational subvariety of groups can be specified with an equation $\delta = e$, where e is the group identity; the term δ and its dual, always with fresh variables, replace the meta-terms. These are similar to the schemas used by B. H. Neumann in [51] and to our inverse loop schema in Sec. 8.1.2; the differences are that our goal is not single axioms, and we use dual equations. The details of our work in this area are presented in [60].

Here we consider bases of cardinalities 2, 3, and 4. These bases were found by considering the 2-, 3-, and 4-bases in the preceding section and conjecturing various positions for the meta-terms α and β. Otter found proofs without much difficulty.

Theorem DUAL-GT-5. An independent self-dual 2-basis schema for GT.

$$\left\{ \begin{array}{l} ((x \cdot y) \cdot (\alpha \cdot z)) \cdot (y \cdot (\alpha \cdot z))' = x \\ ((x \cdot \beta) \cdot y)' \cdot ((x \cdot \beta) \cdot (y \cdot z)) = z \end{array} \right\} \Rightarrow \left\{ \begin{array}{l} x \cdot x' = y \cdot y' \\ x \cdot (y \cdot y') = x \\ (x \cdot y) \cdot z = x \cdot (y \cdot z) \end{array} \right\}.$$

Proof (found by Otter 3.0.4 on gyro at 2.62 seconds).

1	$x = x$	
2	$B \cdot B' = A \cdot A',\ A \cdot (B \cdot B') = A,\ (A \cdot B) \cdot C = A \cdot (B \cdot C) \to \square$	
3	$((x \cdot y) \cdot (\alpha \cdot z)) \cdot (y \cdot (\alpha \cdot z))' = x$	
5	$((x \cdot \beta) \cdot y)' \cdot ((x \cdot \beta) \cdot (y \cdot z)) = z$	
7	$(x \cdot (\alpha \cdot y)) \cdot ((z \cdot (\alpha \cdot u))' \cdot (\alpha \cdot y))' = (x \cdot z) \cdot (\alpha \cdot u)$	$[3 \to 3]$
16,15	$((x \cdot (y \cdot (\alpha \cdot z))') \cdot y) \cdot (\alpha \cdot z) = x$	$[3 \to 7,\ \text{flip}]$
28,27	$(((x \cdot (\beta \cdot (\alpha \cdot y))') \cdot \beta) \cdot \alpha)' \cdot x = y$	$[15 \to 5]$
45	$(x \cdot y)' \cdot (x \cdot (y \cdot z)) = z$	$[27 \to 5\ {:}28]$
99	$((x \cdot y)' \cdot x)' \cdot z = y \cdot z$	$[45 \to 45]$
139,138	$x \cdot ((y \cdot x)' \cdot (y \cdot z)) = z$	$[45 \to 99,\ \text{flip}]$
140	$((x \cdot y)' \cdot x)' = y$	$[99 \to 15\ {:}16,\ \text{flip}]$
144	$(x' \cdot (y \cdot z)')' = y \cdot (z \cdot x)$	$[45 \to 140]$
154	$(x \cdot (y \cdot x)')' = y$	$[140 \to 140]$
170	$x \cdot (y \cdot ((x \cdot y)' \cdot z)) = z$	$[154 \to 45]$
276	$((x \cdot (y \cdot z)') \cdot y) \cdot z = x$	$[138 \to 15\ {:}139]$
279	$((x \cdot y) \cdot z) \cdot (y \cdot z)' = x$	$[138 \to 3\ {:}139]$
309	$((x \cdot y) \cdot (z \cdot y)') \cdot z = x$	$[140 \to 276]$
419	$x' \cdot (((x \cdot y) \cdot (z \cdot y)') \cdot (z \cdot u)) = u$	$[309 \to 45]$

435,434	$x \cdot (y' \cdot y) = x$	[154 → 144, flip]
480	$(x \cdot y') \cdot y = x$	[276 → 434 :435, flip]
489	$((x' \cdot x) \cdot y')' = y$	[434 → 154]
493	$(x \cdot y) \cdot y' = x$	[434 → 279 :435]
521,520	$(x \cdot y) \cdot z = x \cdot (y \cdot z)$	[279 → 480, flip]
553,552	$x \cdot (y \cdot y') = x$	[493 :521]
556	$(x' \cdot (x \cdot y'))' = y$	[489 :521]
580,579	$x' \cdot (x \cdot y) = y$	[419 :521,521,521,139]
634	$B \cdot B' = A \cdot A' \rightarrow \square$	[2 :553,521 :1,1]
635	$x'' = x$	[556 :580]
648,647	$x \cdot (y \cdot x)' = y'$	[154 → 635, flip]
675	$x \cdot x' = y \cdot y'$	[552 → 170 :648]
676	$\square$	[675,634]

Theorem DUAL-GT-6. An independent self-dual 3-basis schema for GT.

$$\left\{ \begin{array}{l} x \cdot (x' \cdot y) = y \\ (x \cdot y') \cdot y = x \\ ((x \cdot \alpha) \cdot y) \cdot (\beta \cdot z) = (x \cdot \alpha) \cdot (y \cdot (\beta \cdot z)) \end{array} \right\} \Rightarrow \left\{ \begin{array}{l} x \cdot x' = y \cdot y' \\ x \cdot (y \cdot y') = x \\ (x \cdot y) \cdot z = x \cdot (y \cdot z) \end{array} \right\}.$$

Proof (found by Otter 3.0.4 on gyro at 0.22 seconds).

1	$x = x$	
2	$B \cdot B' = A \cdot A', \; A \cdot (B \cdot B') = A, \; (A \cdot B) \cdot C = A \cdot (B \cdot C) \rightarrow \square$	
4,3	$x \cdot (x' \cdot y) = y$	
6,5	$(x \cdot y') \cdot y = x$	
7	$((x \cdot \alpha) \cdot y) \cdot (\beta \cdot z) = (x \cdot \alpha) \cdot (y \cdot (\beta \cdot z))$	
9	$x'' \cdot y = x \cdot y$	[3 → 3, flip]
13	$x'' = x$	[9 → 5 :6, flip]
15	$x' \cdot (x \cdot y) = y$	[9 → 3]
17	$(x \cdot y) \cdot y' = x$	[13 → 5]
21	$(x \cdot y) \cdot (\beta \cdot z) = x \cdot (y \cdot (\beta \cdot z))$	[5 → 7 :6]
27	$x \cdot (y \cdot x)' = y'$	[15 → 17]
31	$(x \cdot y)' = y' \cdot x'$	[17 → 27, flip]
35	$(x \cdot (y \cdot z)) \cdot (z' \cdot y') = x$	[31 → 17]
54,53	$(x \cdot y) \cdot z = x \cdot (y \cdot z)$	[3 → 21 :4]
58,57	$x \cdot (y \cdot y') = x$	[35 :54,54,4]
59	$B \cdot B' = A \cdot A' \rightarrow \square$	[2 :58,54 :1,1]
63	$x \cdot x' = y \cdot y'$	[57 → 3]
64	$\square$	[63,59]

Theorem DUAL-GT-7. An independent self-dual 4-basis schema for GT.

$$\left\{ \begin{array}{l} x \cdot (x' \cdot y) = y \\ (x \cdot y') \cdot y = x \\ x' \cdot x = y \cdot y' \\ ((x \cdot (\alpha \cdot y)) \cdot \beta) \cdot z = x \cdot (\alpha \cdot ((y \cdot \beta) \cdot z)) \end{array} \right\} \Rightarrow \left\{ \begin{array}{l} x \cdot x' = y \cdot y' \\ x \cdot (y \cdot y') = x \\ (x \cdot y) \cdot z = x \cdot (y \cdot z) \end{array} \right\}.$$

Proof (found by Otter 3.0.4 on gyro at 3.38 seconds).

1	$x = x$	
2	$B \cdot B' = A \cdot A', \ A \cdot (B \cdot B') = A, \ (A \cdot B) \cdot C = A \cdot (B \cdot C) \ \rightarrow \ \square$	
3	$x \cdot (x' \cdot y) = y$	
5	$(x \cdot y') \cdot y = x$	
7	$x' \cdot x = y \cdot y'$	
8	$((x \cdot (\alpha \cdot y)) \cdot \beta) \cdot z = x \cdot (\alpha \cdot ((y \cdot \beta) \cdot z))$	
15	$x \cdot x' = y \cdot y'$	$[7 \rightarrow 7]$
16	$(x \cdot x') \cdot y = y''$	$[7 \rightarrow 5]$
18,17	$x \cdot (y \cdot y') = x$	$[7 \rightarrow 3]$
19	$x'' = (y \cdot y') \cdot x$	[flip 16]
20	$(A \cdot B) \cdot C = A \cdot (B \cdot C) \ \rightarrow \ \square$	$[2 :18 :15,1]$
25,24	$x'' = x$	$[3 \rightarrow 17]$
28,27	$(x \cdot x') \cdot y = y$	$[19 :25, \text{flip}]$
31	$(x \cdot y) \cdot y' = x$	$[24 \rightarrow 5]$
33	$x' \cdot (x \cdot y) = y$	$[24 \rightarrow 3]$
37	$((x \cdot y) \cdot \beta) \cdot z = x \cdot (\alpha \cdot (((\alpha' \cdot y) \cdot \beta) \cdot z))$	$[3 \rightarrow 8]$
38	$(x \cdot \beta) \cdot y = x \cdot (\alpha \cdot ((\alpha' \cdot \beta) \cdot y))$	$[17 \rightarrow 8]$
48	$x \cdot (\alpha \cdot ((\alpha' \cdot \beta) \cdot y)) = (x \cdot \beta) \cdot y$	[flip 38]
52,51	$(x' \cdot x) \cdot y = y$	$[24 \rightarrow 27]$
61	$(x \cdot y)' \cdot x = y'$	$[31 \rightarrow 33]$
76,75	$(x \cdot y)' = y' \cdot x'$	$[33 \rightarrow 61, \text{flip}]$
94	$(x' \cdot y') \cdot ((y \cdot x) \cdot z) = z$	$[75 \rightarrow 33]$
160,159	$\alpha \cdot (((\alpha' \cdot x) \cdot \beta) \cdot y) = (x \cdot \beta) \cdot y$	$[51 \rightarrow 37 :52, \text{flip}]$
167,166	$x \cdot ((x' \cdot \beta) \cdot y) = \beta \cdot y$	$[15 \rightarrow 37 :28,160, \text{flip}]$
173,172	$x \cdot (((x' \cdot y) \cdot \beta) \cdot z) = (y \cdot \beta) \cdot z$	$[3 \rightarrow 37 :160, \text{flip}]$
181	$x \cdot ((y \cdot \beta) \cdot ((\beta' \cdot (y' \cdot x')) \cdot z)) = z$	$[3 \rightarrow 37 :76,76,173, \text{flip}]$
195,194	$(x \cdot \beta) \cdot y = x \cdot (\beta \cdot y)$	$[48 :167, \text{flip}]$
215	$x \cdot (y \cdot (\beta \cdot ((\beta' \cdot (y' \cdot x')) \cdot z))) = z$	$[181 :195]$
233	$x \cdot (\beta \cdot ((\beta' \cdot x') \cdot y)) = y$	$[3 \rightarrow 194 :76, \text{flip}]$
380	$(x' \cdot y) \cdot ((y' \cdot x) \cdot z) = z$	$[24 \rightarrow 94]$
420,419	$\beta \cdot ((\beta' \cdot x) \cdot y) = x \cdot y$	$[233 \rightarrow 3 :25, \text{flip}]$
423	$x \cdot (y \cdot ((y' \cdot x') \cdot z)) = z$	$[215 :420]$
563	$(x \cdot y) \cdot z = x \cdot (y \cdot z)$	$[380 \rightarrow 423 :25, \text{flip}]$
565	$\square$	$[563,20]$

7.3 Self-Dual Bases for Boolean Algebra

This section is on self-dual bases for Boolean algebras of type $\langle 2,2,1,0,0 \rangle$, with corresponding operations $\langle +, \cdot, ', 0, 1 \rangle$. Here, the dual of an equation is obtained by swapping $+$ with $\cdot$ and 0 with 1.

We present in some detail the sequence of events that led to the results, in order to illustrate some of the ways Otter and MACE are used.

7.3.1 Padmanabhan's 6-Basis

The following theorem is a straightforward verification of Padmanabhan's previously known 6-basis [56].

Theorem DUAL-BA-1. An independent self-dual 6-basis for Boolean algebra.

$$\left\{ \begin{array}{ll} (x+y)\cdot y = y & (x\cdot y)+y = y \\ x\cdot(y+z) = (y\cdot x)+(z\cdot x) & x+(y\cdot z) = (y+x)\cdot(z+x) \\ x+x' = 1 & x\cdot x' = 0 \end{array} \right\}.$$

Independence is easily proved with three MACE runs showing that each equation of the left column is independent of the remaining five equations; a model of size three is quickly found in each case. In order to show that the set is a basis for Boolean algebra, it is sufficient to derive commutativity and associativity of either of the operations.

Proof (found by Otter 3.0.4 on gyro at 28.95 seconds).

1	$x = x$	
2	$B\cdot A = A\cdot B,\ (A\cdot B)\cdot C = A\cdot(B\cdot C) \to \square$	
4,3	$(x+y)\cdot y = y$	
5	$x\cdot(y+z) = (y\cdot x)+(z\cdot x)$	
6	$(x\cdot y)+(z\cdot y) = y\cdot(x+z)$	[flip 5]
9,8	$x+x' = 1$	
11,10	$(x\cdot y)+y = y$	
12	$x+(y\cdot z) = (y+x)\cdot(z+x)$	
13	$x\cdot x' = 0$	
15	$(x+y)\cdot(z+y) = y+(x\cdot z)$	[flip 12]
18	$0+x' = x'$	[13 → 10]
21,20	$x+x = x$	[3 → 10]
23,22	$x\cdot x = x$	[10 → 3]
29	$x\cdot(y+(z+x)) = x$	[3 → 6 :11, flip]
31	$x\cdot y = y\cdot x$	[20 → 6 :21]
36,35	$x\cdot(y+x) = x$	[22 → 6 :11, flip]
38,37	$x+(y\cdot x) = x\cdot(x+y)$	[22 → 6]
56	$x+y = y+x$	[22 → 12 :23]
72	$0' = 1$	[8 → 18, flip]
76	$0\cdot 1 = 0$	[72 → 13]
94,93	$x\cdot(x+y) = x$	[20 → 15 :4,38, flip]
98	$x+(y\cdot(z\cdot x)) = x$	[10 → 15 :4, flip]
115	$x+(y\cdot x) = x$	[37 :94]
130	$x+0 = (0+x)\cdot(1+x)$	[76 → 12]
135	$(0+x)\cdot(1+x) = x+0$	[flip 130]
141	$x'\cdot x = 0$	[13 → 31, flip]
150	$x+0 = (y'+x)\cdot(y+x)$	[141 → 12]

153,152	$0 + x = x$	$[141 \rightarrow 10]$
155,154	$x \cdot (x' + y) = y \cdot x$	$[141 \rightarrow 6 :153, \text{flip}]$
156	$(x' + y) \cdot (x + y) = y + 0$	$[\text{flip } 150]$
161,160	$x + 0 = x$	$[135 :153,36, \text{flip}]$
164	$(x' + y) \cdot (x + y) = y$	$[\text{flip } 156 :161, \text{flip}]$
235,234	$(x \cdot y) + z = (x + z) \cdot (y + z)$	$[12 \rightarrow 56, \text{flip}]$
236	$x' + x = 1$	$[8 \rightarrow 56, \text{flip}]$
276,275	$x \cdot 1 = x$	$[236 \rightarrow 29]$
366	$(x + y) + x = x + y$	$[93 \rightarrow 115]$
1269	$x + (x + y) = x + y$	$[56 \rightarrow 366]$
1623	$(x' + y) \cdot x = y \cdot x$	$[1269 \rightarrow 154 :155, \text{flip}]$
1922	$(x + y') \cdot (y + x) = x$	$[56 \rightarrow 164]$
4171,4170	$((x + y') \cdot (z + y')) \cdot y = (x \cdot z) \cdot y$	$[12 \rightarrow 1623]$
5275,5274	$(x \cdot y) \cdot z = x \cdot (y \cdot z)$	$[98 \rightarrow 1922 :235,235,9,276,4171]$
6098	$\square$	$[2 :5275 :31,1]$

7.3.2 A 2-Basis from Pixley Reduction

It was previously known that one can construct a self-dual 2-basis for Boolean algebra using a reduction schema for Pixley polynomials, but the straightforward method we tried first produced a dependent pair (i.e., either member is a single axiom). After several Otter and MACE experiments with various constructions, several independent pairs emerged. First, we present some background.

A *Pixley polynomial* [64] is a term $p(x, y, z)$ satisfying the so-called Pixley properties

$$p(y, y, x) = p(x, z, z) = p(x, u, x) = x.$$

Boolean algebra admits Pixley polynomials, and we use

$$p(x, y, z) = (x \cdot y') + (x \cdot z) + (y' \cdot z)$$

and its dual $\widetilde{p}(x, y, z)$, which are easily seen to satisfy the Pixley properties.

In [61], Padmanabhan and Quackenbush showed how to construct a single equational axiom for any finitely based equational theory that admits a Pixley polynomial. Briefly, any equation $\alpha = \beta$ can be made into an equivalent (modulo Pixley properties) absorption equation $p(x, \alpha, \beta) = x$, and any pair of absorption equations $g(x) = x$ and $h(y) = y$ can be made into an equivalent (modulo Pixley properties) equation $p(u, g(x), x) = p(y, h(y), y)$. Iterating in this way, one can construct an absorption equation $f(x) = x$ such that

$$\{f(x) = x, \ p(y, y, x) = p(x, z, z) = p(x, u, x) = x\}$$

is a basis for the theory. Finally, this set of identities is collapsed into a single equation with the reduction schema of the following theorem [61], which we verify with Otter.

Theorem PIX-2. A reduction schema for Pixley polynomials.

$$\{p(p(x,x,y),p(f(z),u,z),z)=y\} \Leftrightarrow \left\{ \begin{array}{l} p(x,x,y)=y \\ p(x,y,y)=x \\ p(x,y,x)=x \\ f(x)=x \end{array} \right\}.$$

The proof of ($\Leftarrow$) is obvious. The proof of ($\Rightarrow$) follows.

Proof (found by Otter 3.0.4 on gyro at 0.06 seconds).

1	$x=x$	
2	$p(A,A,B)=B,\ p(A,B,B)=A,\ p(A,B,A)=A,\ f(A)=A \to \square$	
3	$p(p(x,x,y),p(f(z),u,z),z)=y$	
6,5	$p(x,p(f(y),z,y),y)=x$	$[3 \to 3]$
8,7	$p(x,x,y)=y$	$[3 :6]$
9	$p(A,B,B)=A,\ p(A,B,A)=A,\ f(A)=A \to \square$	$[2 :8 :1]$
11,10	$p(x,y,y)=x$	$[7 \to 5]$
12	$p(x,f(y),y)=x$	$[5 \to 5]$
14	$p(f(x),y,x)=x$	$[7 \to 5,\text{ flip}]$
16	$p(A,B,A)=A,\ f(A)=A \to \square$	$[9 :11 :1]$
18,17	$f(x)=x$	$[7 \to 12,\text{ flip}]$
19	$p(A,B,A)=A \to \square$	$[16 :18 :1]$
20	$p(x,y,x)=x$	$[14 :18]$
22	$\square$	$[20,19]$

Our first approach to finding an independent self-dual 2-basis for Boolean algebra was to take Padmanabhan's self-dual 6-basis and turn each "half" into an equation, using the above method. This gives us an equation $F(x)=x$ equivalent to the set

$$\left\{ \begin{array}{ll} (x+y)\cdot y = y & p(y,y,x)=x \\ x\cdot(y+z)=(y\cdot x)+(z\cdot x) & p(x,z,z)=x \\ x+x'=1 & p(x,y,x)=x \end{array} \right\}$$

and the corresponding dual equation $\widetilde{F}(x)=x$ for the other half.

A digression. To analyze independence, we don't need to actually construct $F(x)=x$, but for the curious, we show how Otter can be used as a symbolic calculator to do so. The input file below follows the method given at the beginning of this section. All of the predicate symbols that start with "EQ" can be interpreted as equality; they are distinct so that we have control over the way $F(x)=x$ is constructed. The EQ symbols that end with "a" indicate an absorption equation. (And we use "@" for complement.)

```
op(400, xfx, [*,+,^,v,/,\,#]).   % infix operators
op(300, yf, @).                  % postfix operator

set(hyper_res).   % Hyperresolution inference rule
```

```
list(usable).
-EQ2(x,y) | EQ2a(p(u,x,y), u).
-EQ3(x,y) | EQ3a(p(u,x,y), u).
-EQ1a(x1,y1) | -EQ3a(x2,y2) | EQ5(p(x,x1,y1),p(x,x2,y2)).
-EQ5(x,y) | EQ5a(p(w,x,y),w).
-EQ2a(x1,y1) | -EQ5a(x2,y2) | EQ6(p(x,x1,y1),p(x,x2,y2)).
-EQ6(x,y) | EQ6a(p(w,x,y),w).
-EQ6a(x1,z) | p(p(u,u,x),p(x1,y,z),z)=x.
end_of_list.

list(sos).
EQ1a((x + y) * y, y).
EQ2(x * (y + z), (y * x) + (z * x)).
EQ3(x + x@, 1).
end_of_list.

list(demodulators).
% p(x,y,z) = (x * y@) + ((x * z) + (y@ * z)).
end_of_list.
```

Note that the demodulator at the end is disabled so that the Pixley terms are not expanded. From this input file, Otter produces the equation

$$p(p(x, x, y), p(p(z, p(u, p(v, w \cdot (v_6 + v_7), (v_6 \cdot w) + (v_7 \cdot w)), v),$$
$$p(u, p(v_8, p(v_9, (v_{10} + v_{11}) \cdot v_{11}, v_{11}), p(v_9, p(v_{12}, v_{13} + v_{13}', 1),$$
$$v_{12})), v_8)), v_{14}, z), z) = y.$$

If the Pixley terms are expanded, the resulting equation has length 3183. *End of digression.*

The fact that the two halves of the 6-basis are independent does not help us show independence of our newly constructed pair, because each member of the pair satisfies the Pixley properties as well as half of the 6-basis. In fact, Otter shows in the following theorem that they are *not* independent.

Theorem DUAL-BA-2. A basis for Boolean algebra (to show dependence).

$$\left\{ \begin{array}{l} (x + y) \cdot y = y \\ x \cdot (y + z) = (x \cdot y) + (x \cdot z) \\ x + x' = 1 \\ p(x, y, z) = (x \cdot y') + ((x \cdot z) + (y' \cdot z)) \\ p(x, x, y) = y \\ p(x, y, y) = x \\ p(x, y, x) = x \end{array} \right\}.$$

We derive the "other half" of the 6-basis. The constant 0 is not explicit, so instead of $x \cdot x' = 0$, we show the existence of an element with that property, that is, $x \cdot x' = y \cdot y'$.

(Otter's original proof, which was found with the standard strategy with `max_weight=28`, is 135 steps and was found in 92 seconds. Larry Wos, starting with that proof, found the following 78-step proof by using the resonance strategy [82]).

Proof (found by Otter 3.0.4 on gyro at 3.81 seconds).

1	$x = x$	
2	$A + (B \cdot C) = (A + B) \cdot (A + C)$, $(A \cdot B) + B = B$,	
	$\quad B \cdot B' = A \cdot A' \rightarrow \square$	
3	$(x + y) \cdot y = y$	
5	$x \cdot (y + z) = (x \cdot y) + (x \cdot z)$	
7,6	$(x \cdot y) + (x \cdot z) = x \cdot (y + z)$	[flip 5]
9,8	$x + x' = 1$	
11,10	$p(x, y, z) = (x \cdot y') + ((x \cdot z) + (y' \cdot z))$	
12	$p(x, x, y) = y$	
13	$(x \cdot x') + ((x \cdot y) + (x' \cdot y)) = y$	[copy,12 :11]
15	$p(x, y, y) = x$	
17,16	$(x \cdot y') + ((x \cdot y) + (y' \cdot y)) = x$	[copy,15 :11]
22,21	$1 \cdot x' = x'$	[8 → 3]
23	$x' + (1 \cdot y) = 1 \cdot (x' + y)$	[21 → 6]
26,25	$x + ((y + x) \cdot z) = (y + x) \cdot (x + z)$	[3 → 6]
28,27	$((x + y) \cdot z) + y = (x + y) \cdot (z + y)$	[3 → 6]
29	$(x \cdot (y + z)) \cdot (x \cdot z) = x \cdot z$	[6 → 3]
31	$1 \cdot (x' + y') = x' + y'$	[21 → 23, flip]
33	$(x + y) \cdot (y + y) = y + y$	[3 → 25, flip]
35	$((x + y) \cdot (z + y)) \cdot y = y$	[27 → 3]
38,37	$1 \cdot 1 = 1$	[8 → 31 :9]
43	$((x + y') \cdot 1) \cdot y' = y'$	[8 → 35]
45	$1 \cdot (x' + 1) = x' + 1$	[37 → 23, flip]
48,47	$(1 \cdot (x + 1)) \cdot 1 = 1$	[37 → 29 :38]
49	$(1 \cdot x) + 1 = 1 \cdot (x + 1)$	[37 → 6]
52,51	$1' + (1 + (1' \cdot 1)) = 1$	[37 → 13 :22]
55	$x' + (((y + x') \cdot x) + (x' \cdot x)) = y + x'$	[3 → 16]
57	$x' \cdot (x' + (x + x)) = x'$	[6 → 16 :7]
60,59	$(x \cdot y) + ((x \cdot z) + (z' \cdot z)) = x \cdot (y + ((x \cdot z) + (z' \cdot z)))$	
		[16 → 27 :17]
61	$x \cdot ((x \cdot y) + (y' \cdot y)) = (x \cdot y) + (y' \cdot y)$	[16 → 3]
63	$x \cdot (y' + ((x \cdot y) + (y' \cdot y))) = x$	[16 :60]
71	$(1 \cdot (x + 1)) \cdot (1 + 1) = 1 + 1$	[49 → 33]
74,73	$(x' \cdot y) + x' = x' \cdot (y + (x' + (x + x)))$	[57 → 6]
76,75	$1 \cdot (x + 1) = 1$	[47 → 63 :52,48, flip]
78,77	$1 + 1 = 1$	[71 :76,76, flip]
82,81	$x' + 1 = 1$	[45 :76, flip]
84,83	$1' \cdot 1 = 1'$	[77 → 57 :82]

86,85	$(x + 1') \cdot 1 = x + 1'$	$[83 \to 55 : 28,9,26,82]$
87	$x \cdot (1' + ((x \cdot 1) + 1')) = x$	$[83 \to 63]$
90,89	$x \cdot ((x \cdot 1) + 1') = (x \cdot 1) + 1'$	$[83 \to 61 :84]$
92,91	$1' + 1' = 1'$	$[83 \to 73 :78,82,78,84]$
93	$1' \cdot (x + 1') = 1' \cdot (x + 1)$	$[83 \to 59 :84,92,74,78,82,84,84,92, \text{flip}]$
98,97	$1' + (x + 1') = x + 1'$	$[85 \to 25 :82,86]$
100,99	$(x \cdot 1) + 1' = x$	$[87 :98,90]$
101	$x \cdot x = x$	$[89 :100,100]$
104,103	$x + 1 = 1$	$[3 \to 99 :9, \text{flip}]$
105	$1' \cdot (x + 1') = 1'$	$[93 :104,84]$
108,107	$1' + x = x$	$[99 \to 97 :100]$
110,109	$x \cdot 1 = x$	$[99 \to 85 :100]$
112,111	$x \cdot 1' = 1'$	$[99 \to 43 :110]$
114,113	$x + 1' = x$	$[99 :110]$
116,115	$1' \cdot x = 1'$	$[105 :114]$
118,117	$(x \cdot y) + x = x \cdot (y + x)$	$[101 \to 6]$
120,119	$x + (x \cdot y) = x \cdot (x + y)$	$[101 \to 6]$
126,125	$x \cdot (y + x) = x$	$[109 \to 6 :118,104,110]$
127	$x \cdot (x + y) = x \cdot (1 + y)$	$[109 \to 6 :120]$
129,128	$(x \cdot y) + x = x$	$[117 :126]$
130	$x \cdot x' = 1'$	$[111 \to 13 :112,114,114]$
133	$1' = x \cdot x'$	$[\text{flip } 130]$
134	$x' \cdot x = 1'$	$[61 \to 115 :116,108]$
138	$x \cdot ((y + x) + z) = x \cdot (x + z)$	$[125 \to 6 :120, \text{flip}]$
140	$x \cdot (1 + x') = x$	$[8 \to 127 :110, \text{flip}]$
143,142	$x'' = x$	$[130 \to 13 :114,7,9,110, \text{flip}]$
145,144	$(x \cdot y) + (x' \cdot y) = y$	$[130 \to 13 :108]$
148	$x \cdot x' = y \cdot y'$	$[133 \to 133]$
151	$x' \cdot (y + x) = x' \cdot y$	$[134 \to 6 :114, \text{flip}]$
154,153	$x' \cdot (x + y) = x' \cdot y$	$[134 \to 6 :108, \text{flip}]$
161	$(x \cdot (1 + y)) + (x' \cdot y) = x + y$	$[127 \to 144 :154]$
167,166	$1 + x = 1$	$[140 \to 144 :143,126,9,143, \text{flip}]$
169,168	$x + (x' \cdot y) = x + y$	$[161 :167,110]$
171,170	$x \cdot (x + y) = x$	$[127 :167,110]$
173	$x \cdot ((y + x) + z) = x$	$[138 :171]$
176,175	$x + (x \cdot y) = x$	$[119 :171]$
180,179	$x' \cdot ((y + x) + z) = x' \cdot (y + z)$	$[151 \to 6 :7, \text{flip}]$
189	$x + (x + y) = x + y$	$[153 \to 168 :169, \text{flip}]$
197	$(x + y) + z = y + (x + z)$	$[173 \to 144 :180,169, \text{flip}]$
200,199	$(x \cdot y) + y = y$	$[144 \to 189 :145]$
201	$A + (B \cdot C) = (A + B) \cdot (A + C) \to \square$	$[2 :200 :1,148]$
206	$x + ((y \cdot x) + z) = x + z$	$[199 \to 197, \text{flip}]$
213,212	$x + (y \cdot x) = x$	$[113 \to 206 :114]$
216	$x + (y \cdot (x + z)) = x + (y \cdot z)$	$[6 \to 206]$

218	$(x \cdot y) + (y \cdot z) = y \cdot ((x \cdot y) + z)$	$[212 \rightarrow 25 \, {:}213]$
220	$(x \cdot y) + (z \cdot x) = x \cdot (y + (z \cdot x))$	$[212 \rightarrow 27 \, {:}213]$
223,222	$x + ((x + y) \cdot z) = (x + y) \cdot (x + z)$	$[170 \rightarrow 218 \, {:}171]$
226	$x \cdot y = y \cdot x$	$[218 \rightarrow 220 \, {:}129,176]$
236,235	$x + (y \cdot z) = (x + z) \cdot (x + y)$	$[226 \rightarrow 216 \, {:}223, \text{flip}]$
239	$(A + C) \cdot (A + B) = (A + B) \cdot (A + C) \; \rightarrow \; \square$	$[201 \, {:}236]$
240	$\square$	$[239,226]$

An immediate corollary of the preceding theorem is that the equation $F(x) = x$ (which has length 3183) is a single axiom for Boolean algebra.

Our next approach toward the goal of an independent self-dual 2-basis was to weaken the 6-basis (retaining self-duality) so that when we split it and make it into a pair of equations, they might be independent. Of course, we must prove that adding the Pixley properties and their duals gives us Boolean algebra. Our first attempt was to replace the absorption equations with commutativity of the two operations, and we succeeded. Otter showed that the set

$$\left\{ \begin{array}{ll} x \cdot (y + z) = (x \cdot y) + (x \cdot z) & x + (y \cdot z) = (x + y) \cdot (x + z) \\ x + y = y + x & x \cdot y = y \cdot x \\ x + x' = 1 & x \cdot x' = 0 \\ p(y, y, x) = x & \widetilde{p}(y, y, x) = x \\ p(x, z, z) = x & \widetilde{p}(x, z, z) = x \\ p(x, y, x) = x & \widetilde{p}(x, y, x) = x \end{array} \right\}$$

is a basis for Boolean algebra, and MACE showed that either column by itself is not. Since each column can be replaced with an equivalent single equation, we have an independent self-dual 2-basis. When expanded with a "program" similar to the Otter input file above, each equation has length 4207. We later found that the two commutativity laws can be deleted, giving us equations of length 1103. We give that proof here.

Theorem DUAL-BA-3. A self-dual 2-basis for BA (Pixley reduction).

$$\left\{ \begin{array}{ll} x \cdot (y + z) = (x \cdot y) + (x \cdot z) & x + (y \cdot z) = (x + y) \cdot (x + z) \\ x + x' = 1 & x \cdot x' = 0 \\ (x \cdot y') + (x \cdot x) + (y' \cdot x) = x & (x + y') \cdot (x + x) \cdot (y' + x) = x \\ (x \cdot x') + (x \cdot z) + (x' \cdot z) = z & (x + x') \cdot (x + z) \cdot (x' + z) = z \\ (x \cdot y') + (x \cdot y) + (y' \cdot y) = x & (x + y') \cdot (x + y) \cdot (y' + y) = x \end{array} \right\} .$$

By duality, it is sufficient to derive either of the absorption laws.

(Otter's original proof, which was found with the standard strategy with `max_weight=28` and `max_mem=28000`, is 816 steps and was found in a few hours. We asked Larry Wos to try some of his new methods for shortening proofs by using the resonance strategy [82], and he found the following 99-step proof.)

Proof (found by Otter 3.0.4 on gyro at 8.56 seconds).

1	$(A + B) \cdot B \neq B$	
2	$x \cdot (y + z) = (x \cdot y) + (x \cdot z)$	
3	$x + (y \cdot z) = (x + y) \cdot (x + z)$	
4	$x + x' = 1$	
5	$x \cdot x' = 0$	
7	$x \cdot (y + z) = (x \cdot y) + (x \cdot z)$	
8	$(x \cdot y) + (x \cdot z) = x \cdot (y + z)$	[flip 7]
11,10	$x + (y \cdot z) = (x + y) \cdot (x + z)$	
13,12	$x + x' = 1$	
15,14	$x \cdot x' = 0$	
16	$(x \cdot y') + ((x \cdot x) + (y' \cdot x)) = x$	
17	$((x \cdot y') + ((x \cdot x) + y')) \cdot ((x \cdot y') + ((x \cdot x) + x)) = x$	
		[copy,16 :11,11]
19	$(x \cdot x') + ((x \cdot y) + (x' \cdot y)) = y$	
21,20	$(0 + ((x \cdot y) + x')) \cdot (0 + ((x \cdot y) + y)) = y$	[copy,19 :15,11,11]
22	$(x \cdot y') + ((x \cdot y) + (y' \cdot y)) = x$	
23	$((x \cdot y') + ((x \cdot y) + y')) \cdot ((x \cdot y') + ((x \cdot y) + y)) = x$	
		[copy,22 :11,11]
25	$(x + y') \cdot ((x + x) \cdot (y' + x)) = x$	
27	$(x + x') \cdot ((x + y) \cdot (x' + y)) = y$	
28	$1 \cdot ((x + y) \cdot (x' + y)) = y$	[copy,27 :13]
31,30	$(x + y') \cdot ((x + y) \cdot (y' + y)) = x$	
33,32	$((x \cdot y) + x) \cdot ((x \cdot y) + z) = x \cdot (y + z)$	[8 :11]
35,34	$(x + ((y \cdot z) + y)) \cdot (x + ((y \cdot z) + u)) = (x + y) \cdot (x + (z + u))$	
		[32 → 10 :11, flip]
37,36	$(x + y) \cdot (x + (y' + z)) = (x + (0 + y)) \cdot (x + (0 + z))$	
		[5 → 34 :15, flip]
38	$x \cdot (y + (x' + z)) = ((x \cdot y) + (0 + x)) \cdot ((x \cdot y) + (0 + z))$	
		[32 → 36]
40	$((x \cdot y) + (0 + x)) \cdot ((x \cdot y) + (0 + z)) = x \cdot (y + (x' + z))$	
		[36 → 32]
42,41	$(x + y) \cdot (x + y') = x + 0$	[14 → 10, flip]
44,43	$x \cdot (x' + y) = (0 + x) \cdot (0 + y)$	[14 → 32 :15, flip]
45	$((x + y) \cdot (x + z)) \cdot (x + (y \cdot z)') = x + 0$	[3 → 41]
47	$(x + x) \cdot 1 = x + 0$	[4 → 41]
48	$(x + (y + z)) \cdot (x + (y + z')) = x + (y + 0)$	[41 → 3, flip]
51,50	$(x \cdot y) + 0 = (x \cdot (y + y)) \cdot 1$	[4 → 45 :33, flip]
52	$x + 0 = (x + x) \cdot 1$	[flip 47]
53	$x \cdot (y + x') = (x \cdot (y + y)) \cdot 1$	[32 → 41 :51]
54	$(x \cdot (y + y)) \cdot 1 = x \cdot (y + x')$	[flip 53]
56,55	$(x \cdot (x + x)) \cdot 1 = x \cdot 1$	[4 → 53, flip]
57	$x \cdot ((y \cdot z) + x') = (x \cdot (y \cdot (z + z))) \cdot 1$	[3 → 54 :33, flip]
60	$x \cdot (y + (x' + x')) = (x \cdot y) + (0 + 0)$	[40 → 48]

61	$x + ((y \cdot z) + (0 + 0)) = (x + y) \cdot (x + (z + (y' + y')))$	$[60 \to 3]$
63,62	$(x + (0 + y)) \cdot (x + (0 + (y' + y'))) = x + (0 + (0 + 0))$	
		$[5 \to 61 :37, \text{flip}]$
64	$x' \cdot (x' + (x' + x')) = x'$	$[34 \to 17 :33]$
66	$(x + y') \cdot (x + (y' + (y' + y'))) = x + y'$	$[64 \to 3, \text{flip}]$
68	$1 \cdot (x + (x' + (x' + x'))) = 1$	$[4 \to 66 :13]$
71,70	$(1 \cdot 1) + (0 + (0 + 0)) = 1$	$[38 \to 68 :63]$
72	$(0 + (0 + x')) \cdot (0 + (0 + x')) = x'$	$[14 \to 20 :15]$
74	$0' = (0 + 1) \cdot (0 + 1)$	$[4 \to 72 :13, \text{flip}]$
76	$0 \cdot ((0 + 1) \cdot (0 + 1)) = 0$	$[74 \to 5]$
78	$(x + 0) \cdot ((x + (0 + 1)) \cdot (x + (0 + 1))) = x + 0$	$[76 \to 3 :11, \text{flip}]$
80	$(x + (0 + ((y \cdot z) + y'))) \cdot (x + (0 + ((y \cdot z) + z))) = x + z$	
		$[20 \to 10, \text{flip}]$
83,82	$1 \cdot ((1 \cdot 1) + (0 + 1)) = 1$	$[52 \to 70 :11,11,71]$
84	$(x + 1) \cdot (x + ((1 \cdot 1) + (0 + 1))) = x + 1$	$[82 \to 3, \text{flip}]$
86	$(0 + 1) \cdot (0 + (1 + ((1 \cdot 1) + (0 + 1)))) = (1 \cdot 1) + (0 + 1)$	
		$[82 \to 20 :83,13]$
88	$1 \cdot (x + ((1 \cdot 1) + (0 + 1))) = (1 \cdot x) + 1$	$[32 \to 84]$
89	$x + ((1 \cdot y) + 1) = (x + 1) \cdot (x + (y + ((1 \cdot 1) + (0 + 1))))$	$[88 \to 3]$
90	$(x + 1) \cdot (x + (y + ((1 \cdot 1) + (0 + 1)))) = x + ((1 \cdot y) + 1)$	$[\text{flip } 89]$
92,91	$(1 \cdot 1) + (0 + 1) = 0 + ((1 \cdot 1) + 1)$	$[86 \to 90]$
93	$x' \cdot (x' + (x + x)) = x'$	$[34 \to 23 :33]$
95	$(x + y') \cdot (x + (y' + (y + y))) = x + y'$	$[93 \to 3, \text{flip}]$
97	$1 \cdot (x + (x' + (x + x))) = 1$	$[4 \to 95 :13]$
100,99	$(1 \cdot 1) \cdot ((0 + 1) \cdot (0 + (1 + 1))) = 1 \cdot 1$	$[91 \to 78 :51,56,92,35,51,56]$
102,101	$(0 + ((1 \cdot 1) + 1)) \cdot ((1 \cdot 1) + (0 + (1 + 1))) = 1$	$[38 \to 97 :92]$
103	$(x + (0 + 1)) \cdot (x + (0 + (1 + 1))) = (x + (0 + 1)) \cdot (x + (0 + 1))$	
		$[99 \to 80 :100,13,11,92,102,11, \text{flip}]$
106,105	$(0 + 1) \cdot (0 + (1 + 1)) = (0 + 1) \cdot (0 + 1)$	
		$[99 \to 20 :100,13,11,92,102, \text{flip}]$
108,107	$(0 + 1) \cdot (0 + 1) = 1$	$[91 \to 103 :102,92,92,35,106, \text{flip}]$
109	$(0 + 1) \cdot (0 + (1 + 1)) = 1$	$[105 :108]$
112,111	$0 \cdot 1 = 0$	$[76 :108]$
114,113	$0' = 1$	$[74 :108]$
115	$(x + 0) \cdot (x + 1) = x + 0$	$[111 \to 3, \text{flip}]$
118,117	$0 + 1 = 1$	$[113 \to 4]$
119	$1 \cdot (0 + (1 + 1)) = 1$	$[109 :118]$
122,121	$1 \cdot 1 = 1$	$[107 :118,118]$
124,123	$0 + (1 + 1) = 1 + 1$	$[91 :122,118,122, \text{flip}]$
126,125	$(x + 1) \cdot (x + (1 + 1)) = x + 1$	$[119 \to 3 :124, \text{flip}]$
128,127	$1 + 1 = 1$	$[101 :122,124,122,124,126]$
129	$1 + (0 + (0 + 0)) = 1$	$[70 :122]$
132,131	$x \cdot (0 + x') = (x \cdot 0) \cdot 1$	$[111 \to 57 :128,112]$
134,133	$0 + 0 = 0$	$[111 \to 50 :128,112,112]$

136,135 $1 + 0 = 1$ [129 :134,134]

137     $((x \cdot y) + z') \cdot ((x \cdot (y + y)) \cdot ((z' + x) \cdot (z' + y))) = x \cdot y$

                                                              $[10 \rightarrow 25 :33,11]$

139     $1 \cdot ((x \cdot (y + y)) \cdot (((x \cdot y)' + x) \cdot ((x \cdot y)' + y))) = x \cdot y$ $[4 \rightarrow 137]$

142,141 $1 \cdot 0 = 0$ $[5 \rightarrow 139 :44,42,134,114,15,114,42,136,112,15]$

147     $1 \cdot (1 \cdot (x' + x')) = x'$ $[12 \rightarrow 28]$

150,149 $1' = 0$ $[53 \rightarrow 147 :44,118,132,142,112,112,142, \text{flip}]$

152,151 $(x + 0) \cdot ((x + 1) \cdot 1) = x$ $[149 \rightarrow 30 :150,118]$

154,153 $(x + (y + z')) \cdot ((x + (y + z)) \cdot (x + (z' + z))) = x + y$

                                                              $[30 \rightarrow 10 :11, \text{flip}]$

155     $1 \cdot (((x + y') \cdot ((x + y) \cdot ((y' + y) + (y' + y)))) \cdot (x' + x)) = x$

                                                       $[30 \rightarrow 139 :11,33,31,11,154,31]$

157     $x \cdot (y + x') = (x \cdot y) \cdot 1$ $[151 \rightarrow 57 :11,33,128,152]$

160,159 $x + 0 = x \cdot 1$ $[151 \rightarrow 50 :11,33,128,152]$

162,161 $(x \cdot x) \cdot 1 = x \cdot 1$ $[4 \rightarrow 157, \text{flip}]$

164,163 $(x \cdot 1) \cdot ((x + 1) \cdot 1) = x$ $[151 :160]$

166,165 $(x \cdot 1) \cdot (x + 1) = x \cdot 1$ $[115 :160,160]$

168,167 $(x + x) \cdot 1 = x \cdot 1$ $[52 :160, \text{flip}]$

170,169 $(x + y) \cdot (x + y') = x \cdot 1$ $[41 :160]$

171     $x \cdot (x' \cdot 1) = (0 + x) \cdot 0$ $[159 \rightarrow 43 :160,112]$

173     $(x \cdot 1) \cdot ((x + 1) + y) = (x \cdot 1) \cdot ((x + 1) + y)$

                                                $[165 \rightarrow 32 :166,11,33,128, \text{flip}]$

176,175 $(x \cdot 1) \cdot ((x \cdot 1) \cdot 1) = x$ $[159 \rightarrow 173 :164,160, \text{flip}]$

178,177 $x + x = x$ $[167 \rightarrow 175 :168,176, \text{flip}]$

180,179 $x \cdot x = x$ $[161 \rightarrow 175 :162,176, \text{flip}]$

182,181 $1 \cdot (0 + x) = x$ $[155 :178,31,44,180]$

184,183 $x' \cdot (x' + x) = x'$ $[93 :178]$

188,187 $(x \cdot y) + x = x \cdot (y + x)$ $[32 \rightarrow 179, \text{flip}]$

189     $x' + (x' + x) = x' + x$ $[183 \rightarrow 20 :184,13,118,182]$

192,191 $0 + (0 + x) = 0 + x$ $[189 \rightarrow 38 :44,180,15,15,180, \text{flip}]$

194,193 $0 + x = x$ $[191 \rightarrow 80 :192,21, \text{flip}]$

195     $((x \cdot y) + x') \cdot ((x \cdot y) + y) = y$ $[191 \rightarrow 80 :194,194,194,194]$

198,197 $1 \cdot x = x$ $[191 \rightarrow 181 :194,194]$

201     $(x + y') \cdot (x + y) = x \cdot 1$

                $[191 \rightarrow 153 :194,194,194,194,194,37,194,194,180,160]$

203     $x \cdot (x' \cdot 1) = x \cdot 0$ $[171 :194]$

206,205 $x \cdot 1 = x$ $[4 \rightarrow 201 :178,198, \text{flip}]$

207     $(x + y) \cdot x = x$ $[203 \rightarrow 3 :11,160,206,206,170,206]$

209     $((x + y) \cdot (x + z)) \cdot x = x$ $[3 \rightarrow 207]$

211     $x + 1 = 1$ $[205 \rightarrow 195 :13,206,198]$

214,213 $x \cdot (y + x) = x$ $[211 \rightarrow 2 :206,206,188, \text{flip}]$

219     $x \cdot (y \cdot x) = y \cdot x$ $[195 \rightarrow 209]$

221     $(x + y) \cdot y = y$ $[213 \rightarrow 219 :214]$

223     $\square$ $[221,1]$

For each column of the set listed in the preceding theorem, we can construct an equivalent single equation. The following theorem shows that the left (and therefore the right) column is not a basis for Boolean algebra, so the pair of equations, each of length 1103, is an independent self-dual 2-basis for Boolean algebra.

Theorem DUAL-BA-4. Independence of a Pixley 2-basis for BA.

The following set is not a basis for Boolean algebra.

$$
\left\{
\begin{array}{l}
x \cdot (y + z) = (x \cdot y) + (x \cdot z) \\
x + x' = 1 \\
(x \cdot y') + (x \cdot x) + (y' \cdot x) = x \\
(x \cdot x') + (x \cdot z) + (x' \cdot z) = z \\
(x \cdot y') + (x \cdot y) + (y' \cdot y) = x
\end{array}
\right\} .
$$

The clauses

$$x \cdot (y + z) = (x \cdot y) + (x \cdot z)$$
$$x + x' = 1$$
$$(x \cdot y') + ((x \cdot x) + (y' \cdot x)) = x$$
$$(x \cdot x') + ((x \cdot z) + (x' \cdot z)) = z$$
$$(x \cdot y') + ((x \cdot y) + (y' \cdot y)) = x$$
$$A + A \neq A$$

have the following model (found by MACE 1.2.0 on gyro at 0.03 seconds).

```
 . | 0 1        + | 0 1        '   0 1       A: 1
--+----        --+----        -------
 0 | 0 0        0 | 0 1          1 0
 1 | 0 1        1 | 1 0
```

7.3.3 A 2-Basis from Majority Reduction

While doing the 3-basis work described in the next subsection, we discovered that adding the self-dual equation

$$(x + y') \cdot (x + y) = (x \cdot y') + (x \cdot y) \qquad \text{(SD-cut)}$$

to lattice theory gives us Boolean algebra. This gave us the idea to split this equation in half and use it with the McKenzie lattice theory basis to obtain a 2-basis for Boolean algebra by majority reduction.

Recall (Sec. 6.4.2) that a ternary majority polynomial $m(x, y, z)$ satisfies the equations (we use m here to emphasize the distinction from the Pixley polynomial)

$$m(x, x, y) = m(z, x, x) = m(x, w, x) = x$$

and that a theory that (1) admits a majority polynomial and (2) has a finite basis Σ consisting exclusively of absorption equations is one-based [53]. A single equation can be constructed by first combining the members of Σ into an equivalent (modulo majority properties) pair of absorption equations, $f(y) = y$ and $g(y) = y$, then using the reduction schema (Thm. MAJ-3)

$$m(m(x, y, y), m(x, m(y, z, f(y))), g(y)), u) = y.$$

We use the same majority polynomial we used for lattice theory,

$$m(x, y, z) = (x \cdot z) + (y \cdot (x + z)),$$

along with its dual $\tilde{m}(x, y, z)$.

The McKenzie lattice theory basis $\{$L1, L2, L3, L4$\}$ (p. 137) is self-dual and consists of absorption equations. Splitting (SD-cut) gives us the (stronger) self-dual pair

$$(x + y) \cdot (x + y') = x, \quad \text{(B1)}$$
$$(x \cdot y) + (x \cdot y') = x, \quad \text{(B2)}$$

so we can obtain a self-dual 2-basis for lattice theory by majority reduction from the two sets

$$\left\{ \begin{array}{l} \text{(L1), (L3), (B1),} \\ m(y, y, x) = x \\ m(x, z, z) = x \\ m(x, y, x) = x \end{array} \right\}, \left\{ \begin{array}{l} \text{(L2), (L4), (B2),} \\ \tilde{m}(y, y, x) = x \\ \tilde{m}(x, z, z) = x \\ \tilde{m}(x, y, x) = x \end{array} \right\}.$$

When written in full, each equation has length 91 and 7 variables. We first give an Otter proof that the pair is a basis for BA (note that the majority properties are not necessary), then a MACE model to show its independence.

Theorem DUAL-BA-5. A self-dual 2-basis for BA (majority reduction).

$$\left\{ \begin{array}{ll} y + (x \cdot (y \cdot z)) = y & \text{(L1)}, \quad y \cdot (x + (y + z)) = y \quad \text{(L2)} \\ ((x \cdot y) + (y \cdot z)) + y = y & \text{(L3)}, \quad ((x + y) \cdot (y + z)) \cdot y = y \quad \text{(L4)} \\ (x + y) \cdot (x + y') = x & \text{(B1)}, \quad (x \cdot y) + (x \cdot y') = x \quad \text{(B2)} \end{array} \right\}.$$

Since we already have a basis for lattice theory, we may include associativity and commutativity of the two operations. We derive half of Padmanabhan's 6-basis; the other half follows by duality.

(Our first proof, which was found with the standard strategy with `max_weight=23`, is 119 steps and was found in 80 seconds. While experimenting with the prototype program "eqp", we stumbled across an 18-step proof of distributivity, and we used that proof to guide Otter with the hints strategy [75] to the following short proof.)

Proof (found by Otter 3.0.4 on gyro at 2.51 seconds).

19          $x = x$
20          $(A \cdot B) + (A \cdot C) = A \cdot (B + C), \quad (A + B) \cdot B = B,$
               $B + B' = A + A' \rightarrow \square$

21	$x + (y \cdot (x \cdot z)) = x$	
25	$(x + y) \cdot (x + y') = x$	
27	$x \cdot (y + (x + z)) = x$	
31	$(x \cdot y) + (x \cdot y') = x$	
33	$x + y = y + x$	
34	$x \cdot y = y \cdot x$	
35	$(x + y) + z = x + (y + z)$	
38,37	$(x \cdot y) \cdot z = x \cdot (y \cdot z)$	
51	$x \cdot (y + x) = x$	$[21 \to 27]$
57	$x + (y \cdot x) = x$	$[27 \to 21]$
64,63	$(x + y) \cdot y = y$	$[34 \to 51]$
65	$(A \cdot B) + (A \cdot C) = A \cdot (B + C),\ B + B' = A + A'\ \to\ \square$	
		$[20 :64 :19]$
104	$(x \cdot y) + (y \cdot x') = y$	$[34 \to 31]$
155	$x + ((y \cdot x) + z) = x + z$	$[57 \to 35,\ \text{flip}]$
238	$(x + y) \cdot ((x + y') \cdot z) = x \cdot z$	$[25 \to 37,\ \text{flip}]$
247	$x \cdot (y \cdot z) = y \cdot (z \cdot x)$	$[34 \to 37]$
252	$x \cdot (y \cdot z) = z \cdot (x \cdot y)$	$[\text{flip } 247]$
257	$(x \cdot (y \cdot z)) + (x \cdot (y \cdot z')) = x \cdot y$	$[37 \to 31 :38]$
305	$(x + y) \cdot (z \cdot y) = z \cdot y$	$[51 \to 252,\ \text{flip}]$
362,361	$(x \cdot y) + (z \cdot (x \cdot y')) = (x \cdot y) + (z \cdot x)$	$[257 \to 155,\ \text{flip}]$
408,407	$(x + y) \cdot (z \cdot y') = x \cdot (z \cdot y')$	$[305 \to 238]$
409	$(x \cdot y) + (z \cdot x) = (z + y) \cdot x$	$[305 \to 257 :408,362]$
533,532	$(x \cdot y) + (y \cdot z) = (x + z) \cdot y$	$[33 \to 409]$
554,553	$(x + x') \cdot y = y$	$[104 :533]$
655	$x + x' = y + y'$	$[553 \to 31 :554]$
768	$\square$	$[34 \to 65 :533 :34,655]$

Theorem DUAL-BA-6. Independence of a majority 2-basis for BA.

The following set is not a basis for Boolean algebra.

$$\left\{ \begin{array}{ll} y + (x \cdot (y \cdot z)) = y & \text{(L1)} \\ ((x \cdot y) + (y \cdot z)) + y = y & \text{(L3)} \\ (x + y) \cdot (x + y') = x & \text{(B1)} \\ ((x \cdot z) + x) \cdot (x + z) = x & \\ ((x \cdot x) + y) \cdot (x + x) = x & \\ ((y \cdot x) + x) \cdot (y + x) = x & \end{array} \right\}.$$

The clauses

$$\begin{aligned} &y + (x \cdot (y \cdot z)) = y \\ &((x \cdot y) + (y \cdot z)) + y = y \\ &(x + y) \cdot (x + y') = x \\ &((x \cdot z) + x) \cdot (x + z) = x \\ &((x \cdot x) + y) \cdot (x + x) = x \end{aligned}$$

$$((y \cdot x) + x) \cdot (y + x) = x$$
$$A'' \neq A$$

have the following model (found by MACE 1.2.0 on gyro at 0.03 seconds).

```
+ | 0 1         . | 0 1         '   0 1          A: 0
--+----        --+----        -------
0 | 0 0        0 | 0 1          1 1
1 | 0 1        1 | 1 1
```

For comparison with the preceding model, we present a mathematician's model, found independently, which shows that the set is not a basis for Boolean algebra. Take any lattice with 0 and 1, and define $x' = 0$ for all x. Because we are using a lattice, (L1), (L2), (L3), and (L4) are automatically valid. Moreover, $(x \cdot z) + (y \cdot (x + z))$ is a majority polynomial. Also, $(x + y) \cdot (x + y') = (x + y) \cdot (x + 0) = (x + y) \cdot x = x$, and hence the equation (B1) is also valid. Thus the equation $B(y, x_1, \cdots, x_n) = y$, constructed by majority reduction from (and equivalent to) $\{(L1),(L3),(B1),\text{majority properties}\}$, is valid in this algebra. However, $(x \cdot y) + (x \cdot y') = (x \cdot y) + (x \cdot 0) = (x \cdot y) + 0 = x \cdot y \neq x$, and hence $\widetilde{B}(y, x_1, \cdots, x_n) = y$ is *not* valid.

7.3.4 A 3-Basis from Majority Reduction

A self-dual 3-basis must have either one or three self-dual equations. We searched for a basis with one self-dual equation, focusing mainly on one that gives us distributivity in lattices:

$$(x \cdot y) + (y \cdot z) + (z \cdot x) = (x + y) \cdot (y + z) \cdot (z + x). \quad \text{(SD-dist)}$$

Our first approach was similar to our first 2-basis approach, that is, to use the Pixley polynomial reduction schema to combine equations. It was straightforward to construct a self-dual 3-basis. Consider the three sets

$$\{\text{SD-dist}\}$$

$$\left\{ \begin{array}{l} (x + y) \cdot y = y \\ x + (y + z) = y + (z + x) \\ p(y, y, x) = x \\ p(x, z, z) = x \\ p(x, y, x) = x \end{array} \right\} \left\{ \begin{array}{l} (x \cdot y) + y = y \\ x \cdot (y \cdot z) = y \cdot (z \cdot x) \\ \widetilde{p}(y, y, x) = x \\ \widetilde{p}(x, z, z) = x \\ \widetilde{p}(x, y, x) = x \end{array} \right\}.$$

Otter shows that their union is a basis for Boolean algebra, and each can be made into a single equation with the reduction schema for Pixley polynomials. The resulting 3-basis is self-dual, but Otter also shows that the union of (SD-dist) with either of the other sets is a basis, so the 3-basis is not independent.

We tried replacing the subset $\{(x + y) \cdot y = y, \ x + (y + z) = y + (z + x)\}$ (and its dual) with many other sets (and their duals). Otter shows that each of

$$\{x + y = y + x, \quad (x + y) + z = x + (y + z)\}$$
$$\{x + (y + z) = y + (z + x), \quad x + x' = 1\}$$
$$\{(x + y) + z = x + (y + z), \quad x + x' = 1\}$$

yields Boolean algebra, but we were not able to show dependence or independence of the resulting 3-bases. We then tried replacing the odd equation (SD-dist) with other self-dual equations, including each of

$$x + 0 = x \cdot 1$$
$$(x \cdot y) + (z \cdot (x + y)) = (x + y) \cdot (z + (x \cdot y)) \quad \text{(lattice modularity)}$$
$$(x + y') \cdot (x + y) = (x \cdot y') + (x \cdot y)$$

with various combinations of equations in the Pixley sets; With Otter's help we found that many self-dual 3-bases could be constructed by the Pixley schema reduction, but MACE was not able to show any of them independent.

Our next approach was to use the majority polynomial reduction schema. We tried the following three sets:

$$\{\text{SD-dist}\}$$

$$M = \left\{ \begin{array}{l} y + (x \cdot (y \cdot z)) = y \\ ((x \cdot y) + (y \cdot z)) + y = y \\ (x + x') \cdot y = y \\ m(x, x, y) = x \\ m(y, x, x) = x \\ m(x, y, x) = x \end{array} \right\} \quad \widetilde{M} = \left\{ \begin{array}{l} y \cdot (x + (y + z)) = y \\ ((x + y) \cdot (y + z)) \cdot y = y \\ (x \cdot x') + y = y \\ \widetilde{m}(x, x, y) = x \\ \widetilde{m}(y, x, x) = x \\ \widetilde{m}(x, y, x) = x \end{array} \right\}.$$

Each of the majority sets can be replaced with an equivalent equation (modulo nothing); so, if we can show that the union of the three sets is a basis for Boolean algebra and that no pair of sets is a basis, we will have an independent self-dual 3-basis for Boolean algebra. The rest of the Otter proofs and MACE models in this section accomplish this.

We prove the union to be a basis in two steps, first showing that $x + x' = y + y'$ and $x \cdot x' = y \cdot y'$, which allows us to introduce constants 0 and 1 in the second step. Also, since the McKenzie equations are a basis for lattice theory (Thm. LT-9), we may include associativity of the two operations. Note that we don't need the majority properties to show that the union is a basis for Boolean algebra.

Lemma DUAL-BA-7. Dual BA 3-basis, existence of 0 and 1.

$$\left\{ \begin{array}{l} \{(\text{SD-dist})\} \\ y + (x \cdot (y \cdot z)) = y \\ ((x \cdot y) + (y \cdot z)) + y = y \\ (x + x') \cdot y = y \\ y \cdot (x + (y + z)) = y \\ ((x + y) \cdot (y + z)) \cdot y = y \\ (x \cdot x') + y = y \end{array} \right\} \Rightarrow \left\{ \begin{array}{l} x + x' = y + y' \\ x \cdot x' = y \cdot y' \end{array} \right\}.$$

Proof (found by Otter 3.0.4 on gyro at 2.82 seconds).

2	$B + B' = A + A',\ B \cdot B' = A \cdot A' \ \rightarrow \ \square$	
3	$(x \cdot y) + ((y \cdot z) + (z \cdot x)) = (x + y) \cdot ((y + z) \cdot (z + x))$	
5	$x + (y \cdot (x \cdot z)) = x$	
7	$((x \cdot y) + (y \cdot z)) + y = y$	
10,9	$(x + x') \cdot y = y$	
11	$x \cdot (y + (x + z)) = x$	
15	$(x \cdot x') + y = y$	
22,21	$(x \cdot x')' \cdot y = y$	$[15 \rightarrow 9]$
28,27	$x + (x \cdot y) = x$	$[9 \rightarrow 5]$
37	$x + x = x$	$[9 \rightarrow 7 :28]$
43	$(x \cdot y) + y = y$	$[5 \rightarrow 7]$
49	$x \cdot (y + x) = x$	$[37 \rightarrow 11]$
55	$x \cdot (x + y) = x$	$[43 \rightarrow 11]$
69	$(x \cdot y) \cdot x = x \cdot y$	$[27 \rightarrow 49]$
150,149	$(x \cdot x')' + y = (x \cdot x')'$	$[55 \rightarrow 21,\ \text{flip}]$
152,151	$x + (y \cdot y')' = (y \cdot y')'$	$[49 \rightarrow 21,\ \text{flip}]$
155	$(x \cdot (y \cdot y')') + z = z + x$	$[21 \rightarrow 3 :28,152,150,22,22]$
158	$x + y = (y \cdot (z \cdot z')') + x$	$[\text{flip } 155]$
160,159	$x \cdot (y \cdot y')' = x$	$[21 \rightarrow 69 :22]$
163	$x \cdot (y + y') = x$	$[9 \rightarrow 69 :10]$
165	$x + y = y + x$	$[158 :160]$
182	$x + (y \cdot y') = x$	$[15 \rightarrow 165,\ \text{flip}]$
286	$x + x' = y + y'$	$[9 \rightarrow 163]$
293	$x \cdot x' = y \cdot y'$	$[15 \rightarrow 182]$
726	$\square$	$[2,286,293]$

Theorem DUAL-BA-8. A self-dual 3-basis for BA (majority reduction).

$$
\left\{
\begin{array}{l}
\text{(SD-dist)}\\
y + (x \cdot (y \cdot z)) = y\\
((x \cdot y) + (y \cdot z)) + y = y\\
(x + x') \cdot y = y\\
y \cdot (x + (y + z)) = y\\
((x + y) \cdot (y + z)) \cdot y = y\\
(x \cdot x') + y = y\\
x + x' = 1\\
x \cdot x' = 0\\
(x + y) + z = x + (y + z)\\
(x \cdot y) \cdot z = x \cdot (y \cdot z)
\end{array}
\right\}
\Rightarrow
\left\{
\begin{array}{l}
(x \cdot y) + (x \cdot z) = x \cdot (y + z)\\
(x \cdot y) + y = y
\end{array}
\right\}.
$$

Proof (found by Otter 3.0.4 on gyro at 24.59 seconds).

1	$x = x$
2	$(A \cdot B) + (A \cdot C) = A \cdot (B + C),\ (A \cdot B) + B = B \ \rightarrow \ \square$

3	$(x \cdot y) + ((y \cdot z) + (z \cdot x)) = (x + y) \cdot ((y + z) \cdot (z + x))$	
5	$x + (y \cdot (x \cdot z)) = x$	
7	$((x \cdot y) + (y \cdot z)) + y = y$	
9	$(x + x') \cdot y = y$	
11	$x \cdot (y + (x + z)) = x$	
15	$(x \cdot x') + y = y$	
18,17	$x + x' = 1$	
20,19	$x \cdot x' = 0$	
22,21	$(x + y) + z = x + (y + z)$	
24,23	$(x \cdot y) \cdot z = x \cdot (y \cdot z)$	
26,25	$1 \cdot x = x$	$[9 :18]$
28,27	$0 + x = x$	$[15 :20]$
29	$(x \cdot y) + ((y \cdot z) + y) = y$	$[7 :22]$
35	$(x \cdot 1) + (y + (y \cdot x)) = (x + 1) \cdot ((1 + y) \cdot (y + x))$	$[25 \to 3]$
39	$0' = 1$	$[27 \to 17]$
41	$1 + (x \cdot y) = 1$	$[25 \to 5]$
44,43	$x + (x \cdot y) = x$	$[25 \to 5]$
45	$x \cdot (0 \cdot y) = 0$	$[27 \to 5]$
47	$(x \cdot 1) + y = (x + 1) \cdot ((1 + y) \cdot (y + x))$	$[35 :44]$
49	$0 \cdot 1 = 0$	$[39 \to 19]$
51	$1' = 0$	$[25 \to 19]$
57	$(x \cdot y) + (y' \cdot x) = (x + y) \cdot (y' + x)$	$[19 \to 3 :28,18,26]$
58	$(x' \cdot y) + (y \cdot x) = (x' + y) \cdot (y + x)$	$[19 \to 3 :28,18,26]$
60	$(x + y) \cdot (y' + x) = (x \cdot y) + (y' \cdot x)$	$[\text{flip } 57]$
61	$1 + 0 = 1$	$[51 \to 17]$
67	$(1 + x) \cdot (x + 0) = x$	$[49 \to 3 :26,44,28,28,26, \text{flip}]$
70,69	$x + 1 = 1$	$[61 \to 11 :26]$
74,73	$x \cdot 1 = x$	$[17 \to 11 :70]$
75	$x \cdot (y + x) = x$	$[5 \to 11]$
80,79	$x \cdot (x + y) = x$	$[27 \to 11]$
86	$x + y = (1 + y) \cdot (y + x)$	$[47 :74,70,26]$
93	$x + (y \cdot x) = x$	$[11 \to 5]$
96,95	$1 + x = 1$	$[73 \to 41]$
99	$x + y = y + x$	$[86 :96,26]$
101,100	$x + 0 = x$	$[67 :96,26]$
106	$x + ((y \cdot (x \cdot z)) + u) = x + u$	$[5 \to 21, \text{flip}]$
117,116	$x + x = x$	$[73 \to 43]$
119,118	$0 \cdot x = 0$	$[27 \to 43]$
123,122	$x \cdot 0 = 0$	$[45 :119]$
125,124	$x + ((x \cdot y) + z) = x + z$	$[43 \to 21, \text{flip}]$
134	$x \cdot y = y \cdot x$	$[118 \to 3 :123,28,28,101,28,80]$
137	$x \cdot (x' \cdot y) = 0$	$[19 \to 23 :119, \text{flip}]$
152,151	$x \cdot x = x$	$[116 \to 75]$
160,159	$x \cdot ((y + x) \cdot z) = x \cdot z$	$[75 \to 23, \text{flip}]$

161	$(x \cdot y) + ((y \cdot x) + x) = (x + y) \cdot ((y + x) \cdot x)$	$[151 \to 3 : 117]$
164,163	$(x \cdot y) + y = (x + y) \cdot y$	$[151 \to 3 : 44,117,80]$
165	$x \cdot (x \cdot y) = x \cdot y$	$[151 \to 23, \text{flip}]$
169	$(x \cdot y) + ((y + x) \cdot x) = (x + y) \cdot ((y + x) \cdot x)$	$[161 : 164]$
171	$(A \cdot B) + (A \cdot C) = A \cdot (B + C), \ (A + B) \cdot B = B \ \to \ \Box$	$[2 : 164]$
174	$x \cdot (y \cdot ((y + z) \cdot (z + x))) = x \cdot y$	$[3 \to 79 : 24,160]$
181,180	$x \cdot ((x + y) \cdot z) = x \cdot z$	$[79 \to 23, \text{flip}]$
183,182	$x \cdot (y \cdot (z + x)) = x \cdot y$	$[174 : 181]$
185,184	$(x \cdot y) + x = x$	$[118 \to 29 : 28]$
187,186	$x + (y + x) = y + x$	$[75 \to 29 : 185]$
193,192	$(x + y) \cdot y = y$	$[151 \to 29 : 117,164]$
196	$(A \cdot B) + (A \cdot C) = A \cdot (B + C) \ \to \ \Box$	$[171 : 193 : 1]$
197	$(x + y) \cdot x = x$	$[169 : 193,185,193, \text{flip}]$
200,199	$(x \cdot y) + y = y$	$[163 : 193]$
204	$x + (y + (z \cdot (x + y))) = x + y$	$[21 \to 93]$
206	$x + ((y \cdot x) + z) = x + z$	$[93 \to 21, \text{flip}]$
208	$x + (y + z) = z + (x + y)$	$[21 \to 99]$
210,209	$x' + x = 1$	$[17 \to 99, \text{flip}]$
211	$(x \cdot (y \cdot z)) + y = y$	$[5 \to 99, \text{flip}]$
214	$x + (y + z) = y + (z + x)$	$[\text{flip } 208]$
221	$x \cdot (y \cdot z) = z \cdot (x \cdot y)$	$[23 \to 134]$
224	$(x + (y + z)) \cdot y = y$	$[11 \to 134, \text{flip}]$
227	$(x \cdot y) + ((y \cdot z) + (x \cdot z)) = (x + y) \cdot ((y + z) \cdot (z + x))$	$[134 \to 3]$
239	$x'' + x = x'' \cdot x$	$[19 \to 57 : 28,18,26, \text{flip}]$
243	$(x \cdot y) + (x \cdot y') = (x + y) \cdot (y' + x)$	$[134 \to 57]$
246	$((x + y') \cdot y) + y' = x + y'$	$[75 \to 57 : 22,210,70,187,26]$
249	$(x + y) \cdot (y' + x) = (x \cdot y) + (x \cdot y')$	$[\text{flip } 243]$
281,280	$x + (y \cdot (z + x)) = (z + x) \cdot (x + y)$	$[192 \to 3 : 125,22,117,183]$
287,286	$x + ((x + y) \cdot (y + z)) = x + y$	$[204 : 281]$
322	$(x \cdot y) + (y' \cdot x) = (y' + x) \cdot (x + y)$	$[99 \to 58]$
338	$x \cdot (y \cdot x') = 0$	$[134 \to 137]$
340	$x' \cdot (y \cdot x) = 0$	$[134 \to 137 : 24]$
376,375	$(x \cdot y) + (y' \cdot x) = (x + y) \cdot (x + y')$	$[99 \to 60, \text{flip}]$
386	$(x + y) \cdot (x + y') = (y' + x) \cdot (x + y)$	$[322 : 376]$
388	$(x + y) \cdot (x + y') = (x + y) \cdot (y' + x)$	$[57 : 376]$
491	$(x' + (y \cdot x)) \cdot x = y \cdot x$	$[340 \to 58 : 24,152,28,200, \text{flip}]$
734	$(x \cdot y) + ((x \cdot (y \cdot z)) + (z \cdot x)) = x \cdot ((x \cdot y) + z)$	
		$[165 \to 3 : 24,44,183]$
812,811	$(x + y) \cdot (z \cdot (y \cdot u)) = z \cdot (y \cdot u)$	$[211 \to 224]$
824,823	$(x + y) \cdot (z \cdot (x \cdot u)) = z \cdot (x \cdot u)$	$[106 \to 224]$
828,827	$(x + y) \cdot (z \cdot x) = z \cdot x$	$[3 \to 224 : 24,24,824,812]$
1108,1107	$(x \cdot y) + ((x \cdot (y \cdot z)) + u) = (x \cdot y) + u$	$[23 \to 124]$
1124,1123	$x' + (y \cdot x) = x' + y$	$[58 \to 124 : 287, \text{flip}]$
1130,1129	$(x \cdot y) + (z \cdot x) = x \cdot ((x \cdot y) + z)$	$[734 : 1108]$

1133	$(x' + y) \cdot x = y \cdot x$	[491 :1124]
1139	$x \cdot ((x \cdot y) + y') = (x + y) \cdot (x + y')$	[375 :1130]
1796,1795	$(x + y') \cdot y = x \cdot y$	[99 $\rightarrow$ 1133]
1798,1797	$x'' \cdot x = x$	[17 $\rightarrow$ 1133 :26, flip]
1806,1805	$(x \cdot y) + y' = x + y'$	[246 :1796]
1808,1807	$x'' + x = x$	[239 :1798]
1810,1809	$(x + y) \cdot (x + y') = x$	[1139 :1806,80, flip]
1812,1811	$(x + y) \cdot (y' + x) = x$	[388 :1810, flip]
1813	$(x' + y) \cdot (y + x) = y$	[386 :1810, flip]
1819	$(x \cdot y) + (x \cdot y') = x$	[249 :1812, flip]
1973	$x'' = x$	[1797 $\rightarrow$ 43 :1808, flip]
1976,1975	$(x + y) \cdot x' = y \cdot x'$	[1973 $\rightarrow$ 1133]
2749	$(x' + y) \cdot (x + y) = y$	[99 $\rightarrow$ 1813]
2879,2878	$x + (y \cdot x') = x + y$	[197 $\rightarrow$ 1819 :1976]
3083	$A \cdot ((A \cdot B) + C) = A \cdot (B + C) \rightarrow \square$	[134 $\rightarrow$ 196 :1130]
4371	$(x' + y) \cdot (x + ((z \cdot x') + y)) = (z \cdot x') + y$	[206 $\rightarrow$ 2749]
4412,4411	$x + (y \cdot (z \cdot x')) = x + (y \cdot z)$	[23 $\rightarrow$ 2878]
4429,4428	$x + ((y \cdot x') + z) = x + (y + z)$	[2878 $\rightarrow$ 21 :22, flip]
4432,4431	$(x \cdot y') + z = (y' + z) \cdot (y + (x + z))$	[4371 :4429, flip]
4617,4616	$(x + y) \cdot (z \cdot (x + (y + u))) = (x + y) \cdot z$	[214 $\rightarrow$ 182]
4739,4738	$(x + y) \cdot (z \cdot (y' + x)) = z \cdot x$	[1813 $\rightarrow$ 221, flip]
4862,4861	$(x \cdot y) + (x \cdot z) = x \cdot ((x \cdot y) + z)$	[165 $\rightarrow$ 227 :24,1108,44,183]
4885	$x \cdot ((x \cdot y) + (z \cdot y')) = (y + z) \cdot x$	
		[338 $\rightarrow$ 227 :28,4862,2879,4432,4617,4739]
4908,4907	$x \cdot ((x \cdot y) + z) = (y + (x \cdot z)) \cdot x$	[165 $\rightarrow$ 227 :200,4862,185,828]
4968,4967	$(x + (y \cdot z)) \cdot y = (x + z) \cdot y$	[4885 :4908,4412]
4971	$(B + C) \cdot A = A \cdot (B + C) \rightarrow \square$	[3083 :4908,4968]
4972	$\square$	[4971,134]

Example DUAL-BA-9. Dual BA 3-basis, independence (1).

{SD-dist} is independent of $M \cup \widetilde{M}$. We ask MACE to find a model of $M \cup \widetilde{M}$ that is not a Boolean algebra.

The clauses

$$y + (x \cdot (y \cdot z)) = y$$
$$((x \cdot y) + (y \cdot z)) + y = y$$
$$(x + x') \cdot y = y$$
$$y \cdot (x + (y + z)) = y$$
$$((x + y) \cdot (y + z)) \cdot y = y$$
$$(x \cdot x') + y = y$$
$$((x + z) \cdot x) + (x \cdot z) = x$$
$$((x + x) \cdot y) + (x \cdot x) = x$$
$$((y + x) \cdot x) + (y \cdot x) = x$$
$$((x \cdot z) + x) \cdot (x + z) = x$$
$$((x \cdot x) + y) \cdot (x + x) = x$$

$$((y \cdot x) + x) \cdot (y + x) = x$$
$$A'' \neq A$$

have the following model (found by MACE 1.2.0 on gyro at 2.74 seconds).

```
+ | 0 1 2 3 4            . | 0 1 2 3 4
--+----------           --+----------
0 | 0 0 0 0 0           0 | 0 1 2 3 4
1 | 0 1 2 3 4           1 | 1 1 1 1 1
2 | 0 2 2 0 0           2 | 2 1 2 1 1
3 | 0 3 0 3 0           3 | 3 1 1 3 1
4 | 0 4 0 0 4           4 | 4 1 1 1 4


'   0 1 2 3 4           A: 4
  -------------
    1 0 3 2 2
```

Example DUAL-BA-10. Dual BA 3-basis, independence (2).

The set $\widetilde{M}$ is independent of $M \cup \{\text{SD-dist}\}$. We ask MACE to find a model of $M \cup \{\text{SD-dist}\}$ that is not a Boolean algebra.

The clauses

$$(x \cdot y) + ((y \cdot z) + (z \cdot x)) = (x + y) \cdot ((y + z) \cdot (z + x))$$
$$y + (x \cdot (y \cdot z)) = y$$
$$((x \cdot y) + (y \cdot z)) + y = y$$
$$(x + x') \cdot y = y$$
$$((x \cdot z) + x) \cdot (x + z) = x$$
$$((x \cdot x) + y) \cdot (x + x) = x$$
$$((y \cdot x) + x) \cdot (y + x) = x$$
$$A'' \neq A$$

have the following model (found by MACE 1.2.0 on gyro at 0.63 seconds).

```
+ | 0 1        . | 0 1        '   0 1        A: 1
--+----        --+----        --------
0 | 0 0        0 | 0 1          0 0
1 | 0 1        1 | 1 1
```

8. Miscellaneous Topics

This chapter contains new bases for inverse and Moufang loops, some previously known theorems on other subvarieties of quasigroups, and several previously known theorems on algebras of set difference.

8.1 Inverse Loops and Moufang Loops

A quasigroup can be defined as a set with a binary operation, having unique left and right solutions, that is, satisfying the laws $\forall x \forall y \exists! z, z \cdot x = y$ and $\forall x \forall y \exists! z, x \cdot z = y$. However, we define quasigroups as algebras of type $\langle 2, 2, 2 \rangle$ satisfying the four equations

$$x \cdot (x \backslash y) = y, \quad (x/y) \cdot y = x,$$
$$x \backslash (x \cdot y) = y, \quad (x \cdot y)/y = x,$$

in which $x \cdot y$ is the quasigroup operation, $x \backslash y$ is the right solution of x and y (i.e., the unique element z such that $x \cdot z = y$), and x/y is the left solution of x and y. (In fact, if $\langle S; \cdot \rangle$ is a quasigroup, then $\langle S; \backslash \rangle$ and $\langle S; / \rangle$ are quasigroups as well.)

A *loop* is a quasigroup containing an element, say 1, that is a left and right identity: $1 \cdot x = x \cdot 1 = x$. If we define $L(x)$ as $1/x$ and $R(x)$ as $x \backslash 1$, then it follows immediately that all elements (of loops) have unique left and right inverses, $L(x) \cdot x = x \cdot R(x) = 1$. We can therefore use the following set as an equational basis for the variety of loops,

$$x \cdot (x \backslash y) = y \quad (x/y) \cdot y = x,$$
$$x \backslash (x \cdot y) = y \quad (x \cdot y)/y = x,$$
$$1 \cdot x = x \quad x \cdot 1 = x,$$
$$L(x) \cdot x = 1 \quad x \cdot R(x) = 1,$$

where the equations for left and right inverse are optional. An *inverse loop* is a loop in which $L(x) = R(x)$. An independent basis for inverse loops is the set

$$x' \cdot x = 1,$$
$$x' \cdot (x \cdot y) = y,$$
$$(x \cdot y) \cdot y' = x,$$

in which $'$ denotes inverse. The quasigroup properties follow, and the set is easily proved independent with three MACE runs.

8.1.1 Bases for Moufang Loops

A *Moufang loop* [8] is a loop that satisfies *any* of the Moufang equations:[1]

$$(x \cdot (y \cdot z)) \cdot x = (x \cdot y) \cdot (z \cdot x), \qquad \text{(Moufang-1)}$$
$$((x \cdot y) \cdot z) \cdot y = x \cdot (y \cdot (z \cdot y)), \qquad \text{(Moufang-2)}$$
$$((x \cdot y) \cdot x) \cdot z = x \cdot (y \cdot (x \cdot z)). \qquad \text{(Moufang-3)}$$

That is, given a loop, the three Moufang equations are equivalent. We show this in the following three theorems as

$$\text{Moufang-1} \Rightarrow \text{Moufang-2} \Rightarrow \text{Moufang-3} \Rightarrow \text{Moufang-1}.$$

Theorem MFL-1. Moufang-1 $\Rightarrow$ Moufang-2 in loops.

$$\left\{ \begin{array}{l} \text{loop} \\ (x \cdot (y \cdot z)) \cdot x = (x \cdot y) \cdot (z \cdot x) \end{array} \right\} \Rightarrow \{((x \cdot y) \cdot z) \cdot y = x \cdot (y \cdot (z \cdot y))\}.$$

Proof (found by Otter 3.0.4 on gyro at 11.55 seconds).

3,2	$1 \cdot x = x$	
5,4	$x \cdot 1 = x$	
6	$x \cdot (x \backslash y) = y$	
8	$x \backslash (x \cdot y) = y$	
10	$(x/y) \cdot y = x$	
13,12	$(x \cdot y)/y = x$	
14	$x \cdot R(x) = 1$	
18	$(x \cdot (y \cdot z)) \cdot x = (x \cdot y) \cdot (z \cdot x)$	
20	$((A \cdot B) \cdot C) \cdot B \neq A \cdot (B \cdot (C \cdot B))$	
30,29	$x \backslash 1 = R(x)$	$[14 \to 8]$
32,31	$x \backslash x = 1$	$[4 \to 8]$
51	$(x \cdot (y/z)) \cdot (z \cdot x) = (x \cdot y) \cdot x$	$[10 \to 18, \text{flip}]$
53	$(x \cdot y) \cdot x = (x \cdot z) \cdot ((z \backslash y) \cdot x)$	$[6 \to 18]$
55,54	$(x \cdot y) \cdot x = x \cdot (y \cdot x)$	$[2 \to 18 :5]$
56	$(x \cdot y) \cdot ((y \backslash z) \cdot x) = x \cdot (z \cdot x)$	$[\text{flip } 53 :55]$
58	$(x \cdot (y/z)) \cdot (z \cdot x) = x \cdot (y \cdot x)$	$[51 :55]$
60	$x \cdot ((y \cdot z) \cdot x) = (x \cdot y) \cdot (z \cdot x)$	$[18 :55]$
64	$x \cdot (R(x) \cdot x) = x$	$[14 \to 54 :3, \text{flip}]$
68	$x \cdot ((x \backslash y) \cdot x) = y \cdot x$	$[6 \to 54, \text{flip}]$
70	$(x \cdot (y \cdot x))/x = x \cdot y$	$[54 \to 12]$

---

[1] K. Kunen points out that Moufang-2 and Moufang-3 are duals (mirror images after renaming variables), so it would be natural to also consider the dual of Moufang-1 as well; it is obviously equivalent (modulo loop axioms) to the other three.

78	$R(x) \cdot x = 1$	$[64 \to 8 : 32, \text{flip}]$
85,84	$R(R(x)) = x$	$[78 \to 8 : 30]$
104	$x \backslash (y \cdot x) = (x \backslash y) \cdot x$	$[68 \to 8]$
106	$(x \backslash y) \cdot R(x) = R(x) \cdot (y \cdot R(x))$	$[78 \to 56 : 3]$
119,118	$(x \cdot y)/x = x \cdot (y/x)$	$[10 \to 70]$
149,148	$R(x) \cdot (y/x) = R(x) \cdot (y \cdot R(x))$	$[14 \to 58 : 5]$
150	$(x \cdot (y/(z/x))) \cdot z = x \cdot (y \cdot x)$	$[10 \to 58]$
188	$x \backslash ((x \cdot y) \cdot (z \cdot x)) = (y \cdot z) \cdot x$	$[60 \to 8]$
236	$(x \backslash (y/x)) \cdot x = x \backslash y$	$[10 \to 104, \text{flip}]$
255	$x \cdot ((x \backslash y)/x) = y/x$	$[6 \to 118, \text{flip}]$
286	$x \backslash (y/x) = (x \backslash y)/x$	$[236 \to 12, \text{flip}]$
493,492	$x \backslash y = R(x) \cdot y$	$[106 \to 12 : 119,13, \text{flip}]$
551,550	$(R(x) \cdot y)/x = R(x) \cdot (y \cdot R(x))$	$[286 : 493,149,493, \text{flip}]$
559,558	$x/y = y \cdot (R(y) \cdot (x \cdot R(y)))$	$[255 : 493,551, \text{flip}]$
566	$R(x) \cdot ((x \cdot y) \cdot (z \cdot x)) = (y \cdot z) \cdot x$	$[188 : 493]$
585,584	$R(x) \cdot (x \cdot y) = y$	$[8 : 493]$
587,586	$x \cdot (R(x) \cdot y) = y$	$[6 : 493]$
630	$(x \cdot (y \cdot R(z \cdot R(x)))) \cdot z = x \cdot (y \cdot x)$	$[150 : 559,587,559,587]$
634	$(x \cdot R(y)) \cdot y = x$	$[10 : 559,587]$
658	$(x \cdot y) \cdot R(y) = x$	$[84 \to 634]$
672	$x \cdot R(y \cdot x) = R(y)$	$[584 \to 658]$
687,686	$R(x \cdot y) = R(y) \cdot R(x)$	$[658 \to 672, \text{flip}]$
692	$(x \cdot (y \cdot (x \cdot R(z)))) \cdot z = x \cdot (y \cdot x)$	$[630 : 687,85]$
946	$(x \cdot (y \cdot x)) \cdot z = x \cdot (y \cdot (x \cdot z))$	$[692 \to 634 : 85]$
954	$((x \cdot y) \cdot z) \cdot y = x \cdot (y \cdot (z \cdot y))$	$[946 \to 566 : 585, \text{flip}]$
956	$\square$	$[954,20]$

Theorem MFL-2. Moufang-2 $\Rightarrow$ Moufang-3 in loops.

$$\left\{ \begin{array}{l} \text{loop} \\ ((x \cdot y) \cdot z) \cdot y = x \cdot (y \cdot (z \cdot y)) \end{array} \right\} \Rightarrow \{((x \cdot y) \cdot x) \cdot z = x \cdot (y \cdot (x \cdot z))\}.$$

Proof (found by Otter 3.0.4 on gyro at 13.45 seconds).

3,2	$1 \cdot x = x$	
4	$x \cdot 1 = x$	
8	$x \backslash (x \cdot y) = y$	
11,10	$(x/y) \cdot y = x$	
12	$(x \cdot y)/y = x$	
14	$x \cdot R(x) = 1$	
18	$((x \cdot y) \cdot z) \cdot y = x \cdot (y \cdot (z \cdot y))$	
20	$((A \cdot B) \cdot A) \cdot C \neq A \cdot (B \cdot (A \cdot C))$	
30,29	$x \backslash 1 = R(x)$	$[14 \to 8]$
32,31	$x \backslash x = 1$	$[4 \to 8]$
37	$(x/y) \backslash x = y$	$[10 \to 8]$
49	$x \cdot (R(x) \cdot (y \cdot R(x))) = y \cdot R(x)$	$[14 \to 18 : 3, \text{flip}]$

53,52	$(x \cdot y) \cdot x = x \cdot (y \cdot x)$	$[2 \to 18 :3]$
54	$x \cdot (y \cdot (R(x \cdot y) \cdot y)) = y$	$[14 \to 18 :3, \text{flip}]$
61	$(A \cdot (B \cdot A)) \cdot C \neq A \cdot (B \cdot (A \cdot C))$	$[20 :53]$
68	$x \cdot (R(x) \cdot x) = x$	$[14 \to 52 :3, \text{flip}]$
77,76	$(x \cdot (y \cdot x))/x = x \cdot y$	$[52 \to 12]$
84	$R(x) \cdot x = 1$	$[68 \to 8 :32, \text{flip}]$
93,92	$R(R(x)) = x$	$[84 \to 8 :30]$
96	$x \cdot (R(x) \cdot y) = y$	$[10 \to 49 :11]$
98	$R(x) \cdot (x \cdot y) = y$	$[92 \to 96]$
102	$x \cdot (y \cdot ((R(x \cdot y) \cdot z) \cdot y)) = z \cdot y$	$[96 \to 18, \text{flip}]$
109,108	$x \backslash y = R(x) \cdot y$	$[96 \to 8]$
114	$R(x/y) \cdot x = y$	$[37 :109]$
118	$x/(y \cdot x) = R(y)$	$[98 \to 12]$
136	$(x \cdot y) \cdot z = R(z/x) \cdot (z \cdot (y \cdot z))$	$[114 \to 18]$
138	$R(x/y) \cdot (x \cdot (z \cdot x)) = (y \cdot z) \cdot x$	$[\text{flip } 136]$
154	$x \cdot R(y \cdot x) = R(y)$	$[54 \to 118 :77]$
165,164	$x/y = x \cdot R(y)$	$[114 \to 154 :93, \text{flip}]$
172,171	$R(x \cdot R(y)) = y \cdot R(x)$	$[10 \to 154 :165, \text{flip}]$
179	$(x \cdot R(y)) \cdot (y \cdot (z \cdot y)) = (x \cdot z) \cdot y$	$[138 :165,172]$
203,202	$R(x \cdot y) = R(y) \cdot R(x)$	$[154 \to 98, \text{flip}]$
204	$x \cdot (y \cdot (((R(y) \cdot R(x)) \cdot z) \cdot y)) = z \cdot y$	$[102 :203]$
236	$(x \cdot y) \cdot ((R(y) \cdot R(x)) \cdot z) = z$	$[202 \to 96]$
374	$(x \cdot y) \cdot ((R(y) \cdot z) \cdot x) = x \cdot (z \cdot x)$	$[179 \to 236]$
463,462	$x \cdot (((R(x) \cdot y) \cdot z) \cdot x) = y \cdot (z \cdot x)$	$[204 \to 96 :93, \text{flip}]$
506	$(x \cdot (y \cdot x)) \cdot z = x \cdot (y \cdot (x \cdot z))$	$[374 \to 18 :463]$
508	$\square$	$[506,61]$

Theorem MFL-3. Moufang-3 $\Rightarrow$ Moufang-1 in loops.
$$\left\{ \begin{array}{l} \text{loop} \\ ((x \cdot y) \cdot x) \cdot z = x \cdot (y \cdot (x \cdot z)) \end{array} \right\} \Rightarrow \{(x \cdot (y \cdot z)) \cdot x = (x \cdot y) \cdot (z \cdot x)\}.$$

Proof (found by Otter 3.0.4 on gyro at 14.58 seconds).

3,2	$1 \cdot x = x$	
5,4	$x \cdot 1 = x$	
7,6	$x \cdot (x \backslash y) = y$	
12	$(x \cdot y)/y = x$	
14	$x \cdot R(x) = 1$	
16	$L(x) \cdot x = 1$	
18	$((x \cdot y) \cdot x) \cdot z = x \cdot (y \cdot (x \cdot z))$	
20	$(A \cdot (B \cdot C)) \cdot A \neq (A \cdot B) \cdot (C \cdot A)$	
40,39	$1/x = L(x)$	$[16 \to 12]$
41	$L(R(x)) = x$	$[14 \to 12 :40]$
47	$L(x) \cdot (x \cdot (L(x) \cdot y)) = L(x) \cdot y$	$[16 \to 18 :3, \text{flip}]$
60,59	$(x \cdot y) \cdot x = x \cdot (y \cdot x)$	$[4 \to 18 :5]$

65	$A \cdot ((B \cdot C) \cdot A) \neq (A \cdot B) \cdot (C \cdot A)$	[20 :60]
66	$(x \cdot (y \cdot x)) \cdot z = x \cdot (y \cdot (x \cdot z))$	[18 :60]
92	$L(x) \cdot (x \cdot y) = y$	[6 → 47 :7]
95,94	$x \cdot (R(x) \cdot y) = y$	[41 → 92]
102	$L(L(x)) = x$	[16 → 92 :5]
105,104	$L(x) = R(x)$	[14 → 92 :5]
111,110	$R(x) \cdot x = 1$	[4 → 92 :105]
115,114	$R(R(x)) = x$	[102 :105,105]
120	$R(x) \cdot (x \cdot y) = y$	[92 :105]
136	$x \cdot ((R(x) \cdot y) \cdot x) = y \cdot x$	[94 → 59, flip]
156	$x/(y \cdot x) = R(y)$	[120 → 12]
190	$(x \cdot (y \cdot (x \cdot z)))/z = x \cdot (y \cdot x)$	[66 → 12]
204	$(x \cdot R(y)) \cdot y = x$	[66 → 136 :111,5,95, flip]
222	$(x \cdot y) \cdot R(y) = x$	[114 → 204]
233,232	$x/y = R(y \cdot R(x))$	[204 → 156]
238,237	$R(x \cdot R(y)) = y \cdot R(x)$	[204 → 12 :233]
243	$(x \cdot (y \cdot (x \cdot z))) \cdot R(z) = x \cdot (y \cdot x)$	[190 :233,238]
256	$x \cdot R(y \cdot x) = R(y)$	[120 → 222]
267,266	$R(x \cdot y) = R(y) \cdot R(x)$	[222 → 256, flip]
440	$(x \cdot (y \cdot z)) \cdot (R(z) \cdot x) = x \cdot (y \cdot x)$	[94 → 243 :267,115]
552	$(x \cdot y) \cdot (z \cdot x) = x \cdot ((y \cdot z) \cdot x)$	[222 → 440 :115]
559	$x \cdot ((y \cdot z) \cdot x) = (x \cdot y) \cdot (z \cdot x)$	[flip 552]
560	□	[559,65]

In the preceding proof we see that Moufang loops are in fact inverse loops (i.e., $L(x) = R(x)$), and we henceforth write the inverse of x as x'.

It turns out that we can simplify our equational basis for Moufang loops, but Moufang-1, Moufang-2, and Moufang-3 are no longer "equivalent" in the simplified basis. In particular, the set

$$\left\{ \begin{array}{l} 1 \cdot x = x \\ x' \cdot x = 1 \\ ((x \cdot y) \cdot z) \cdot y = x \cdot (y \cdot (z \cdot y)) \quad \text{(Moufang-2)} \end{array} \right\}$$

is a basis for Moufang loops, and we can replace Moufang-2 with Moufang-3, but we cannot replace Moufang-2 with Moufang-1.

To prove that the preceding set is a basis, we show that 1 is also a right identity, that x' is also a right inverse of x, and that $\cdot$ is a quasigroup (by showing that left and right solutions exist and that the left and right cancellation laws hold). The following three theorems accomplish this task.

Theorem MFL-4. Simple basis with Moufang-2 (part 1).

$$\left\{ \begin{array}{l} 1 \cdot x = x \\ x' \cdot x = 1 \\ \text{Moufang-2} \end{array} \right\} \Rightarrow \left\{ \begin{array}{l} x \cdot 1 = x \\ x \cdot x' = 1 \\ \forall x \forall y \exists z \ (x \cdot z = y) \\ \forall x \forall y \exists z \ (z \cdot x = y) \end{array} \right\}.$$

Proof (found by Otter 3.0.4 on gyro at 7.45 seconds).

1	$x = x$	
3,2	$1 \cdot x = x$	
5,4	$x' \cdot x = 1$	
7,6	$((x \cdot y) \cdot z) \cdot y = x \cdot (y \cdot (z \cdot y))$	
8	$A \cdot 1 = A, \ B \cdot B' = 1, \ C \cdot x = D, \ y \cdot E = F \ \rightarrow \ \square$	
9	$x' \cdot (x \cdot (y \cdot x)) = y \cdot x$	$[4 \rightarrow 6 :3, \text{flip}]$
12,11	$(x \cdot y) \cdot x = x \cdot (y \cdot x)$	$[2 \rightarrow 6 :3]$
13	$(x \cdot (y \cdot (z \cdot y))) \cdot z = (x \cdot y) \cdot (z \cdot (y \cdot z))$	$[6 \rightarrow 6]$
17	$x' \cdot (x \cdot x') = x'$	$[4 \rightarrow 11 :3, \text{flip}]$
19	$(x \cdot (y \cdot x)) \cdot y = x \cdot (y \cdot (x \cdot y))$	$[11 \rightarrow 6]$
33	$x' \cdot (x \cdot (x \cdot (y \cdot x))) = x \cdot (y \cdot x)$	$[11 \rightarrow 9 :12]$
37	$x' \cdot (x \cdot (y \cdot (x \cdot (z \cdot x)))) = y \cdot (x \cdot (z \cdot x))$	$[6 \rightarrow 9 :7]$
42,41	$x' \cdot (x \cdot x) = x$	$[2 \rightarrow 9 :3]$
44,43	$x \cdot x' = 1$	$[17 \rightarrow 9 :5, \text{flip}]$
47	$A \cdot 1 = A, \ C \cdot x = D, \ y \cdot E = F \ \rightarrow \ \square$	$[8 :44 :1]$
50	$x \cdot (y \cdot ((x \cdot y)' \cdot y)) = y$	$[43 \rightarrow 6 :3, \text{flip}]$
53,52	$x \cdot 1 = x$	$[43 \rightarrow 11 :3,5, \text{flip}]$
54	$x \cdot (x' \cdot (y \cdot x')) = y \cdot x'$	$[43 \rightarrow 6 :3, \text{flip}]$
56	$C \cdot x = D, \ y \cdot E = F \ \rightarrow \ \square$	$[47 :53 :1]$
58,57	$(x \cdot y) \cdot y = x \cdot (y \cdot y)$	$[2 \rightarrow 13 :53,53,3, \text{flip}]$
68,67	$x' \cdot ((x \cdot x) \cdot x') = 1$	$[41 \rightarrow 11 :44, \text{flip}]$
73	$x \cdot (x' \cdot x') = x'$	$[43 \rightarrow 57 :3, \text{flip}]$
95	$x \cdot ((x' \cdot x') \cdot x) = 1$	$[73 \rightarrow 11 :5, \text{flip}]$
98,97	$x'' = (x \cdot x) \cdot x'$	$[67 \rightarrow 9 :53]$
99	$(x \cdot y') \cdot (y \cdot (y \cdot y)) = x \cdot (y \cdot y)$	$[67 \rightarrow 13 :53,42,12, \text{flip}]$
112,111	$(x \cdot x) \cdot (x' \cdot x') = 1$	$[97 \rightarrow 4 :58]$
113	$(x' \cdot x') \cdot x = x'$	$[95 \rightarrow 9 :53, \text{flip}]$
149	$x \cdot ((y \cdot ((x \cdot y)' \cdot y)) \cdot x) = y \cdot x$	$[50 \rightarrow 11, \text{flip}]$
168,167	$(x \cdot x) \cdot x' = x$	$[113 \rightarrow 54 :98,68,53,98,98,7,58,112,53, \text{flip}]$
176,175	$x'' = x$	$[97 :168]$
183	$x \cdot ((x \cdot x)' \cdot x) = 1$	$[50 \rightarrow 33 :5, \text{flip}]$
192,191	$(x \cdot x)' \cdot x = x'$	$[183 \rightarrow 9 :53, \text{flip}]$
206,205	$x' \cdot (x \cdot y) = y$	$[191 \rightarrow 37 :44,53,192,44,53]$
208,207	$x \cdot (x' \cdot y) = y$	$[167 \rightarrow 37 :176,5,53,168,5,53]$
220,219	$x \cdot ((y \cdot x)' \cdot x) = y' \cdot x$	$[50 \rightarrow 205, \text{flip}]$
235	$x \cdot ((x' \cdot y) \cdot x) = y \cdot x$	$[149 :220]$
316,315	$(x \cdot y) \cdot x' = x \cdot (y \cdot x')$	$[235 \rightarrow 207 :176, \text{flip}]$
317	$(x \cdot y) \cdot (y' \cdot x) = x \cdot x$	$[235 \rightarrow 19 :208,58,208]$
321,320	$(x' \cdot y) \cdot x = x' \cdot (y \cdot x)$	$[235 \rightarrow 205, \text{flip}]$
354,353	$(x \cdot x) \cdot y = x \cdot (x \cdot y)$	$[317 \rightarrow 6 :321,208]$
887	$(x \cdot (y \cdot y)) \cdot y' = x \cdot y$	$[99 \rightarrow 6 :316,354,44,53,206]$
909	$(x \cdot y') \cdot y = x$	$[6 \rightarrow 887 :354,44,53,5,53, \text{flip}]$
995	$\square$	$[56,207,909]$

Theorem MFL-5. Simple basis with Moufang-2 (part 2).

$$\left\{ \begin{array}{l} 1 \cdot x = x \\ x' \cdot x = 1 \\ \text{Moufang-2} \end{array} \right\} \Rightarrow \{\text{left cancellation}\}.$$

Proof (found by Otter 3.0.4 on gyro at 1.52 seconds).

3,2	$1 \cdot x = x$	
5,4	$x' \cdot x = 1$	
7,6	$((x \cdot y) \cdot z) \cdot y = x \cdot (y \cdot (z \cdot y))$	
8	$A \cdot C = A \cdot B$	
10	$C \neq B$	
13	$x' \cdot (x \cdot (y \cdot x)) = y \cdot x$	$[4 \to 6 : 3, \text{flip}]$
16,15	$(x \cdot y) \cdot x = x \cdot (y \cdot x)$	$[2 \to 6 : 3]$
23	$x' \cdot (x \cdot x') = x'$	$[4 \to 15 : 3, \text{flip}]$
45	$x' \cdot (x \cdot (x \cdot (y \cdot x))) = x \cdot (y \cdot x)$	$[15 \to 13 : 16]$
53	$x' \cdot (x \cdot (y \cdot (x \cdot (z \cdot x)))) = y \cdot (x \cdot (z \cdot x))$	$[6 \to 13 : 7]$
60,59	$x \cdot x' = 1$	$[23 \to 13 : 5, \text{flip}]$
69	$x \cdot (y \cdot ((x \cdot y)' \cdot y)) = y$	$[59 \to 6 : 3, \text{flip}]$
72,71	$x \cdot 1 = x$	$[59 \to 15 : 3,5, \text{flip}]$
267	$x \cdot ((x \cdot x)' \cdot x) = 1$	$[69 \to 45 : 5, \text{flip}]$
276,275	$(x \cdot x)' \cdot x = x'$	$[267 \to 13 : 72, \text{flip}]$
294,293	$x' \cdot (x \cdot y) = y$	$[275 \to 53 : 60,72,276,60,72]$
333	$C = B$	$[8 \to 293 : 294, \text{flip}]$
335	$\square$	$[333,10]$

Theorem MFL-6. Simple basis with Moufang-2 (part 3).

$$\left\{ \begin{array}{l} 1 \cdot x = x \\ x' \cdot x = 1 \\ \text{Moufang-2} \end{array} \right\} \Rightarrow \{\text{right cancellation}\}.$$

Proof (found by Otter 3.0.4 on gyro at 0.49 seconds).

3,2	$1 \cdot x = x$	
5,4	$x' \cdot x = 1$	
7,6	$((x \cdot y) \cdot z) \cdot y = x \cdot (y \cdot (z \cdot y))$	
8	$C \cdot A = B \cdot A$	
10	$C \neq B$	
11	$C \cdot (A \cdot (x \cdot A)) = B \cdot (A \cdot (x \cdot A))$	$[8 \to 6 : 7, \text{flip}]$
13	$x' \cdot (x \cdot (y \cdot x)) = y \cdot x$	$[4 \to 6 : 3, \text{flip}]$
15	$(x \cdot y) \cdot x = x \cdot (y \cdot x)$	$[2 \to 6 : 3]$
29	$x' \cdot (x \cdot x') = x'$	$[4 \to 15 : 3, \text{flip}]$
60,59	$x \cdot x' = 1$	$[29 \to 13 : 5, \text{flip}]$
68,67	$x \cdot 1 = x$	$[59 \to 15 : 3,5, \text{flip}]$

71	$(x \cdot y) \cdot y = x \cdot (y \cdot y)$	$[67 \to 6 :3]$
95	$x \cdot (x' \cdot x') = x'$	$[59 \to 71 :3, \text{flip}]$
121	$x \cdot ((x' \cdot x') \cdot x) = 1$	$[95 \to 15 :5, \text{flip}]$
142,141	$(x' \cdot x') \cdot x = x'$	$[121 \to 13 :68, \text{flip}]$
143	$C = B$	$[121 \to 11 :68,142,60,68]$
145	$\square$	$[143,10]$

That completes the proof that the simplified set is a basis for Moufang loops. The following problem shows that we can obtain an alternative basis by replacing Moufang-2 with Moufang-3.

Theorem MFL-7. Simple basis with Moufang-3.

$$\left\{ \begin{array}{l} 1 \cdot x = x \\ x' \cdot x = 1 \\ \text{Moufang-3} \end{array} \right\} \Rightarrow \{\text{Moufang-2}\}.$$

Proof (found by Otter 3.0.4 on gyro at 33.60 seconds).

1	$x = x$	
3,2	$1 \cdot x = x$	
5,4	$x' \cdot x = 1$	
7,6	$((x \cdot y) \cdot x) \cdot z = x \cdot (y \cdot (x \cdot z))$	
8	$((A \cdot B) \cdot C) \cdot B \neq A \cdot (B \cdot (C \cdot B))$	
9	$x' \cdot (x \cdot (x' \cdot y)) = x' \cdot y$	$[4 \to 6 :3, \text{flip}]$
12,11	$(x \cdot 1) \cdot y = x \cdot y$	$[2 \to 6 :3,3]$
13	$(x \cdot (y \cdot (x \cdot 1))) \cdot z = x \cdot (y \cdot (x \cdot z))$	$[6 \to 11 :7]$
17	$(x \cdot x) \cdot y = x \cdot (x \cdot y)$	$[11 \to 6 :3]$
19	$((x \cdot y) \cdot (x \cdot 1)) \cdot z = x \cdot (y \cdot (x \cdot z))$	$[11 \to 6 :12,12]$
34,33	$(x \cdot (x \cdot 1)) \cdot y = x \cdot (x \cdot y)$	$[11 \to 17 :12,12]$
38,37	$x' \cdot (x \cdot 1) = 1$	$[4 \to 9 :5]$
43	$((x \cdot y) \cdot x)' \cdot (x \cdot (y \cdot (x \cdot 1))) = 1$	$[6 \to 37]$
51	$x \cdot (x' \cdot (x \cdot y)) = x \cdot y$	$[37 \to 13 :12, \text{flip}]$
53	$(x \cdot (y \cdot (y \cdot (x \cdot 1)))) \cdot z = x \cdot (y \cdot (y \cdot (x \cdot z)))$	$[33 \to 13 :34]$
61	$x' \cdot (x'' \cdot 1) = 1$	$[37 \to 51 :38]$
128,127	$x'' \cdot (x' \cdot (x' \cdot (x'' \cdot y))) = y$	$[61 \to 53 :38,3, \text{flip}]$
144,143	$x'' \cdot (x' \cdot (x' \cdot 1)) = x' \cdot 1$	$[37 \to 127]$
146,145	$x' \cdot (x'' \cdot y) = y$	$[9 \to 127 :128, \text{flip}]$
148,147	$x' \cdot 1 = x'$	$[4 \to 127 :144]$
182,181	$x' \cdot (x \cdot y) = y$	$[145 \to 9 :146]$
185	$(x \cdot y') \cdot z = y' \cdot ((y'' \cdot x) \cdot (y' \cdot z))$	$[145 \to 19 :148]$
188	$x' \cdot ((x'' \cdot y) \cdot (x' \cdot z)) = (y \cdot x') \cdot z$	$[\text{flip } 185]$
205	$((x \cdot y) \cdot x)'' = x \cdot (y \cdot (x \cdot 1))$	$[43 \to 181 :148]$
224,223	$x'' = x$	$[4 \to 181 :148]$
225	$x' \cdot ((x \cdot y) \cdot (x' \cdot z)) = (y \cdot x') \cdot z$	$[188 :224]$
237	$(x \cdot y) \cdot x = x \cdot (y \cdot (x \cdot 1))$	$[205 :224]$

245	$x \cdot (x' \cdot y) = y$	$[223 \to 181]$
248,247	$x \cdot 1 = x$	$[223 \to 147 :224]$
249	$x \cdot x' = 1$	$[223 \to 4]$
252,251	$(x \cdot y) \cdot x = x \cdot (y \cdot x)$	$[237 :248]$
261	$(x \cdot (y \cdot x)) \cdot z = x \cdot (y \cdot (x \cdot z))$	$[13 :248]$
285	$x \cdot ((x' \cdot y) \cdot (x \cdot z)) = (y \cdot x) \cdot z$	$[223 \to 225 :224,224]$
286	$(x \cdot y') \cdot y = x$	$[249 \to 225 :248,182,224, \text{flip}]$
291	$(x \cdot y) \cdot z = y \cdot ((y' \cdot x) \cdot (y \cdot z))$	$[\text{flip } 285]$
295	$(x \cdot y) \cdot y' = x$	$[223 \to 286]$
307	$x \cdot (y \cdot x)' = y'$	$[181 \to 295]$
314,313	$(x \cdot y)' = y' \cdot x'$	$[295 \to 307, \text{flip}]$
562	$(x \cdot (y \cdot (x \cdot z))) \cdot z' = x \cdot (y \cdot x)$	$[261 \to 295]$
843	$B \cdot (((B' \cdot A) \cdot (B \cdot C)) \cdot B) \neq A \cdot (B \cdot (C \cdot B))$	$[291 \to 8 :252]$
1465	$(x \cdot (y \cdot z)) \cdot (z' \cdot x) = x \cdot (y \cdot x)$	$[245 \to 562 :314,224]$
1791	$(x \cdot y) \cdot (z \cdot x) = x \cdot ((y \cdot z) \cdot x)$	$[295 \to 1465 :224]$
1824,1823	$x \cdot (((x' \cdot y) \cdot z) \cdot x) = y \cdot (z \cdot x)$	$[245 \to 1791, \text{flip}]$
1837	$A \cdot (B \cdot (C \cdot B)) \neq A \cdot (B \cdot (C \cdot B))$	$[843 :1824,252]$
1838	$\square$	$[1837,1]$

To show that one does *not* obtain a basis for Moufang loops by replacing Moufang-2 with Moufang-1, we use MACE to find a counterexample.

Example MFL-8. Simple basis does not work with Moufang-1.

$$\left\{ \begin{array}{l} 1 \cdot x = x \\ x' \cdot x = 1 \\ \text{Moufang-1} \end{array} \right\} \not\Rightarrow \{\text{Moufang-2}\}.$$

The clauses

$$1 \cdot x = x$$
$$x' \cdot x = 1$$
$$(x \cdot (y \cdot z)) \cdot x = (x \cdot y) \cdot (z \cdot x)$$
$$((A \cdot B) \cdot C) \cdot B \neq A \cdot (B \cdot (C \cdot B))$$

have the following model (found by MACE 1.2.0 on gyro at 0.78 seconds).

```
. | 0 1 2           '   0 1 2
--+------          ----------
0 | 1 0 1              0 1 0
1 | 0 1 2
2 | 1 2 1          A: 0,  B: 2,  C: 1
```

8.1.2 Single Axioms for Inverse Loops and Moufang Loops

In [52], Padmanabhan presents the following single axiom, in terms of division $(x/y = x \cdot y')$, for inverse loops:

$$(u/u)/((x/y)/(z/(y/x))) = z.$$

From this, he derives the following single axiom schema for subvarieties of inverse loops:

$$((u/u)/\mathcal{W})/((x/y)/(z/(y/x))) = z.$$

The term $\mathcal{W}$, containing fresh variables, represents the equation $\mathcal{W} = 1$, which specifies the properties of the subvariety. For example, a single axiom for Moufang loops (in terms of division) can be easily built from Padmanabhan's schema.

Padmanabhan's results are a generalization of the important work of Higman and Neumann [19] for groups and subvarieties in terms of division. Neumann much later obtained similar results for groups in terms of product and inverse [51], but the schema is more complicated, with two meta-terms, because there is no single axiom for groups in terms of product, inverse, and identity [72].

Here, in a generalization of Neumann's results, we solve the problem of finding a single axiom and a schema for inverse loops in terms of product and inverse. To find the axioms, we used Otter with methods similar to those presented in [35].

Theorem IL-1. A single axiom for inverse loops.

$$\{x \cdot ((((x \cdot y') \cdot y)' \cdot z) \cdot (u' \cdot u)) = z\} \Leftrightarrow \left\{ \begin{array}{c} x' \cdot x = y' \cdot y \\ x' \cdot (x \cdot y) = y \\ (x \cdot y) \cdot y' = x \end{array} \right\}.$$

We prove ($\Rightarrow$) in two steps, first showing that $x' \cdot x$ is a constant, then including $x' \cdot x = 1$ to derive the right-hand side. The third Otter proof is ($\Leftarrow$).

Proof (found by Otter 3.0.4 on gyro at 0.46 seconds).

1	$A' \cdot A = B' \cdot B \rightarrow \square$	
4,3	$x \cdot ((((x \cdot y') \cdot y)' \cdot z) \cdot (u' \cdot u)) = z$	
5	$x \cdot (y \cdot (z' \cdot z)) = (((((x \cdot u') \cdot u)' \cdot v') \cdot v) \cdot y) \cdot (w' \cdot w)$	$[3 \rightarrow 3]$
6	$(((((x \cdot y') \cdot y)' \cdot z') \cdot z) \cdot u) \cdot (v' \cdot v) = x \cdot (u \cdot (w' \cdot w))$	$[\text{flip } 5]$
28	$x \cdot (y \cdot (z' \cdot z)) = x \cdot (y \cdot (u' \cdot u))$	$[6 \rightarrow 6]$
64	$x \cdot (y' \cdot y) = x \cdot (z' \cdot z)$	$[28 \rightarrow 3 : 4]$
84	$x' \cdot x = y' \cdot y$	$[64 \rightarrow 3 : 4]$
85	$\square$	$[84,1]$

Proof ($\Rightarrow$) found by Otter 3.0.4 on gyro at 0.66 seconds.

1	$x = x$
2	$B' \cdot B = A' \cdot A,\ A' \cdot (A \cdot B) = B,\ (A \cdot B) \cdot B' = A \rightarrow \square$
3	$x \cdot ((((x \cdot y') \cdot y)' \cdot z) \cdot (u' \cdot u)) = z$
6,5	$x' \cdot x = 1$

7	$x \cdot ((((x \cdot y') \cdot y)' \cdot z) \cdot 1) = z$	[3 :6]
9	$A' \cdot (A \cdot B) = B, \ (A \cdot B) \cdot B' = A \ \rightarrow \ \square$	[2 :6,6 :1]
10	$x'' \cdot (((1 \cdot x)' \cdot y) \cdot 1) = y$	[5 → 7]
12	$(((((x \cdot y') \cdot y)' \cdot z') \cdot z)' \cdot u) \cdot 1 = x \cdot (u \cdot 1)$	[7 → 7, flip]
15,14	$(x \cdot y') \cdot y = x \cdot (1 \cdot 1)$	[5 → 7, flip]
17,16	$(((x \cdot (1 \cdot 1))' \cdot (1 \cdot 1))' \cdot y) \cdot 1 = x \cdot (y \cdot 1)$	[12 :15,15]
18	$x \cdot (((x \cdot (1 \cdot 1))' \cdot y) \cdot 1) = y$	[7 :15]
23,22	$x'' \cdot (1 \cdot 1) = 1 \cdot x$	[5 → 14, flip]
24	$x'' \cdot ((1 \cdot x)' \cdot (1 \cdot 1)) = 1'$	[14 → 10]
28	$(1 \cdot 1)' \cdot ((1' \cdot x) \cdot 1) = x$	[5 → 18]
36	$((1' \cdot (1 \cdot 1))' \cdot x) \cdot 1 = (1 \cdot 1)' \cdot (x \cdot 1)$	[18 → 28, flip]
39,38	$(1 \cdot 1)' \cdot (1' \cdot (1 \cdot 1)) = 1'$	[14 → 28]
40	$1'' \cdot (x \cdot 1) = (1' \cdot x) \cdot 1$	[28 → 10]
42	$1'' \cdot 1 = 1' \cdot (1 \cdot 1)$	[38 → 10 :6]
44	$(x \cdot (1 \cdot 1)') \cdot (y \cdot 1) = x \cdot (y \cdot 1)$	[14 → 16 :17, flip]
48	$(x \cdot (1 \cdot 1))' \cdot (x \cdot (y \cdot 1)) = y$	[16 → 18]
55	$(1' \cdot ((1 \cdot 1)' \cdot x)) \cdot 1 = x$	[18 → 40 :23, flip]
66	$1' \cdot ((1 \cdot 1)' \cdot (x \cdot 1)) = x$	[5 → 48]
69,68	$(x \cdot (1 \cdot 1))' \cdot (x \cdot y) = 1' \cdot ((1 \cdot 1)' \cdot y)$	[55 → 48]
71,70	$1'' = 1' \cdot 1'$	[42 → 48 :69,39, flip]
75	$1' \cdot ((1 \cdot 1)' \cdot (x \cdot (1 \cdot 1))) = x \cdot 1'$	[14 → 48 :69]
82	$((x \cdot 1)' \cdot (1 \cdot 1))' \cdot 1 = x$	[5 → 48]
104	$(1' \cdot 1') \cdot (1 \cdot 1) = 1 \cdot 1$	[70 → 22]
110	$(1' \cdot 1')' \cdot (((1 \cdot 1')' \cdot x) \cdot 1) = x$	[70 → 10]
121,120	$1' \cdot (1 \cdot 1) = 1'$	[70 → 24 :6,15]
124	$(1 \cdot 1)' \cdot 1' = 1'$	[38 :121]
126	$((1' \cdot 1') \cdot x) \cdot 1 = (1 \cdot 1)' \cdot (x \cdot 1)$	[36 :121,71]
127	$(1 \cdot 1)' \cdot (x \cdot 1) = ((1' \cdot 1') \cdot x) \cdot 1$	[flip 126]
128	$(1 \cdot 1)' \cdot 1 = 1 \cdot 1$	[120 → 28 :6]
137,136	$(1 \cdot 1)' = 1$	[128 → 66 :6,6, flip]
142,141	$((1' \cdot 1') \cdot x) \cdot 1 = 1 \cdot (x \cdot 1)$	[127 :137, flip]
144,143	$1 \cdot 1' = 1'$	[124 :137]
149	$1' \cdot (1 \cdot (x \cdot (1 \cdot 1))) = x \cdot 1'$	[75 :137]
151	$(x \cdot (1 \cdot 1))' \cdot (x \cdot y) = 1' \cdot (1 \cdot y)$	[68 :137]
158,157	$(x \cdot 1) \cdot (y \cdot 1) = x \cdot (y \cdot 1)$	[44 :137]
162	$(1' \cdot 1')' \cdot (1 \cdot (x \cdot 1)) = x$	[110 :144,71,142]
173,172	$1' = 1 \cdot (1 \cdot 1)$	[136 → 22 :121]
177,176	$1 \cdot (1 \cdot 1) = 1$	[136 → 5]
187,186	$1 \cdot (1 \cdot (x \cdot 1)) = x$	[162 :173,177,173,177,137]
195,194	$1 \cdot 1 = 1$	[104 :173,177,173,177,158,177, flip]
202	$(x \cdot 1)' \cdot (x \cdot y) = 1 \cdot (1 \cdot y)$	[151 :195,173,195,195]
206,205	$x \cdot 1 = x$	[149 :173,195,195,195,187,173,195,195, flip]
226,225	$x'' = x$	[82 :206,206,206,206]
232,231	$1 \cdot x = x$	[22 :226,206,206, flip]

235	$(x \cdot y') \cdot y = x$	[14 :206,206]
238,237	$x' \cdot (x \cdot y) = y$	[202 :206,232,232]
239	$(A \cdot B) \cdot B' = A \rightarrow \square$	[9 :238 :1]
244	$(x \cdot y) \cdot y' = x$	[225 $\rightarrow$ 235]
246	$\square$	[244,239]

Proof ($\Leftarrow$) found by Otter 3.0.4 on gyro at 0.08 seconds.

1	$x = x$	
2	$A \cdot ((((A \cdot B') \cdot B)' \cdot C) \cdot (D' \cdot D)) = C \rightarrow \square$	
3	$x' \cdot x = y' \cdot y$	
4	$x' \cdot (x \cdot y) = y$	
6	$(x \cdot y) \cdot y' = x$	
8	$A \cdot ((((A \cdot B') \cdot B)' \cdot C) \cdot (x' \cdot x)) = C \rightarrow \square$	[3 $\rightarrow$ 2]
10,9	$x'' \cdot y = x \cdot y$	[4 $\rightarrow$ 4]
12,11	$x \cdot (y' \cdot y) = x$	[3 $\rightarrow$ 4 :10]
13	$A \cdot (((A \cdot B') \cdot B)' \cdot C) = C \rightarrow \square$	[8 :12]
24	$x'' = x$	[4 $\rightarrow$ 11, flip]
31,30	$(x \cdot y') \cdot y = x$	[24 $\rightarrow$ 6]
33,32	$x \cdot (x' \cdot y) = y$	[24 $\rightarrow$ 4]
35	$C = C \rightarrow \square$	[13 :31,33]
36	$\square$	[35,1]

In the following two lemmas, α and β are meta-terms specifying a subvariety of inverse loops; here we can treat them as constants.

Lemma IL-2. Inverse loop schema gives inverse loop basis.

$$\{\beta = \alpha\} \Rightarrow$$

$$\left(\{x \cdot ((((x \cdot y') \cdot y)' \cdot z) \cdot ((\alpha \cdot u)' \cdot (\beta \cdot u))) = z\} \Leftrightarrow \left\{ \begin{array}{l} x' \cdot x = y' \cdot y \\ x' \cdot (x \cdot y) = y \\ (x \cdot y) \cdot y' = x \end{array} \right\} \right).$$

As for Thm. IL-1, we prove ($\Rightarrow$) in two steps, first showing that $x' \cdot x$ is a constant, then including $x' \cdot x = 1$ to derive the right-hand side. The third Otter proof is ($\Leftarrow$).

Proof (found by Otter 3.0.4 on gyro at 1.28 seconds).

1	$A' \cdot A = B' \cdot B \rightarrow \square$	
4,3	$\beta = \alpha$	
5	$x \cdot ((((x \cdot y') \cdot y)' \cdot z) \cdot ((\alpha \cdot u)' \cdot (\beta \cdot u))) = z$	
7,6	$x \cdot ((((x \cdot y') \cdot y)' \cdot z) \cdot ((\alpha \cdot u)' \cdot (\alpha \cdot u))) = z$	[copy,5 :4]
10,9	$x \cdot ((((x \cdot y') \cdot y)' \cdot z) \cdot (u' \cdot u)) = z$	[6 $\rightarrow$ 6 :7]
12	$x \cdot (y \cdot (z' \cdot z)) = (((((x \cdot u') \cdot u)' \cdot v') \cdot v)' \cdot y) \cdot (w' \cdot w)$	[9 $\rightarrow$ 9]
13	$(((((x \cdot y') \cdot y)' \cdot z') \cdot z)' \cdot u) \cdot (v' \cdot v) = x \cdot (u \cdot (w' \cdot w))$	[flip 12]
35	$x \cdot (y \cdot (z' \cdot z)) = x \cdot (y \cdot (u' \cdot u))$	[13 $\rightarrow$ 13]

141	$x \cdot (y' \cdot y) = x \cdot (z' \cdot z)$	$[35 \to 9 : 10]$
174	$x' \cdot x = y' \cdot y$	$[141 \to 9 : 10]$
175	$\square$	$[174, 1]$

Proof ($\Rightarrow$) found by Otter 3.0.4 on gyro at 0.66 seconds.

1	$x = x$	
2	$B' \cdot B = A' \cdot A, \ A' \cdot (A \cdot B) = B, \ (A \cdot B) \cdot B' = A \ \to \ \square$	
4,3	$\beta = \alpha$	
6,5	$x' \cdot x = 1$	
7	$x \cdot ((((x \cdot y') \cdot y)' \cdot z) \cdot ((\alpha \cdot u)' \cdot (\beta \cdot u))) = z$	
8	$x \cdot ((((x \cdot y') \cdot y)' \cdot z) \cdot 1) = z$	[copy,7 :4,6]
10	$A' \cdot (A \cdot B) = B, \ (A \cdot B) \cdot B' = A \ \to \ \square$	$[2 : 6,6 : 1]$
11	$x'' \cdot (((1 \cdot x)' \cdot y) \cdot 1) = y$	$[5 \to 8]$
13	$(((((x \cdot y') \cdot y)' \cdot z') \cdot z)' \cdot u) \cdot 1 = x \cdot (u \cdot 1)$	$[8 \to 8, \text{flip}]$
16,15	$(x \cdot y') \cdot y = x \cdot (1 \cdot 1)$	$[5 \to 8, \text{flip}]$
18,17	$(((x \cdot (1 \cdot 1))' \cdot (1 \cdot 1))' \cdot y) \cdot 1 = x \cdot (y \cdot 1)$	$[13 : 16,16]$
19	$x \cdot (((x \cdot (1 \cdot 1))' \cdot y) \cdot 1) = y$	$[8 : 16]$
24,23	$x'' \cdot (1 \cdot 1) = 1 \cdot x$	$[5 \to 15, \text{flip}]$
25	$x'' \cdot ((1 \cdot x)' \cdot (1 \cdot 1)) = 1'$	$[15 \to 11]$
29	$(1 \cdot 1)' \cdot ((1' \cdot x) \cdot 1) = x$	$[5 \to 19]$
37	$((1' \cdot (1 \cdot 1))' \cdot x) \cdot 1 = (1 \cdot 1)' \cdot (x \cdot 1)$	$[19 \to 29, \text{flip}]$
40,39	$(1 \cdot 1)' \cdot (1' \cdot (1 \cdot 1)) = 1'$	$[15 \to 29]$
41	$1'' \cdot (x \cdot 1) = (1' \cdot x) \cdot 1$	$[29 \to 11]$
43	$1'' \cdot 1 = 1' \cdot (1 \cdot 1)$	$[39 \to 11 : 6]$
45	$(x \cdot (1 \cdot 1)') \cdot (y \cdot 1) = x \cdot (y \cdot 1)$	$[15 \to 17 : 18, \text{flip}]$
49	$(x \cdot (1 \cdot 1))' \cdot (x \cdot (y \cdot 1)) = y$	$[17 \to 19]$
56	$(1' \cdot ((1 \cdot 1)' \cdot x)) \cdot 1 = x$	$[19 \to 41 : 24, \text{flip}]$
73	$1' \cdot ((1 \cdot 1)' \cdot (x \cdot 1)) = x$	$[5 \to 49]$
76,75	$(x \cdot (1 \cdot 1))' \cdot (x \cdot y) = 1' \cdot ((1 \cdot 1)' \cdot y)$	$[56 \to 49]$
78,77	$1'' = 1' \cdot 1'$	$[43 \to 49 : 76,40, \text{flip}]$
82	$1' \cdot ((1 \cdot 1)' \cdot (x \cdot (1 \cdot 1))) = x \cdot 1'$	$[15 \to 49 : 76]$
89	$((x \cdot 1)' \cdot (1 \cdot 1))' \cdot 1 = x$	$[5 \to 49]$
104	$x \cdot 1' = 1' \cdot ((1 \cdot 1)' \cdot (x \cdot (1 \cdot 1)))$	$[\text{flip } 82]$
107	$(1' \cdot 1') \cdot (1 \cdot 1) = 1 \cdot 1$	$[77 \to 23]$
117	$(1' \cdot 1')' \cdot (((1 \cdot 1')' \cdot x) \cdot 1) = x$	$[77 \to 11]$
124,123	$1' \cdot (1 \cdot 1) = 1'$	$[77 \to 25 : 6,16]$
127	$(1 \cdot 1)' \cdot 1' = 1'$	$[39 : 124]$
129	$((1' \cdot 1') \cdot x) \cdot 1 = (1 \cdot 1)' \cdot (x \cdot 1)$	$[37 : 124,78]$
130	$(1 \cdot 1)' \cdot (x \cdot 1) = ((1' \cdot 1') \cdot x) \cdot 1$	$[\text{flip } 129]$
131	$(1 \cdot 1)' \cdot 1 = 1 \cdot 1$	$[123 \to 29 : 6]$
142,141	$(1 \cdot 1)' = 1$	$[131 \to 73 : 6,6, \text{flip}]$
149,148	$((1' \cdot 1') \cdot x) \cdot 1 = 1 \cdot (x \cdot 1)$	$[130 : 142, \text{flip}]$
151,150	$1 \cdot 1' = 1'$	$[127 : 142]$
152	$x \cdot 1' = 1' \cdot (1 \cdot (x \cdot (1 \cdot 1)))$	$[104 : 142]$

156	$(x \cdot (1 \cdot 1))' \cdot (x \cdot y) = 1' \cdot (1 \cdot y)$	[75 :142]
163,162	$(x \cdot 1) \cdot (y \cdot 1) = x \cdot (y \cdot 1)$	[45 :142]
171	$(1' \cdot 1')' \cdot (1 \cdot (x \cdot 1)) = x$	[117 :151,78,149]
192,191	$1' = 1 \cdot (1 \cdot 1)$	[141 $\to$ 23 :124]
196,195	$1 \cdot (1 \cdot 1) = 1$	[141 $\to$ 5]
200,199	$1 \cdot 1 = 1$	[107 :192,196,192,196,163,196, flip]
204,203	$1 \cdot (1 \cdot (x \cdot 1)) = x$ [171 :192,200,200,192,200,200,200,192,200,200]
219	$(x \cdot 1)' \cdot (x \cdot y) = 1 \cdot (1 \cdot y)$ [156 :200,192,200,200]
223,222	$x \cdot 1 = x$ [152 :192,200,200,192,200,200,200,204]
237,236	$x'' = x$ [89 :223,223,223,223]
247,246	$1 \cdot x = x$ [23 :237,223,223, flip]
250	$(x \cdot y') \cdot y = x$ [15 :223,223]
253,252	$x' \cdot (x \cdot y) = y$ [219 :223,247,247]
254	$(A \cdot B) \cdot B' = A \to \square$ [10 :253 :1]
255	$(x \cdot y) \cdot y' = x$ [236 $\to$ 250]
257	$\square$ [255,254]

Proof ($\Leftarrow$) found by Otter 3.0.4 on gyro at 0.12 seconds.

2	$A \cdot ((((A \cdot B') \cdot B)' \cdot C) \cdot ((\alpha \cdot D)' \cdot (\beta \cdot D))) = C \to \square$	
4,3	$\beta = \alpha$	
5	$x' \cdot x = y' \cdot y$	
7,6	$x' \cdot (x \cdot y) = y$	
8	$(x \cdot y) \cdot y' = x$	
10	$A \cdot ((((A \cdot B') \cdot B)' \cdot C) \cdot ((\alpha \cdot D)' \cdot (\alpha \cdot D))) = C \to \square$	[2 :4]
12,11	$x'' \cdot y = x \cdot y$	[6 $\to$ 6]
15	$x \cdot (y \cdot x)' = y'$	[6 $\to$ 8]
22,21	$x \cdot (y' \cdot y) = x$	[5 $\to$ 6 :12]
23	$A \cdot (((A \cdot B') \cdot B)' \cdot C) = C \to \square$	[10 :22]
27,26	$x'' = x$	[6 $\to$ 21, flip]
31,30	$(x \cdot y)' = y' \cdot x'$	[8 $\to$ 15, flip]
32	$A \cdot (A' \cdot C) = C \to \square$	[23 :31,31,27,7]
41	$x \cdot (x' \cdot y) = y$	[26 $\to$ 6]
43	$\square$	[41,32]

Lemma IL-3. Inverse loop schema and basis imply $\beta = \alpha$.

$$\left.\begin{array}{l} x \cdot ((((x \cdot y') \cdot y)' \cdot z) \cdot ((\alpha \cdot u)' \cdot (\beta \cdot u))) = z \\ x' \cdot x = y' \cdot y \\ x' \cdot (x \cdot y) = y \\ (x \cdot y) \cdot y' = x \end{array}\right\} \Rightarrow \{\beta = \alpha\}.$$

Proof (found by Otter 3.0.4 on gyro at 0.41 seconds).

1	$\beta = \alpha \to \square$
3	$x \cdot ((((x \cdot y') \cdot y)' \cdot z) \cdot ((\alpha \cdot u)' \cdot (\beta \cdot u))) = z$
5	$x' \cdot x = y' \cdot y$

6	$x' \cdot (x \cdot y) = y$	
8	$(x \cdot y) \cdot y' = x$	
11	$x \cdot ((((x \cdot y') \cdot y)' \cdot z) \cdot (u' \cdot (\beta \cdot ((((\alpha \cdot v') \cdot v)' \cdot u) \cdot$	
	$((\alpha \cdot w)' \cdot (\beta \cdot w)))))) = z$	$[3 \to 3]$
19,18	$(((x \cdot y') \cdot y)' \cdot z) \cdot ((\alpha \cdot u)' \cdot (\beta \cdot u)) = x' \cdot z$	$[3 \to 6, \text{flip}]$
23	$x \cdot ((((x \cdot y') \cdot y)' \cdot z) \cdot (u' \cdot (\beta \cdot (\alpha' \cdot u)))) = z$	$[11 : 19]$
26	$x \cdot (x' \cdot y) = y$	$[3 : 19]$
31,30	$x \cdot (y \cdot x)' = y'$	$[6 \to 8]$
35,34	$x'' = x$	$[26 \to 8 : 31]$
37,36	$(x \cdot y') \cdot y = x$	$[34 \to 8]$
40	$x \cdot ((x' \cdot y) \cdot (z' \cdot (\beta \cdot (\alpha' \cdot z)))) = y$	$[23 : 37]$
48	$x \cdot (y' \cdot y) = x$	$[5 \to 26]$
52	$(x' \cdot x) \cdot y = y$	$[5 \to 36 : 35]$
55,54	$x \cdot (y \cdot y') = x$	$[34 \to 48]$
61,60	$(x \cdot x') \cdot y = y$	$[34 \to 52]$
70	$x \cdot x' = y \cdot y'$	$[54 \to 26]$
84	$x \cdot (y' \cdot (\beta \cdot (\alpha' \cdot y))) = x$	$[70 \to 40 : 61, 35]$
126	$x \cdot (\alpha' \cdot \beta) = x$	$[70 \to 84 : 35, 55]$
143	$\beta = \alpha$	$[26 \to 126]$
145	$\square$	$[143, 1]$

Theorem IL-4. A single axiom schema for inverse loops.

The schema

$$x \cdot ((((x \cdot y') \cdot y)' \cdot z) \cdot ((\alpha \cdot u)' \cdot (\beta \cdot u))) = z$$

can be used to construct a single axiom for any finitely defined equational subvariety of inverse loops.

Proof. First, the equations Σ that specify the subvariety can be bundled together into an equivalent (modulo inverse loops) single equation $\delta = \gamma$; see [19, 52, 51] for the method. Second, $\delta = \gamma$ is transformed to the equivalent (modulo inverse loops) formula $\delta_1 \gamma_1' = \delta_2 \gamma_2'$, in which the two sides share no variables and are unifiable; call this equation $\alpha = \beta$. Third, α and β are plugged into the axiom. The terms α and β are unifiable without instantiating anything else in the axiom, so by Lem. IL-2, we can derive a basis for inverse loops. Given the basis for inverse loops, by Lem. IL-3, we can derive $\alpha = \beta$, so we have the theory of the subvariety and nothing more.

Corollary MFL-9. A single axiom for Moufang loops.

The axiom schema of Thm. IL-4 gives us a single axiom for the variety of Moufang loops, which is the subvariety of inverse loops satisfying one of the Moufang equations, for example, Moufang-3:

$$((x \cdot y) \cdot x) \cdot z = x \cdot (y \cdot (x \cdot z)).$$

We can transform Moufang-3, call it $\delta = \gamma$, into the appropriate form for the schema by writing it as $\delta_1\gamma_1' = \delta_2\gamma_2'$, in which the two sides of the equation have disjoint sets of variables. When we plug these terms into the schema, we get

$$x \cdot ((((x \cdot y') \cdot y)' \cdot z) \cdot ((\\
((((x_1 \cdot y_1) \cdot x_1) \cdot z_1) \cdot (x_1 \cdot (y_1 \cdot (x_1 \cdot z_1)))') \cdot u)' \cdot (\\
((((x_2 \cdot y_2) \cdot x_2) \cdot z_2) \cdot (x_2 \cdot (y_2 \cdot (x_2 \cdot z_2)))') \cdot u))) = z,$$

which is a single axiom, in terms of product and inverse, for the variety of Moufang loops.

8.2 Quasigroups

This section contains previously known but unpublished theorems on bases for subvarieties of quasigroups. The results are due to B. Wolk, N. S. Mendelsohn, and Padmanabhan. Here we simply verify the theorems with Otter.

Theorem QGT-3. A basis for Stein quasigroups.

$$\left\{ \begin{array}{c} \text{quasigroup} \\ x(y(yx)) = yx \end{array} \right\} \Rightarrow \{x(xy) = yx\}.$$

Proof (found by Otter 3.0.4 on gyro at 0.47 seconds).

4	$x \backslash (x \cdot y) = y$	
7,6	$(x/y) \cdot y = x$	
8	$(x \cdot y)/y = x$	
10	$x \cdot (y \cdot (y \cdot x)) = y \cdot x$	
12	$A \cdot (A \cdot B) \neq B \cdot A$	
13	$B \cdot A \neq A \cdot (A \cdot B)$	[flip 12]
14	$(x/y) \backslash x = y$	[6 → 4]
21,20	$x \cdot ((y/x) \cdot y) = y$	[6 → 10 :7]
33,32	$((x/y) \cdot x) \cdot (y \cdot x) = x$	[20 → 10 :21]
34	$x/((x/y) \cdot x) = y$	[20 → 8]
37,36	$x \backslash y = (y/x) \cdot y$	[20 → 4]
38	$(x/(x/y)) \cdot x = y$	[14 :37]
49,48	$x/(y \cdot x) = (x/y) \cdot x$	[34 → 34]
54	$x/y = y/(y/x)$	[38 → 8]
55	$x/(x/y) = y/x$	[flip 54]
68	$(x \cdot y)/x = (y/x) \cdot y$	[8 → 55 :49]
74	$(x/y) \cdot (y/x) = y$	[55 → 6]
80	$((x/y) \cdot x) \cdot y = y \cdot x$	[8 → 74 :49]
132	$x \cdot y = y \cdot (y \cdot x)$	[68 → 80 :33]
133	$\square$	[132,13]

Theorem STN-5. A 1-basis for totally symmetric quasigroups (SQUAGS) (1).

$$\{((xx)y)(z(xy)) = z\} \Leftrightarrow \left\{ \begin{array}{l} xx = x \\ (xy)x = y \end{array} \right\}.$$

Proof ($\Rightarrow$) found by Otter 3.0.4 on gyro at 0.49 seconds.

1	$x = x$	
3,2	$((x \cdot x) \cdot y) \cdot (z \cdot (x \cdot y)) = z$	
4	$A \cdot A = A, \ (A \cdot B) \cdot A = B \ \rightarrow \ \square$	
9	$((x \cdot x) \cdot (y \cdot z)) \cdot x = (y \cdot y) \cdot z$	$[2 \rightarrow 2]$
15	$(x \cdot x) \cdot (y \cdot (y \cdot y)) = x \cdot (y \cdot y)$	$[2 \rightarrow 9, \text{flip}]$
57,56	$(x \cdot (y \cdot y)) \cdot x = (y \cdot y) \cdot (y \cdot y)$	$[15 \rightarrow 9]$
58	$((x \cdot x) \cdot (x \cdot x)) \cdot (y \cdot (x \cdot x)) = y \cdot y$	$[15 \rightarrow 2]$
61	$x \cdot x = ((y \cdot y) \cdot (y \cdot y)) \cdot (x \cdot (y \cdot y))$	$[\text{flip } 58]$
119,118	$(x \cdot x) \cdot (x \cdot x) = (x \cdot x) \cdot x$	$[56 \rightarrow 9 :57]$
138,137	$x \cdot x = x$	$[61 :119,3]$
142,141	$(x \cdot y) \cdot x = y$	$[56 :138,138,138,138]$
143	$\square$	$[4 :138,142 :1,1]$

Proof ($\Leftarrow$) found by Otter 3.0.4 on gyro at 0.06 seconds.

3,2	$x \cdot x = x$	
4	$(x \cdot y) \cdot x = y$	
6	$((A \cdot A) \cdot B) \cdot (C \cdot (A \cdot B)) \neq C$	
7	$(A \cdot B) \cdot (C \cdot (A \cdot B)) \neq C$	$[\text{copy},6 :3]$
8	$x \cdot (y \cdot x) = y$	$[4 \rightarrow 4]$
10	$\square$	$[8,7]$

Theorem STN-6. A 1-basis for SQUAGS (2).

$$\{(((xx)y)x)(zy) = z\} \Leftrightarrow \left\{ \begin{array}{l} xx = x \\ (xy)x = y \end{array} \right\}.$$

Proof ($\Rightarrow$) found by Otter 3.0.4 on gyro at 0.41 seconds.

1	$x = x$	
3,2	$(((x \cdot x) \cdot y) \cdot x) \cdot (z \cdot y) = z$	
4	$A \cdot A = A, \ (A \cdot B) \cdot A = B \ \rightarrow \ \square$	
5	$((((x \cdot x) \cdot x) \cdot y) \cdot (((x \cdot x) \cdot x) \cdot x)) \cdot (z \cdot y) = z$	$[2 \rightarrow 2]$
7	$(((x \cdot x) \cdot (y \cdot z)) \cdot x) \cdot y = ((u \cdot u) \cdot z) \cdot u$	$[2 \rightarrow 2]$
11	$(((x \cdot x) \cdot y) \cdot x) \cdot (((z \cdot z) \cdot u) \cdot z) = ((v \cdot v) \cdot (y \cdot u)) \cdot v$	$[2 \rightarrow 7]$
12	$((x \cdot x) \cdot y) \cdot x = ((z \cdot z) \cdot y) \cdot z$	$[7 \rightarrow 7]$
15	$((x \cdot x) \cdot (y \cdot z)) \cdot x = (((u \cdot u) \cdot y) \cdot u) \cdot (((v \cdot v) \cdot z) \cdot v)$	$[\text{flip } 11]$
23,22	$(((x \cdot x) \cdot y) \cdot x) \cdot (((z \cdot z) \cdot u) \cdot z) = (y \cdot y) \cdot u$	$[12 \rightarrow 2]$
28	$((x \cdot x) \cdot (y \cdot z)) \cdot x = (y \cdot y) \cdot z$	$[15 :23]$

36	$(((x \cdot y) \cdot (x \cdot y)) \cdot (x \cdot y)) \cdot ((x \cdot y) \cdot (x \cdot y)) = (x \cdot x) \cdot y$	$[28 \to 28]$
44	$(((x \cdot x) \cdot ((y \cdot y) \cdot z)) \cdot x) \cdot ((u \cdot u) \cdot (y \cdot z)) = ((v \cdot v) \cdot u) \cdot v$	
		$[28 \to 7]$
46,45	$(((x \cdot x) \cdot y) \cdot x) \cdot ((z \cdot z) \cdot u) = (y \cdot y) \cdot (z \cdot u)$	$[28 \to 2]$
52	$(((x \cdot x) \cdot y) \cdot ((x \cdot x) \cdot y)) \cdot (z \cdot (x \cdot y)) = ((u \cdot u) \cdot z) \cdot u$	$[44\ {:}46]$
62,61	$((x \cdot x) \cdot x) \cdot (y \cdot x) = y$	$[12 \to 5\ {:}3]$
79,78	$(x \cdot x) \cdot y = x \cdot y$	$[36\ {:}62,\ \text{flip}]$
105,104	$x \cdot (y \cdot x) = y$	$[61\ {:}79,79]$
107,106	$(x \cdot y) \cdot x = y$	$[52\ {:}79,79,79,105,79,\ \text{flip}]$
112	$A \cdot A = A \; \to \; \square$	$[4\ {:}107\ {:}1]$
113	$x \cdot x = x$	$[106 \to 78,\ \text{flip}]$
115	$\square$	$[113,112]$

Proof ($\Leftarrow$) found by Otter 3.0.4 on gyro at 0.03 seconds.

3,2	$x \cdot x = x$	
5,4	$(x \cdot y) \cdot x = y$	
6	$(((A \cdot A) \cdot B) \cdot A) \cdot (C \cdot B) \neq C$	
7	$B \cdot (C \cdot B) \neq C$	$[\text{copy},6\ {:}3,5]$
8	$x \cdot (y \cdot x) = y$	$[4 \to 4]$
10	$\square$	$[8,7]$

Theorem STN-7. A 1-basis for commutative SQUAGS.

$$\{((xx)y)(z(yx)) = z\} \Leftrightarrow \left\{ \begin{array}{l} xy = yx \\ xx = x \\ (xy)x = y \end{array} \right\}.$$

Proof ($\Rightarrow$) found by Otter 3.0.4 on gyro at 0.22 seconds.

1	$x = x$	
2	$((x \cdot x) \cdot y) \cdot (z \cdot (y \cdot x)) = z$	
4	$B \cdot A = A \cdot B, \; A \cdot A = A, \; (A \cdot B) \cdot A = B \; \to \; \square$	
5	$x \cdot (y \cdot ((x \cdot ((z \cdot z) \cdot z)) \cdot (z \cdot z))) = y$	$[2 \to 2]$
9	$(((x \cdot y) \cdot (x \cdot y)) \cdot z) \cdot z = (y \cdot y) \cdot x$	$[2 \to 2]$
15	$x \cdot (x \cdot ((y \cdot y) \cdot y)) = (y \cdot y) \cdot y$	$[2 \to 5]$
26,25	$(x \cdot x) \cdot (x \cdot x) = (x \cdot x) \cdot x$	$[2 \to 15]$
29	$((((x \cdot x) \cdot x) \cdot ((x \cdot x) \cdot x)) \cdot y) \cdot ((x \cdot x) \cdot x) = y$	$[15 \to 2]$
36,35	$((x \cdot x) \cdot x) \cdot ((x \cdot x) \cdot x) = ((x \cdot x) \cdot x) \cdot (x \cdot x)$	$[25 \to 25\ {:}26,26]$
37	$(((((x \cdot x) \cdot x) \cdot (x \cdot x)) \cdot y) \cdot ((x \cdot x) \cdot x) = y$	$[29\ {:}36]$
41	$x \cdot (x \cdot (((y \cdot y) \cdot y) \cdot (y \cdot y))) = ((y \cdot y) \cdot y) \cdot (y \cdot y)$	$[25 \to 15\ {:}26]$
46,45	$((x \cdot x) \cdot x) \cdot (x \cdot x) = x \cdot x$	$[25 \to 2\ {:}36]$
55	$x \cdot (x \cdot (y \cdot y)) = y \cdot y$	$[41\ {:}46,46]$
58,57	$((x \cdot x) \cdot y) \cdot ((x \cdot x) \cdot x) = y$	$[37\ {:}46]$
60,59	$x \cdot x = x$	$[35\ {:}58,46,\ \text{flip}]$
62,61	$(x \cdot y) \cdot x = y$	$[57\ {:}60,60,60]$

63	$x \cdot (x \cdot y) = y$	[55 :60,60]
75	$((x \cdot y) \cdot z) \cdot z = y \cdot x$	[9 :60,60]
79	$B \cdot A = A \cdot B \;\rightarrow\; \square$	[4 :60,62 :1,1]
81,80	$(x \cdot y) \cdot y = x$	[61 → 63]
82	$x \cdot y = y \cdot x$	[75 :81]
83	$\square$	[82,79]

Proof ($\Leftarrow$) found by Otter 3.0.4 on gyro at 0.06 seconds.

1	$x = x$	
3,2	$x \cdot x = x$	
4	$(x \cdot y) \cdot x = y$	
6	$x \cdot y = y \cdot x$	
7	$((A \cdot A) \cdot B) \cdot (C \cdot (B \cdot A)) \neq C$	
8	$(A \cdot B) \cdot (C \cdot (B \cdot A)) \neq C$	[copy,7 :3]
10,9	$x \cdot (y \cdot x) = y$	[4 → 4]
15	$C \neq C$	[6 → 8 :10]
16	$\square$	[15,1]

Theorem STN-8. An axiom schema for commutative Steiner quasigroups.

$$\{(g(x) \cdot y) \cdot (z \cdot (y \cdot x)) = z\} \Leftrightarrow \left\{ \begin{array}{l} x \cdot y = y \cdot x \\ (x \cdot y) \cdot x = y \\ g(x) = x \end{array} \right\}.$$

This shows that the equational theory of any finitely based subvariety of Steiner quasigroups is one based. The unary $g(x)$ packs all the remaining equations true in the subvariety in question. For example, we obtain Thm. STN-7 if $g(x) = x \cdot x$,

Proof ($\Rightarrow$) found by Otter 3.0.4 on gyro at 1.89 seconds.

1	$x = x$	
2	$(g(x) \cdot y) \cdot (z \cdot (y \cdot x)) = z$	
4	$B \cdot A = A \cdot B, \; g(A) = A, \; (A \cdot B) \cdot A = B \;\rightarrow\; \square$	
5	$(g(x \cdot (y \cdot z)) \cdot (g(z) \cdot y)) \cdot (u \cdot x) = u$	[2 → 2]
8,7	$(g(x \cdot y) \cdot z) \cdot z = g(y) \cdot x$	[2 → 2]
10,9	$g(x) \cdot (g(y \cdot z) \cdot x) = g(y) \cdot g(z)$	[7 → 7 :8, flip]
13	$(g(x) \cdot y) \cdot (g(z) \cdot u) = g(u \cdot z) \cdot (y \cdot x)$	[7 → 2]
24	$(g(x) \cdot g(y \cdot z)) \cdot (g(y) \cdot g(z)) = g(x)$	[9 → 2]
59,58	$g(x \cdot y) \cdot x = g(y)$	[7 → 5]
69	$g(g(x)) \cdot g(y \cdot x) = g(y)$	[58 → 58]
77,76	$g(x \cdot (y \cdot z)) = g(x) \cdot (g(z) \cdot y)$	[2 → 58, flip]
121	$g(x) \cdot (y \cdot (z \cdot (z \cdot x))) = y$	[58 → 2]
131,130	$g(g(x \cdot y)) = g(g(x)) \cdot g(g(y))$	[58 → 69, flip]
148	$g(x) \cdot (g(x) \cdot g(y)) = g(g(y))$	[69 → 24]
155	$g(x) \cdot (y \cdot (g(x \cdot z) \cdot g(z))) = y$	[69 → 2]

174	$(g(x) \cdot y) \cdot y = (g(x) \cdot g(z)) \cdot g(z)$	[121 → 7 :77,59]
186	$(g(x) \cdot g(y)) \cdot g(y) = (g(x) \cdot z) \cdot z$	[flip 174]
197	$g(x) \cdot g(y) = g(y) \cdot x$	[148 → 7 :10]
233	$((g(g(x)) \cdot g(g(y))) \cdot (g(z) \cdot u)) \cdot (g(z) \cdot u) = g(g(y) \cdot x)$	
		[13 → 58 :77,131]
240	$g(g(x) \cdot y) = ((g(g(y)) \cdot g(g(x))) \cdot (g(z) \cdot u)) \cdot (g(z) \cdot u)$	[flip 233]
245,244	$g(x \cdot y) \cdot g(y) = g(x)$	[69 → 197, flip]
249	$g(x) \cdot (y \cdot g(x)) = y$	[155 :245]
276,275	$g(g(x) \cdot y) = g(x) \cdot g(y)$	[58 → 249, flip]
288	$((g(g(x)) \cdot g(g(y))) \cdot (g(z) \cdot u)) \cdot (g(z) \cdot u) = g(y) \cdot g(x)$	
		[240 :276, flip]
335,334	$g(g(x)) \cdot (y \cdot g(x)) = y$	[249 → 121]
340,339	$g(g(x)) \cdot g(g(y)) = g(x) \cdot g(y)$	[249 → 9 :59,131]
344,343	$g(x \cdot g(y)) = g(x) \cdot g(y)$	[249 → 58, flip]
346,345	$(g(x) \cdot g(y)) \cdot g(y) = g(x)$	[249 → 24 :344,335]
353,352	$g(x) \cdot (y \cdot x) = y$	[249 → 2 :344,346]
370	$((g(x) \cdot g(y)) \cdot (g(z) \cdot u)) \cdot (g(z) \cdot u) = g(y) \cdot g(x)$	[288 :340]
392	$(g(x) \cdot y) \cdot y = g(x)$	[186 :346, flip]
403,402	$g(x \cdot y) = g(x) \cdot g(y)$	[9 :353]
418,417	$((g(x) \cdot g(y)) \cdot z) \cdot z = g(y) \cdot x$	[7 :403]
472,471	$g(x) \cdot g(y) = g(x) \cdot y$	[370 :418, flip]
530,529	$g(x \cdot y) = g(x) \cdot y$	[402 :472]
570	$x \cdot (x \cdot y) = g(y)$	[352 → 392]
585,584	$g(x) = x$	[570 → 121 :530,353]
591,590	$(x \cdot y) \cdot x = y$	[570 → 2 :585,585]
594	$x \cdot y = y \cdot x$	[570 → 2 :585,585]
619	$\square$	[4 :585,591 :594,1,1]

Proof ($\Leftarrow$) found by Otter 3.0.4 on gyro at 0.09 seconds.

1	$x = x$	
2	$x \cdot y = y \cdot x$	
4,3	$g(x) = x$	
5	$(x \cdot y) \cdot x = y$	
7	$(g(A) \cdot B) \cdot (C \cdot (B \cdot A)) \neq C$	
8	$(A \cdot B) \cdot (C \cdot (B \cdot A)) \neq C$	[copy,7 :4]
10,9	$x \cdot (y \cdot x) = y$	[5 → 5]
15	$C \neq C$	[2 → 8 :10]
16	$\square$	[15,1]

8.3 Algebras of Set Difference

The theorems in this section are inspired by J. A. Kalman's work on set difference [21, 12], where he shows that the equations

$$x - (y - x) = x$$
$$x - (x - y) = y - (y - x)$$
$$(x - y) - z = (x - z) - (y - z)$$

form a basis for families of sets closed under set difference.

Theorem SD-2 is a check that intersection, when defined in terms of set difference, is associative and commutative. Theorem SD-3 shows that the third equation of Kalman's basis can be replaced with a simpler one [23].

Theorem SD-2. Intersection in terms of set difference.

$$\left\{ \begin{array}{l} x - (y - x) = x \\ x - (x - y) = y - (y - x) \\ (x - y) - z = (x - z) - (y - z) \\ x \cdot y = x - (x - y) \end{array} \right\} \Rightarrow \left\{ \begin{array}{l} (x \cdot y) \cdot z = x \cdot (y \cdot z) \\ x \cdot y = y \cdot x \end{array} \right\}.$$

Proof (found by Otter 3.0.4 on gyro at 4.82 seconds).

3,2	$x - (y - x) = x$	
4	$x - (x - y) = y - (y - x)$	
5	$(x - y) - z = (x - z) - (y - z)$	
7,6	$x \cdot y = x - (x - y)$	
8	$(A \cdot B) \cdot C = A \cdot (B \cdot C),\ A \cdot B = B \cdot A \to \square$	
9	$(A - (A - B)) - ((A - (A - B)) - C) =$ $A - (A - (B - (B - C))) \to \square$	[copy,8 :7,7,7,7,7,7 :4]
10	$(x - y) - (z - y) = (x - z) - y$	[flip 5]
11	$(x - y) - y = x - y$	[2 → 2]
14,13	$(x - y) - (x - y) = y - y$	[11 → 4 :3]
15	$(x - y) - ((x - y) - x) = x - (y - (y - x))$	[4 → 4, flip]
22,21	$(x - (x - y)) - (y - (y - x)) = x - x$	[4 → 13 :14]
23	$x - x = y - y$	[4 → 13 :22,14]
24	$((x - y) - z) - (y - z) = (x - y) - z$	[11 → 5, flip]
28	$(x - y) - ((z - x) - y) = x - y$	[2 → 5, flip]
44	$(x - (x - y)) - y = z - z$	[5 → 23 :3]
48,47	$x - (y - y) = x$	[23 → 4 :3]
56,55	$(x - x) - y = x - x$	[47 → 2]
76	$(A - (A - B)) - ((B - (B - A)) - C) =$ $A - (A - (B - (B - C))) \to \square$	[4 → 9]
103	$x - x = ((y - (y - z)) - u) - z$	[44 → 10 :56]
107,106	$((x - y) - z) - y = (x - y) - (z - y)$	[11 → 10, flip]
112,111	$(x - y) - (z - x) = x - (y - (z - x))$	[2 → 10, flip]
126,125	$(x - y) - x = x - x$	[55 → 10, flip]
128	$((x - (x - y)) - z) - y = u - u$	[flip 103]
132	$x - (y - (y - x)) = x - y$	[15 :126,48, flip]
148	$(x - (y - z)) - y = x - y$	[125 → 10 :48, flip]
153,152	$x - (x - (x - y)) = x - y$	[125 → 4 :48, flip]
178	$(x - y) - (x - (y - z)) = z - z$	[148 → 125 :14,14]

227,226	$((x - (x - y)) - z) - ((y - x) - z) = (x - (x - y)) - z$	
		$[132 \rightarrow 28]$
294	$(x - (x - y)) - z = (y - (y - x)) - z$	$[4 \rightarrow 24 :227]$
313	$x - (x - ((y - (y - x)) - z)) = (y - (y - x)) - z$	
		$[128 \rightarrow 4 :48, \text{flip}]$
403	$x - (x - ((y - x) - z)) = z - z$	$[2 \rightarrow 178]$
521	$x - (x - ((y - x) - z)) = x - x$	$[132 \rightarrow 403 :14,14,14]$
615,614	$(x - y) - (z - y) = (x - y) - z$	$[521 \rightarrow 132 :48,107, \text{flip}]$
770	$x - (y - (z - x)) = x - y$	$[28 :615,112]$
774	$(x - y) - z = (x - z) - y$	$[10 :615]$
842,841	$x - ((y - (z - x)) - u) = x - (y - u)$	$[774 \rightarrow 770]$
847,846	$(x - (x - y)) - z = y - (y - (x - z))$	$[313 :842, \text{flip}]$
861	$x - (x - (y - z)) = y - (y - (x - z))$	$[294 :847,847]$
862	$B - (B - (A - (B - C))) = A - (A - (B - (B - C))) \rightarrow \square$	
		$[76 :847,847,153]$
863	$\square$	$[862,861]$

Theorem SD-3. A simpler basis for set difference.

$$\left\{ \begin{array}{l} x - (y - x) = x \\ x - (x - y) = y - (y - x) \end{array} \right\} \Rightarrow$$

$$\{(x - y) - z = (x - z) - (y - z) \Leftrightarrow (x - y) - z = (x - z) - y\}.$$

Proof ($\Rightarrow$) found by Otter 3.0.4 on gyro at 2.72 seconds.

3,2	$x - (y - x) = x$	
4	$x - (x - y) = y - (y - x)$	
5	$(x - y) - z = (x - z) - (y - z)$	
6	$(A - C) - B \neq (A - B) - C$	
7	$(x - y) - (z - y) = (x - z) - y$	$[\text{flip } 5]$
8	$(x - y) - y = x - y$	$[2 \rightarrow 2]$
11,10	$(x - y) - (x - y) = y - y$	$[8 \rightarrow 4 :3]$
12	$(x - y) - ((x - y) - x) = x - (y - (y - x))$	$[4 \rightarrow 4, \text{flip}]$
19,18	$(x - (x - y)) - (y - (y - x)) = x - x$	$[4 \rightarrow 10 :11]$
20	$x - x = y - y$	$[4 \rightarrow 10 :19,11]$
35	$(A - B) - (C - B) \neq (A - B) - C$	$[5 \rightarrow 6]$
46,45	$x - (y - y) = x$	$[20 \rightarrow 4 :3]$
53	$(x - x) - y = x - x$	$[45 \rightarrow 2]$
83,82	$((x - y) - z) - y = (x - y) - (z - y)$	$[8 \rightarrow 7, \text{flip}]$
98,97	$(x - y) - x = x - x$	$[53 \rightarrow 7, \text{flip}]$
104	$x - (y - (y - x)) = x - y$	$[12 :98,46, \text{flip}]$
120	$(x - (y - z)) - y = x - y$	$[97 \rightarrow 7 :46, \text{flip}]$
154	$(x - y) - (x - (y - z)) = z - z$	$[120 \rightarrow 97 :11,11]$
359	$x - (x - ((y - x) - z)) = z - z$	$[2 \rightarrow 154]$
569	$x - (x - ((y - x) - z)) = x - x$	$[104 \rightarrow 359 :11,11,11]$

627          $(x - y) - (z - y) = (x - y) - z$ [569 → 104 :46,83, flip]
629   □   [627,35]

Proof ($\Leftarrow$) found by Otter 3.0.4 on gyro at 20.60 seconds.

3,2 $x - (y - x) = x$
4            $x - (x - y) = y - (y - x)$
5            $(x - y) - z = (x - z) - y$
6            $(A - B) - C \neq (A - C) - (B - C)$

7            $(A - C) - (B - C) \neq (A - B) - C$ [flip 6]
8            $(x - y) - y = x - y$ [2 → 2]
11,10 $(x - y) - (x - y) = y - y$ [8 → 4 :3]
19,18 $((x - y) - z) - y = (x - y) - z$ [8 → 5, flip]
21           $(x - (x - y)) - z = (y - z) - (y - x)$ [4 → 5]
22           $(x - y) - (z - x) = x - y$ [2 → 5, flip]
29           $(x - y) - (x - z) = (z - (z - x)) - y$ [flip 21]
33,32 $(x - y) - ((x - z) - y) = z - (z - (x - y))$ [5 → 4]
34           $x - ((y - x) - z) = x$ [5 → 2]
50           $x - x = y - y$ [34 → 4 :19,11]
55,54 $x - (y - y) = x$ [50 → 34]
57           $(x - x) - y = (z - y) - z$ [50 → 5]
61           $(x - y) - x = (z - z) - y$ [flip 57]
67,66 $(x - x) - y = x - x$ [54 → 2]
68           $(x - y) - x = z - z$ [61 :67]
75           $x - (x - (x - y)) = x - y$ [68 → 4 :55, flip]
103,102 $((x - y) - z) - (u - (x - z)) = (x - z) - y$ [5 → 22]
767          $(x - y) - ((z - (z - x)) - y) = (x - z) - ((x - z) - (x - y))$
                                                                  [29 → 4]
841          $(x - (x - y)) - ((x - (x - y)) - (x - z)) = y - (y - (x - z))$
                                                                  [75 → 32 :33, flip]
2094,2093 $(x - y) - (x - (x - z)) = (x - y) - z$ [75 → 102 :103, flip]
2125,2124 $(x - y) - ((x - y) - (x - z)) = (x - y) - z$ [29 → 102 :19]
2130,2129 $(x - (x - y)) - z = y - (y - (x - z))$ [841 :2125]
2132 $(x - y) - (z - y) = (x - z) - y$ [767 :2130,2094,2125]
2133   □   [2132,7]

A. Theorems Proved

We list here the results that are new and interesting to us. Some are entirely new, and some are new kinds of proof for previously known theorems. We start the section numbers with A.3 so that they correspond to the chapter numbers.

A.3 Algebras over Algebraic Curves

Theorem MED-2 (p. 36). Median law for Steiner quasigroups.

Theorem MED-4 (p. 47). Median law for chord-tangent construction (3).

Theorem MED-5 (p. 47). Median law for four group operations.

Theorem ABGT-2 (p. 49). Identity with (gL) is a commutative monoid.

Theorem ABGT-3 (p. 50). Existence of inverses under (gL).

Corollary ABGT-3a (p. 51). Mumford-Ramanujam theorem for elliptic curves.

Theorem UAL-1 (p. 51). Uniqueness of inversive groupoids under (gL).

Theorem MCV-1 (p. 53). Associativity of Mal'cev polynomial under (gL).

Theorem MCV-2 (p. 53). Mal'cev polynomial under (gL).

Corollary UAL-3 (p. 54). Uniqueness of Mal'cev laws under (gL).

Theorem UAL-4 (p. 54). Uniqueness of binary Steiner law under (gL).

Theorem UAL-6 (p. 57). Uniqueness of 5-ary Steiner law under (gL).

Corollary UAL-7 (p. 59). A ruler construction for cubic and conic.

A.4 Other (gL)-Algebras

Theorem ABGT-4 (p. 64). A (gL)-basis for right division in Abelian groups (1).

Theorem ABGT-5 (p. 64). A (gL)-basis for right division in Abelian groups (2).

Theorem ABGT-6 (p. 65). A (gL)-basis for left and right division in Abelian groups.

Theorem ABGT-7 (p. 66). A (gL)-basis for Abelian groups with double inversion (1).

Theorem ABGT-8 (p. 67). A (gL)-basis for Abelian groups with double inversion (2).

Theorem QGT-2 (p. 71). A (gL)-basis for generalized division in Abelian groups.

Theorem CS-GL-1 (p. 74). Cancellative (gL)-semigroups are commutative.

Theorem CS-GL-2 (p. 74). Diassociative cancellative (gL)-groupoids are commutative.

Theorem CS-GL-3 (p. 75). Nearly (1) associative cancellative (gL)-groupoids are commutative.

Theorem OC-1 (p. 80). Validity of binary overlay for quasigroups.

Theorem OC-3 (p. 81). Validity of ternary overlay for quasigroups.

Theorem TC-6 (p. 85). Inconsistency of TC with semilattices.

Theorem RC-1 (p. 86). Inconsistency of RC with semilattices.

Theorem TC-7 (p. 86). Associativity of Mal'cev polynomial under TC.

Theorem TC-8 (p. 87). TC Steiner quasigroups are medial.

Theorem RC-2 (p. 87). Commutative RC Steiner quasigroups are medial.

Theorem RC-4 (p. 88). RC basis for right division in Abelian groups.

Theorem TC-9 (p. 89). TC basis for double inversion in Abelian groups (1).

Theorem TC-10 (p. 89). TC basis for double inversion in Abelian groups (2).

Theorem TC-11 (p. 90). TC groupoids with identity are commutative semigroups.

Theorem TC-12 (p. 90). TC Steiner quasigroups with $xe = ex$ are Abelian groups.

Theorem TC-13 (p. 91). Cancellative medial algebras satisfy TC.

Theorem RC-6 (p. 91). Set difference is inconsistent with RC.

Theorem TC-14 (p. 92). Set difference is inconsistent with TC.

A.5 Semigroups

Theorem CS-1 (p. 96). Support (1) for the CS conjecture.

Theorem CS-2 (p. 96). Support (2) for the CS conjecture.

Theorem CS-3 (p. 97). Support (3) for the CS conjecture.

Theorem CS-4 (p. 99). Support (4) for the CS conjecture.

Theorem CS-5 (p. 100). Support (5) for the CS conjecture.

Theorem CS-6 (p. 100). Support (6) for the CS conjecture.

Theorem CS-7 (p. 101). Support (7) for the CS conjecture.

Theorem CS-8 (p. 102). Support (8) for the CS conjecture.

Theorem CS-9 (p. 103). Support (9) for the CS conjecture.

Theorem CS-13 (p. 104). Nilpotent CS satisfy the quotient condition.

A.6 Lattice-like Algebras

Theorem WAL-2 (p. 135). Uniqueness of the meet operation in WAL.

Theorem LT-10 (p. 139). An absorption 3-basis for LT.

Theorem MAJ-3 (p. 142). A majority schema for two absorption equations.

Theorem LT-12 (p. 144). A short single axiom for LT.

Theorem WAL-4 (p. 146). A short single axiom for WAL.

Theorem TBA-1 (p. 152). A short single axiom for TBA.

A.7 Independent Self-Dual Bases

Theorem DUAL-GT-1 (p. 156). An independent self-dual 2-basis for GT.

Theorem DUAL-GT-2 (p. 157). An independent self-dual 2-basis for Abelian GT.

Theorem DUAL-GT-3 (p. 158). An independent self-dual 3-basis for GT.

Theorem DUAL-GT-4 (p. 159). An independent self-dual 4-basis for GT.

Theorem DUAL-GT-5 (p. 160). An independent self-dual 2-basis schema for GT.

Theorem DUAL-GT-6 (p. 161). An independent self-dual 3-basis schema for GT.

Theorem DUAL-GT-7 (p. 161). An independent self-dual 4-basis schema for GT.

Theorem DUAL-BA-3 (p. 169). A self-dual 2-basis for BA (Pixley reduction).

Theorem DUAL-BA-5 (p. 174). A self-dual 2-basis for BA (majority reduction).

Theorem DUAL-BA-8 (p. 178). A self-dual 3-basis for BA (majority reduction).

A.8 Miscellaneous Topics

Theorem MFL-4 (p. 187). Simple basis with Moufang-2 (part 1).

Theorem MFL-5 (p. 189). Simple basis with Moufang-2 (part 2).

Theorem MFL-6 (p. 189). Simple basis with Moufang-2 (part 3).

Theorem MFL-7 (p. 190). Simple basis with Moufang-3.

Example MFL-8 (p. 191). Simple basis does not work with Moufang-1.

Theorem IL-1 (p. 192). A single axiom for inverse loops.

Theorem IL-4 (p. 197). A single axiom schema for inverse loops.

Corollary MFL-9 (p. 197). A single axiom for Moufang loops.

B. Open Questions

We start the section numbers with B.3 so that they correspond to the chapter numbers.

B.3 Algebras over Algebraic Curves

B.4 Other (gL)-Algebras

Problem TBG-2. Ternary Boolean group associativity.

We have the following (gL)-basis for Boolean groups in terms of a ternary operation (Thm. TBG-1).

$$\left\{ \begin{array}{l} p(x, x, x) = x \\ p(x, y, p(z, u, v)) = p(y, z, p(u, v, x)) \end{array} \right\}.$$

Can the associative law be replaced with an equation with fewer variables?

Problem CS-GL-8. A (gL)-basis for identities common to $x + y$ and $x - y$ (1).

$$\left\{ \begin{array}{l} \text{cancellation} \\ x(ey) = y(ex) \\ xe = x \end{array} \right\} \overset{?}{=\!(gL)\!\Rightarrow} \ \{x(y(zu)) = (x(yz))u\}$$

Problem CS-GL-9. A (gL)-basis for identities common to $x + y$ and $x - y$ (2).

$$\left\{ \begin{array}{l} x(ey) = y(ex) \\ xe = x \end{array} \right\} \overset{?}{=\!(gL)\!\Rightarrow} \ \{x(y(zu)) = (x(yz))u\}.$$

Problem CS-GL-10. Nearly (6) associative cancellative (gL)-groupoids.

$$\left\{ \begin{array}{l} \text{cancellation} \\ x((yz)u) = (x(yz))u \end{array} \right\} \overset{?}{=\!(gL)\!\Rightarrow} \ \{xy = yx\}.$$

B.5 Semigroups

Problem CS-14. On cancellative semigroups.

$$\{CS, (xy)^n = (yx)^n\} \overset{?}{\Rightarrow} \{xy^n = y^n x\}.$$

B.6 Lattice-like Algebras

Problem NL-1. Is the variety of near lattices 1-based?
See Sec. 6.1.3.

Problem TNL-1. Is the variety of transitive near lattices 1-based?
See Sec. 6.1.3.

Problem RBA-1. The Robbins question.

$$\left.\begin{cases} x + y = y + x \\ (x + y) + z = x + (y + z) \\ \text{Robbins axiom} \end{cases}\right\} \overset{?}{\Rightarrow} \{\text{Huntington axiom}\}.$$

Problem TRI-1. A 2-basis for the triangle algebra.

Let T (for triangle) denote the 3-element algebra of type $\langle 2, 2 \rangle$ with two binary operations of join and meet defined on the set $T = \{0, 1, 2\}$ with $0 < 1 < 2 < 0$. The binary relation $<$ is reflexive and antisymmetric but not transitive. Define $x \vee y = l.u.b.(x, y)$ and $x \wedge y = g.l.b.(x, y)$. Thus both $\vee$ and $\wedge$ are idempotent and commutative but not associative. If one draws the Hasse diagram of the algebra, it will be a triangle with $0 \vee 1 = 1$, $1 \vee 2 = 2$, $2 \vee 0 = 0$, etc. It is known that T is not 1-based. Find a simple 2-basis for the equational theory of T.

B.7 Independent Self-Dual Bases

Problem QLT-8. Independent self-dual 5-basis for QLT.

It is easy to show that quasilattice theory (QLT) cannot be defined by any self-dual independent set of identities with fewer that four identities. Does there exist an independent self-dual 5-basis for QLT?

Problem DUAL-1. Equational theory with no independent self-dual basis.

Is there an example of a finitely based equational theory of algebras admitting a duality (i.e., an arity preserving automorphism of period 2) but with no independent self-dual basis?

B.8 Miscellaneous Topics

Problem MFL-10. A short single axiom for Moufang loops.

Is there a single axiom for Moufang loops, in terms of product and inverse, that is simpler than the one given in Cor. MFL-9 (p. 197)? That axiom has 10 variables, 24 occurrences of variables, and 5 occurrences of inverse; using Otter's measure, it has length 52.

Problem ABGT-9. A simple basis for $A2$.

The type is $\langle 2, 0 \rangle$ with constant e. Find the smallest possible equational basis for $A2$.

$$A2 = \left\{ \begin{array}{l} (xe)e = x \\ e(ex) = x \\ ee = e \\ e(xe) = (ex)e \\ ((xx)e)(xx) = e \end{array} \right\}.$$

Problem ABGT-10. A (gL)-schema for Abelian groups in terms of double inversion.

Find an equation A containing a term $f(x)$ or $f(y)$ such that

$$\{A\} \quad \begin{array}{c} =\!(gL)\!\Rightarrow \\ \Longleftarrow \end{array} \quad \left\{ \begin{array}{l} ((x|(((x|y)|z)|(y|e)))|(e|e)) = z \\ f(x) = x \end{array} \right\}.$$

Problem STN-9. Single identity for ternary SQUAGs.

Find a single axiom for the equational theory given by the following set.

$$\left\{ \begin{array}{l} g(x) = x \\ f(x, x, x) = x \\ f(x, y, f(z, x, y)) = z \end{array} \right\}.$$

Problem HBCK-1. A first-order proof of a theorem in HBCK.

The type is $\langle 2, 0 \rangle$ with constant 1. The quasivariety HBCK is defined by

$$\left\{ \begin{array}{ll} x \cdot 1 = 1 & \text{(M3)} \\ 1 \cdot x = x & \text{(M4)} \\ (x \cdot y) \cdot ((z \cdot x) \cdot (z \cdot y)) = 1 & \text{(M5)} \\ x \cdot y = 1, y \cdot x = 1 \rightarrow x = y & \text{(M7)} \\ x \cdot x = 1 & \text{(M8)} \\ x \cdot (y \cdot z) = y \cdot (x \cdot z) & \text{(M9)} \\ (x \cdot y) \cdot (x \cdot z) = (y \cdot x) \cdot (y \cdot z) & \text{(H)} \end{array} \right\}.$$

Theorem (Blok and Ferreirim [9]). HBCK is an equational class of algebras defined by the equations (M3-5,8-9), (H), and (J):

$$(((x \cdot y) \cdot y) \cdot x) \cdot x = (((y \cdot x) \cdot x) \cdot y) \cdot y. \quad \text{(J)}$$

Find a first-order derivation of (J) from the axioms of HBCK:

$$\{M3,M4,M5,M7,M8,M9,H\} \rightarrow \{J\}.$$

The known proof is model theoretic.

C. The Autonomous Mode

Otter has an autonomous mode, which allows the user to give just the denial of the conjecture. When the autonomous mode is specified, Otter sets its own flags and parameters based on a very simple syntactic analysis of the input. We do not ordinarily use the autonomous mode, and it was not used for the main body of this work. However, it is useful, especially for beginners; and in an afterthought, we reran, in the autonomous mode, all of the theorems not involving the inference rule (gL).

Each job was run for at most half an hour, and the results are listed in Table C.1 along with the "tuned" results from the main body of this work. (Recall from Sec. 2.2.6 that the tuning usually involved changing just a few parameters from our basic strategy.) The number of seconds to proof, the proof length, and the memory used (megabytes) are listed; "—" means that no proof was found within the 1800-second time limit.

The autonomous mode for these problems is quite similar to the basic starting strategy we used (Sec. 2.2.6). The main difference is the parameter max_weight, the weight threshold for retained clauses. In the autonomous mode, it is adjusted dynamically based on the amount of memory available, whereas in the basic strategy, it is static for each search and adjusted by the user between runs.

Our interpretation of the table is that even the automatic Otter is useful for many interesting equational problems. Note that the autonomous mode is substantially better for Thms. CS-2 and QLT-6.

Table C.1. Tuned vs. Autonomous Searches on Non-(gL) Theorems

Theorem	Tuned			Autonomous		
	Time (sec.)	Length	Mem. (MB)	Time (sec.)	Length	Mem. (MB)
ABGT-7 (part 2)	0.56	32	0.4	4.80	41	0.9
ABGT-8 (part 2)	0.40	18	0.3	4.15	67	0.9
BA-1	7.84	45	1.0	—	—	4.7
CS-1	7.14	6	1.1	850.06	6	4.2
CS-2	187.41	8	4.3	5.15	6	1.1
CS-3	203.10	24	3.5	—	—	7.2
CS-4	4.67	2	1.7	54.66	2	1.1
CS-5	4.89	7	0.8	39.87	8	1.8
CS-6	269.12	8	4.0	—	—	5.7
CS-7	172.79	31	2.9	—	—	4.2
CS-8	955.82	32	8.4	—	—	4.3
CS-13	25288.37	58	19.9	—	—	4.5
DUAL-BA-1	28.95	33	4.6	59.40	39	4.1
DUAL-BA-2	3.81	78	0.8	—	—	5.6
DUAL-BA-3	8.56	99	1.3	—	—	6.2
DUAL-BA-5	2.51	17	0.7	—	—	4.2
DUAL-BA-7	2.82	17	0.7	25.46	17	3.6
DUAL-BA-8	24.59	103	3.6	—	—	4.3
DUAL-GT-1	0.53	22	0.4	1.25	23	0.4
DUAL-GT-2	1.47	29	0.7	4.08	28	0.7
DUAL-GT-3	0.52	17	0.4	0.42	20	0.2
DUAL-GT-4	0.17	10	0.2	1.14	11	0.3
DUAL-GT-5	2.62	26	0.9	3.28	26	0.8
DUAL-GT-6	0.22	12	0.2	0.26	12	0.1
DUAL-GT-7	3.38	27	0.7	3.45	34	0.5
GT-1	0.35	10	0.2	0.33	10	0.2
IL-1 (part 1)	0.46	5	0.4	1.25	6	0.6
IL-1 (part 2)	0.66	48	0.5	0.78	50	0.3
IL-1 (part 3)	0.08	8	0.2	0.08	8	0.1
IL-2 (part 1)	1.28	7	0.7	0.47	8	0.3
IL-2 (part 2)	0.66	49	0.6	0.69	50	0.3
IL-2 (part 3)	0.12	9	0.2	0.09	8	0.1
IL-3	0.41	16	0.3	0.44	14	0.2
LT-2	23.51	35	3.2	88.70	43	4.5
LT-3	71.49	76	4.2	—	—	4.2
LT-4	4.08	13	0.7	3.57	13	0.7
LT-5	11.74	40	2.0	77.53	71	4.1
LT-6	101.45	97	9.4	—	—	4.6
LT-8	79.56	20	5.2	76.01	20	4.3
LT-9	392.74	50	14.5	889.26	61	4.3
LT-10	7.85	18	1.9	6.33	19	1.5
LT-11	1.64	20	0.6	1.81	21	0.5
MAJ-2	0.09	10	0.2	0.06	11	0.1
MAJ-3	0.78	23	0.9	0.97	24	0.8
MED-1	142.15	17	0.8	—	—	4.6
MED-7	9.76	10	0.6	—	—	4.5
MFL-1	11.55	36	0.9	—	—	4.9

(continued)	Tuned			Autonomous		
	Time (sec.)	Length	Mem. (MB)	Time (sec.)	Length	Mem. (MB)
MFL-2	13.45	29	0.6	—	—	5.0
MFL-3	14.58	26	0.6	—	—	4.7
MFL-4	7.45	37	1.1	38.88	39	4.7
MFL-5	1.52	12	0.4	7.49	25	1.9
MFL-6	0.49	11	0.3	1.05	13	0.4
MFL-7	33.60	39	1.6	—	—	5.4
PIX-2	0.06	10	0.2	0.07	11	0.1
QGT-3	0.47	14	0.3	0.68	14	0.3
QLT-1	4.96	47	1.2	7.72	20	1.4
QLT-2	7.28	19	1.2	5.80	37	1.1
QLT-3	165.96	113	2.0	—	—	4.6
QLT-4	8.18	49	1.5	16.35	49	2.7
QLT-5	12.04	30	0.7	866.29	127	4.4
QLT-6	475.97	61	6.1	265.12	32	4.3
RBA-2	21.99	37	1.9	48.99	37	4.4
RBA-3	1506.42	40	8.2	—	—	4.9
RC-1	0.55	3	0.3	1.36	3	0.1
RC-2	10.48	3	1.1	14.34	3	0.9
RC-3	0.14	1	0.2	0.28	1	0.1
RC-4	8.41	8	0.5	21.48	8	0.3
RC-6	0.79	5	0.3	1.09	6	0.1
SD-2	4.82	34	0.9	35.45	83	4.1
SD-3 (part 1)	2.72	17	0.8	2.77	17	0.6
SD-3 (part 2)	20.60	23	1.8	59.12	35	4.0
STN-5 (part 1)	0.49	8	0.3	0.54	9	0.3
STN-5 (part 2)	0.06	2	0.1	0.04	2	0.1
STN-6 (part 1)	0.41	17	0.4	0.46	18	0.2
STN-6 (part 2)	0.03	2	0.1	0.04	2	0.1
STN-7 (part 1)	0.22	18	0.3	0.45	18	0.2
STN-7 (part 2)	0.06	3	0.1	0.06	3	0.1
STN-8 (part 1)	1.89	36	0.7	2.49	37	0.5
STN-8 (part 2)	0.09	3	0.1	0.05	3	0.1
TBA-1 (part 1)	3.78	27	1.1	933.82	29	4.8
TBA-1 (part 2)	1.73	39	0.7	—	—	4.6
TC-1	5.22	3	1.0	18.32	4	1.4
TC-2	2.23	4	0.3	5.52	4	0.3
TC-4	55.76	5	0.3	376.16	8	2.9
TC-5	74.62	8	0.9	118.84	7	2.7
TC-6	0.41	4	0.3	0.67	4	0.2
TC-7	15.38	7	2.2	—	—	7.5
TC-8	0.09	2	0.2	0.17	2	0.1
TC-9	1.04	7	0.4	—	—	4.3
TC-10	0.64	7	0.3	0.51	6	0.2
TC-11	0.55	3	0.2	—	—	4.1
TC-12	2.15	9	0.5	—	—	7.5
TC-13	1.09	5	0.4	3.98	5	0.4
TC-14	0.75	7	0.4	2.93	10	0.2
WAL-1	267.46	51	5.5	369.58	83	4.3
WAL-2	13.99	18	1.1	18.59	18	3.4
WAL-3	2.69	45	0.6	4.92	62	1.1

Bibliography

1. A. A. Albert. Quasigroups I & II. *Trans. Amer. Math. Soc.*, 54, 55, 1943.

2. Anonymous. The QED Manifesto. In A. Bundy, editor, *Proceedings of CADE-12, Springer-Verlag Lecture Notes in Artificial Intelligence, Vol. 814*, pages 238–251, Berlin, 1994. Springer-Verlag. Also ftp://info.mcs.anl.gov/pub/qed/manifesto.

3. F. Bennett, H. Zhang, and L. Zhu. Self-orthogonal Mendelsohn triple systems. *J. Combinatoric Theory*, 73:207–218, 1996.

4. G. Birkhoff. On the structure of abstract algebras. *Proc. of the Camb. Philos. Soc.*, 29:433–454, 1935.

5. G. Birkhoff. *Lattice Theory*. AMS Colloquium Publications, Providence, R.I., 1967.

6. R. S. Boyer and J S. Moore. *A Computational Logic*. Academic Press, New York, 1979.

7. R. S. Boyer and J S. Moore. *A Computational Logic Handbook*. Academic Press, New York, 1988.

8. R. H. Bruck. *A Survey of Binary Systems*. Springer-Verlag, Berlin, 1958.

9. S. Burris, Jan. 1995. Correspondence by electronic mail.

10. O. Chein, H. O. Pflugfelder, and J. D. H. Smith. *Quasigroups and Loops, Theory and Applications*. Haldermann Verlag, Berlin, 1990.

11. I. M. S. Etherington. Quasigroups and cubic curves. *Proc. Edinburgh Math. Soc.*, 14:273–291, 1965.

12. H. G. Forder and J. A. Kalman. Implication in equational logic. *Math. Gazette*, 46:122–126, 1962.

13. K. Gödel. *The Consistency of the Axiom of Choice and of the Generalized Continuum-Hypothesis with the Axioms of Set Theory*. Princeton University Press, 1940.

14. G. Grätzer. *General Lattice Theory*. Academic Press, New York, 1978.

15. G. Grätzer and R. Padmanabhan. Symmetric difference in Abelian groups. *Pac. J. Math.*, 74(2):339–347, 1978.

16. J. Guard, F. Oglesby, J. Bennett, and L. Settle. Semi-Automated Mathematics. *J. ACM*, 16(1):49–62, 1969.

17. N. Gupta and A. Rhemtulla. A note on centre-by-finite-exponent varieties of groups. *J. Austral. Math. Soc.*, 11:33–36, 1970.

18. J. Hart and K. Kunen. Single axioms for odd exponent groups. *J. Automated Reasoning*, 14(3):383–412, 1995.

19. G. Higman and B. H. Neumann. Groups as groupoids with one law. *Publicationes Mathematicae Debrecen*, 2:215–227, 1952.

20. B. Jónsson. Algebras whose congruence lattices are distributive. *Math. Scand.*, 21:110–121, 1967.

21. J. A. Kalman. Equational completeness and families of sets closed under subtraction. *Indag. Math.*, 22:402–405, 1960.

22. J. A. Kalman. A shortest single axiom for the classical equivalential calculus. *Notre Dame J. Formal Logic*, 19:141–144, 1978.

23. J. A. Kalman, Sept.–Oct. 1994. Correspondence by electronic mail.

24. D. Kelly and R. Padmanabhan. Self-dual equational bases for lattice varieties. To appear.

25. D. Kelly and R. Padmanabhan. Identities common to four Abelian group operations with zero. *Algebra Universalis*, 21:1–24, 1985.

26. A. Knapp, editor. *Elliptic Curves*. Princeton University Press, 1993.

27. D. Knuth and P. Bendix. Simple word problems in universal algebras. In J. Leech, editor, *Computational Problems in Abstract Algebras*, pages 263–297. Pergamon Press, Oxford, U.K., 1970.

28. K. Kunen. Single axioms for groups. *J. Automated Reasoning*, 9(3):291–308, 1992.

29. H. Lakser, R. Padmanabhan, and C. R. Platt. Subdirect decomposition of Płonka sums. *Duke Math. J.*, 39(3):485–488, 1972.

30. A. I. Mal'cev. Über die Einbettung von assoziativen Systemen Gruppen I. *Mat. Sbornik*, 6(48):331–336, 1939.

31. B. Mazur. Arithmetic on curves. *Bull. AMS*, 14:207–259, 1986.

32. J. McCharen, R. Overbeek, and L. Wos. Complexity and related enhancements for automated theorem-proving programs. *Computers and Math. Applic.*, 2:1–16, 1976.

33. J. McCharen, R. Overbeek, and L. Wos. Problems and experiments for and with automated theorem-proving programs. *IEEE Trans. on Computers*, C-25(8):773–782, August 1976.

34. W. McCune. Automated discovery of new axiomatizations of the left group and right group calculi. *J. Automated Reasoning*, 9(1):1–24, 1992.

35. W. McCune. Single axioms for groups and Abelian groups with various operations. *J. Automated Reasoning*, 10(1):1–13, 1993.

36. W. McCune. Single axioms for the left group and right group calculi. *Notre Dame J. Formal Logic*, 34(1):132–139, 1993.

37. W. McCune. A Davis-Putnam program and its application to finite first-order model search: Quasigroup existence problems. Tech. Memo. ANL/MCS-TM-194, Argonne National Laboratory, Argonne, Ill., 1994.

38. W. McCune. Models And Counter-Examples (MACE). http://www.mcs.anl.gov/home/mccune/ar/mace/, 1994.

39. W. McCune. OTTER 3.0 Reference Manual and Guide. Tech. Report ANL-94/6, Argonne National Laboratory, Argonne, Ill., 1994.

40. W. McCune and R. Padmanabhan. Single identities for lattice theory and weakly associative lattices. *Algebra Universalis*, 1996. To appear.

41. W. McCune and L. Wos. The absence and the presence of fixed point combinators. *Theoretical Computer Science*, 87:221–228, 1991.

42. W. McCune and L. Wos. Application of automated deduction to the search for single axioms for exponent groups. In A. Voronkov, editor, *Logic Programming and Automated Reasoning, Springer-Verlag Lecture Notes in Artificial Intelligence, Vol. 624*, pages 131–136, Berlin, July 1992. Springer-Verlag.

43. R. N. McKenzie. Equational bases for lattice theories. *Math. Scand.*, 27:24–38, 1970.

44. R. N. McKenzie, G. F. McNulty, and W. F. Taylor. *Algebras, Lattices, Varieties*, volume I. Wadsworth & Brookes/Cole, California, 1987.

45. G. McNulty. The decision problem for equational bases of algebras. *Ann. Math. Logic*, 11:193–259, 1976.

46. N. S. Mendelsohn, R. Padmanabhan, and B. Wolk. Straight edge constructions on cubic curves. *C. R. Math. Rep. Acad. Sci. Canada*, 10:77–82, 1988.

47. N. S. Mendelsohn, R. Padmanabhan, and B. Wolk. Placement of the Desargues configuration on a cubic curve. *Geometriae Dedicata*, 40:165–170, 1991.

48. J. S. Milne. Abelian varieties. In *Arithmetic, Geometry*, pages 103–150. Springer-Verlag, New York, 1986.

49. D. Mumford. *Abelian Varieties*, 2nd Ed. Oxford University Press, 1974.

50. D. Mumford. *Algebraic Geometry I, Complex Projective Varieties*. Springer-Verlag, 1976.

51. B. H. Neumann. Another single law for groups. *Bull. Australian Math. Soc.*, 23:81–102, 1981.

52. R. Padmanabhan. Inverse loops as groupoids with one law. *J. London Math. Soc.*, 2(1 44):203–206, 1969.

53. R. Padmanabhan. Equational theory of algebras with a majority polynomial. *Algebra Universalis*, 7(2):273–275, 1977.

54. R. Padmanabhan. A first-order proof of a theorem of Frink. *Algebra Universalis*, 13(3):397–400, 1981.

55. R. Padmanabhan. Logic of equality in geometry. *Discrete Mathematics*, 15:319–331, 1982.

56. R. Padmanabhan. A self-dual equational basis for Boolean algebras. *Canad. Math. Bull.*, 26(1):9–12, 1983.

57. R. Padmanabhan and W. McCune. Automated reasoning about cubic curves. *Computers and Math. Applic.*, 29(2):17–26, 1995.

58. R. Padmanabhan and W. McCune. An equational characterization of the conic construction on cubic curves. Preprint MCS-P517-0595, Argonne National Laboratory, Argonne, Ill., 1995.

59. R. Padmanabhan and W. McCune. Single identities for ternary Boolean algebras. *Computers and Math. Applic.*, 29(2):13–16, 1995.

60. R. Padmanabhan and W. McCune. On n-ary Steiner laws and self-dual axiomatizations of group theory. In preparation, 1996.

61. R. Padmanabhan and R. W. Quackenbush. Equational theories of algebras with distributive congruences. *Proc. of AMS*, 41(2):373–377, 1973.

62. J. G. Peterson. The possible shortest single axioms for EC-tautologies. Report 105, Dept. of Mathematics, University of Auckland, 1977.

63. G. W. Pieper. The QED Workshop. Tech. memo., Argonne National Laboratory, Argonne, Ill., May 1994.

64. A. F. Pixley. Distributivity and permutability of congruence relations in equational classes of algebras. *Proc. Amer. Math. Soc.*, 14:105–109, 1963.

65. R. W. Quackenhush. Quasi-affine algebras. *Algebra Universalis*, 20:318–327, 1985.

66. M. Reid. *Undergraduate Algebraic Geometry*. Cambridge University Press, 1988.

67. G. Robinson and L. Wos. Paramodulation and theorem-proving in first-order theories with equality. In D. Michie and R. Meltzer, editors, *Machine Intelligence, Vol. IV*, pages 135–150. Edinburgh University Press, 1969.

68. I. R. Shafarevich. *Basic Algebraic Geometry*. Springer-Verlag, Berlin, 1977.

69. M. Sholander. On the existence of inverse operation in alternation groupoids. *Bull. Amer. Math. Soc.*, 55:746–757, 1949.

70. J. Slaney, M. Fujita, and M. Stickel. Automated reasoning and exhaustive search: Quasigroup existence problems. *Computers and Math. Applic.*, 29:115–132, 1995.

71. M. Stickel and H. Zhang. Studying quasigroup identities by rewriting techniques: Problems and first results. In *Proc. of RTA-95*, 1995.

72. A. Tarski. Equational logic and equational theories of algebras. In K. Schütte, editor, *Contributions to Mathematical Logic*, pages 275–288. North-Holland, Amsterdam, 1968.

73. A. Tarski. An interpolation theorem. *Discrete Math*, 12:185–192, 1975.

74. A. Tarski and S. Givant. *A Formalization of Set Theory without Variables*, volume 41 of *Colloquium Publications*. American Mathematical Society, 1987.

75. R. Veroff. Using hints to increase the effectiveness of an automated reasoning program: Case studies. *J. Automated Reasoning*, 1996. To appear.

76. T.-C. Wang and R. Stevens. Solving open problems in right alternative rings with Z-module reasoning. *J. Automated Reasoning*, 5(2):141–165, 1989.

77. F. J. Wicklin. Pisces: A platform for implicit surfaces and curves and the exploration of singularities. Technical Report GCG #89, The Geometry Center, 1995. Also http://www.geom.umn.edu/locate/pisces/.

78. S. Winker. Generation and verification of finite models and counterexamples using an automated theorem prover answering two open questions. *J. ACM*, 29:273–284, 1982.

79. S. Winker. Robbins algebra: Conditions that make a near-Boolean algebra Boolean. *J. Automated Reasoning*, 6(4):465–489, 1990.

80. L. Wos. Automated reasoning answers open questions. *Notices of the AMS*, 5(1):15–26, January 1993.

81. L. Wos. The kernel strategy and its use for the study of combinatory logic. *J. Automated Reasoning*, 10(3):287–343, 1993.

82. L. Wos. *The Automation of Reasoning: An Experimenter's Notebook with Otter Tutorial.* Academic Press, New York, 1996. To appear.

83. L. Wos, R. Overbeek, E. Lusk, and J. Boyle. *Automated Reasoning: Introduction and Applications,* 2nd edition. McGraw-Hill, New York, 1992.

84. L. Wos, G. Robinson, and D. Carson. Efficiency and completeness of the set of support strategy in theorem proving. *J. ACM*, 12(4):536–541, 1965.

85. L. Wos, S. Winker, B. Smith, R. Veroff, and L. Henschen. A new use of an automated reasoning assistant: Open questions in equivalential calculus and the study of infinite domains. *Artificial Intelligence*, 22:303–356, 1984.

Index

Lecture Notes in Artificial Intelligence (LNAI)

Lecture Notes in Computer Science